MATHEMATICAL STRUCTURES FOR COMPUTER SCIENCE

MATHEMATICAL STRUCTURES
FOR COMPUTER SCIENCE

JUDITH L. GERSTING

Indiana University–
Purdue University at Indianapolis

W. H. Freeman and Company
San Francisco

Project Editor: Judith Wilson
Copy Editor: Stephen McElroy
Designer: Nancy Benedict
Production Coordinator: Sarah Segal
Illustration Coordinator: Richard Quiñones
Artist: John Cordes
Compositor: Syntax International
Printer and Binder: R. R. Donnelley and Sons Company

Cover design adapted from a graphic by Benoit B.
Mandelbrot and reproduced by his kind permission. From
The Fractal Geometry of Nature by Benoit B. Mandelbrot.
W. H. Freeman and Company. Copyright © 1982.

Library of Congress Cataloging in Publication Data

Gersting, Judith L.
  Mathematical structures for computer science.

  Includes bibliographies and index.
  1. Mathematics—1961–   . 2. Mathematical models.
3. Electronic data processing—Mathematics. I. Title.
QA39.2.G47   510   82-2550
ISBN 0-7167-1305-5   AACR2

Printed in the United States of America

1 2 3 4 5 6 7 8 9 0 DO 0 8 9 8 7 6 5 4 3 2

To Adam and Jason,
two beloved discrete structures

CONTENTS

PREFACE

A course in discrete structures is an integral part of the curriculum for most undergraduate computer science students.* Nonetheless, there have been variations of opinion on the content and even the purpose of the course. My thoughts on these issues are as follows.

1. The course in discrete structures should be a survey, not an in-depth study of any particular topic. When taken as an undergraduate terminal theoretical course, it must provide the student with a broad overview. When taken as a prerequisite for further study, it must prepare the student for courses in computability theory, formal language theory, analysis of algorithms, coding theory, and switching theory.

2. The course in discrete structures should introduce the rigorous theoretical frameworks within which ideas about computer science can be expressed. The method of introduction should ensure that students will, at the least, become comfortable with these frameworks and appreciate their value, although not every student will become fluent in their use.

Naturally these opinions, and the anticipated student audience, have influenced the choice and arrangement of topics, as well as the level of presentation. Most students will be sophomores or juniors in computer science with a mathematical background of calculus. This book is appropriate for them. It is also ideally suited for the applied algebra course recommended for the mathematical sciences curriculum.†

Listed below are five unique and helpful features of this book.

1. The writing style is clear, conversational, and informal, avoiding the theorem–proof approach that intimidates so many students. Providing motivation and developing results before giving formal statements seems more natural to the student, and, indeed, parallels the historical process.

2. There are numerous Practice Problems provided throughout the text. These problems are simple, and the student should work them as

---

* Discrete structures was described as course B3 in the Curriculum 68 report of the ACM (*Communications of the ACM*, vol. 11, no. 3, March 1968), and as course MA 4 in the Curriculum 78 report (*Communications of the ACM*, vol. 22, no. 3, March 1979). The report *Curriculum in Computer Science and Engineering* (IEEE Computer Society, 1977) describes discrete structures as course TC-1.
† See *Recommendations for a General Mathematical Sciences Program*, Committee on the Undergraduate Program in Mathematics of the MAA, 1981.

they are encountered. Their purpose is to provide immediate reinforcement of some idea or notation just introduced. My students have found this feature very helpful. Answers to all Practice Problems are provided at the very back of the book. In addition, there are, of course, many exercises at the end of each section; most are straightforward, but others provide extensions of the text material. Answers to selected exercises, denoted in the text by ★, are given at the back of the book. Occasionally, an exercise requires the student to write a computer program; at the instructor's option, such an exercise may entail anything from writing a pseudocode algorithm to actually writing and running a program in some language.

3. Students taking the discrete structures course will generally fail to be moved by the inherent beauty of a theoretical structure. They want to know what it's good for. I have tried to tie material to computer applications, bearing in mind that the students' exposure to computer science may not yet be extensive.

4. The last two chapters cover theory of computation and formal languages much more extensively than most texts for this course. This coverage provides a strong background for students who will later go on to specialized courses in these areas.

5. Because such a variety of topics is covered, I have emphasized the unifying basic theme of simulation, in fact, two kinds of simulation. Simulation I is the concept of a mathematical model, or structure, and Simulation II refers to the simulation of one instance of a structure by another. The theme is first presented in Chapter 3 and reappears in Chapters 5, 7, and 9.

Chapters 1 and 2 provide coverage of logic, sets, combinatorics, relations, functions, and graphs. Although this material is essential to the rest of the text, it can be treated as quickly as the instructor feels the students can proceed. Much of it will be review to most of them. I have found, however, that students may not be as well versed in this area as we expect they are, and that a thorough presentation of these topics is warranted.

Chapter 3 discusses notions of models and homomorphisms, using Boolean algebras as a starting point. Chapter 4 presents material on gating networks.

Chapter 5 considers algebraic structures and homomorphisms. Here again, the depth of coverage can be varied to suit the class. Much of the spirit of this chapter carries over into Chapter 7, but no specific information from Chapter 5 is required in Chapter 7. However, Chapter 5 is a prerequisite for Chapter 6, an aside dealing with coding theory.

Chapter 7 introduces finite-state machines, their homomorphisms and their recognition capabilities (Kleene's Theorem). Chapter 8 is a look at finite-state machine "hardware" and, of course, depends upon Chapter 7; in addition, Chapter 4 is a prerequisite for Section 8.2, and Chapter 5 is a prerequisite for Section 8.4.

Chapter 9 covers Turing machines, unsolvability results, and a bit of computational complexity. Chapter 10 introduces formal languages and relates finite-state machines and Turing machines to languages; Section 10.1 is independent, but Section 10.2 requires Sections 7.3, 9.1, and 9.2.

There are various choices of topics for a course taught from this material. A one-semester course with an engineering emphasis could cover Chapters 1–3, 5, 7, 8, and either 4 or 6. A computer science course might include Chapters 1–3, 5, 7, 9, and 10. A two-quarter sequence could proceed sequentially—Chapters 1–5 and then 6–10—or it could follow a more mathematical–computer science split and treat Chapters 1, 2, 3, 5, and 6 during the first quarter, and then Chapters 4, 7, 8, 9, and 10 during the second quarter. A course for the mathematical sciences might cover Chapters 1–5, 7, and 10.1.

I have included brief lists of suggested readings at the end of each chapter. These lists are by no means comprehensive, nor do they imply that any book omitted is not worthwhile; they are simply sources for the interested student.

Some of the Practice Problems and exercises in this book previously appeared in *Abstract Algebra, A First Look* (Joseph E. Kuczkowski and Judith L. Gersting, New York: Marcel Dekker, Inc., 1977) and are used here with permission of Marcel Dekker, Inc.

I want to express my appreciation to reviewers Allen Acree, William Dorn, Victor Klee, Yale Patt, and Charles Swart for their helpful suggestions. Linda Knabe did a marvelous job in transforming my handwritten notes into flawless typescript. My department chairman, Michael Gemignani, strongly supported my efforts in writing this book, and I owe a great deal to him for his encouragement.

Of course, my special thanks go to my family. My husband John graciously relinquished more and more of his share of the desk as the manuscript grew, and our two small children never ran their toy cars on my papers and always did their best to be quiet whenever I told them "Mommy has to work now."

March 1982 Judith L. Gersting

NOTE TO THE STUDENT

Each chapter in this book begins with a short paragraph or two giving an overview of the chapter. Each section ends with a Checklist, a list of the definitions and main ideas presented in the section as well as the techniques you should have learned. These Checklists will help you summarize the sections and review for exams.

This book contains many Practice Problems. These problems are generally not difficult and are meant to be worked as soon as you get to them. Answers are given at the very back of the book. You'll find learning much easier if you give these problems your best effort, so get your pencil and paper ready. . . .

Judith L. Gersting

MATHEMATICAL STRUCTURES FOR COMPUTER SCIENCE

Chapter 1

HOW TO SPEAK MATHEMATICS: BASIC VOCABULARY I

It is interesting to consider how much thinking and learning a person could do if he or she had no language capability. Language provides both a vehicle for communication and a framework within which to think. Words and sentences in the English language are building blocks for the formation of ideas. There are many words in the English language that we use with a fairly good intuitive understanding of their meaning; yet we would be hard pressed to give a precise dictionary definition for them. If we are to study computer science as a science, then we cannot allow its terminology to remain vague. Precise definitions are required. In this chapter and the next, we will try to give precise meaning to words and concepts you probably already understand in an intuitive way. This process is not a redundancy, but a necessary first step in establishing our tools of communication. When you ask that ever-popular question, "What is this stuff good for?" keep in mind the title of these two chapters. They are chapters on "vocabulary." Throughout the rest of this book, we will use the terminology established here to express ideas having applications in computer science.

Section 1 of this chapter deals primarily with statement logic, a part of first order logic. First order logic formalizes the organized and precise method of thinking that characterizes scientific investigation.

Aside from these generalities, the topics covered in Chapters 1 and 2 have concrete applications to computer science. Computer logic, or circuit logic, is a direct analogue of statement logic (see Chapter 4). First order logic has been used in program verification, in proving that the output of a given computer program will always comply with certain predetermined conditions. Combinatorial techniques are used in algorithm development and analysis, and graphs are used to structure data for efficient computer storage and manipulation.

Section 1.1   LOGIC

For purposes of general communication we often express ourselves in English language questions, exclamations, and so forth, but to communicate facts or information of some sort, we use statements. Technically, a **statement** is a sentence that is either true or false.

1.1 Example   Consider the following:

> (a) Two is a prime number.
> (b) How are you?
> (c) He is certainly tall.
> (d) There is life as we know it on Saturn.

Sentence (a) is a statement because it is true. Item (b) is a question and has no truth value associated with it, so it is not a statement. In (c), the word "he" is a variable in the sentence, and the sentence is neither true nor false until "he" is specified. Therefore (c) is not a statement. Sentence (d) is a statement because it is either true or false; we do not have to be able to decide which.   □

CONNECTIVES AND TRUTH VALUES

To vary our conversations, we do not confine ourselves to simple statements. We combine simple statements with connecting words to make compound statements. The truth value of a compound statement depends on the truth values of its components and the connecting words used. A common connective is the word "and." (Words like "but" and "also" express different shades of meaning but produce the same effect as far as truth values are concerned.) If we combine the two true statements "grass is green" and "horses like oats," we would consider the resulting statement, "grass is green and horses like oats," to be true. We use the symbol $\wedge$ to denote "and" and capital letters to denote statements. We agree, then, that if A is true and B is true, $A \wedge B$ (read "A and B") is also true.

1.2 Practice   If A is true and B is false, what truth value would you assign to $A \wedge B$? If A is false and B is true, what truth value would you assign to $A \wedge B$? If A and B are both false, what truth value would you assign to $A \wedge B$?

(Answer, page A-1)

The statement "$A \wedge B$" is called the **conjunction** of A and B. We can summarize the effects of conjunction by the truth table presented in Fig 1.1.

Another connective is the word "or." The statement "*A* or *B*," symbolized $A \vee B$, is called the **disjunction** of statements *A* and *B*. If *A* and *B* are both true statements, then $A \vee B$ would be considered true, giving us the first line of the truth table for disjunction (see Figure 1.2).

A	B	$A \wedge B$
T	T	T
T	F	F
F	T	F
F	F	F

A	B	$A \vee B$
T	T	T
T	F	
F	T	
F	F	

Figure 1.1   **Figure 1.2**

1.3 Practice   Use your understanding of the word "or" to complete the truth table for disjunction.   *(Answer, page A-1)*

Statements *A* and *B* may be combined in the form "if *A*, then *B*," symbolized by $A \Rightarrow B$. The connective here is **implication**. $A \Rightarrow B$ may also be read as "*A* implies *B*", "*A* is a sufficient condition for *B*," "*A* only if *B*," "*B* follows from *A*," or "*B* is a necessary condition for *A*." In the compound statement $A \Rightarrow B$, *A* is called the **antecedent** and *B* the **consequent**.

1.4 Example   The statement "Thunder is a necessary condition for lightning" can be rewritten as "If there is lightning, then there is thunder." The antecedent is "there is lightning," and the consequent is "there is thunder."   □

1.5 Practice   Name the antecedent and consequent in each of the following statements. (Hint: rewrite each statement in if–then form.)

(a) If *a* is a positive number, then $2a$ is a positive number.
(b) A sufficient condition for using 6 storage locations is that a 2×3 array is to be stored.
(c) Susan will pass her physics course only if she is bright and studies hard.
(d) Good combustion is a necessary condition for high gasoline mileage.   *(Answer, page A-1)*

The truth table for implication is less obvious than that for conjunction or disjunction. To understand its definition, let's suppose you hear your roommate remark, "If I graduate this spring, then I'll work for the summer in Yosemite Park." If he graduates and goes to work in Yosemite, his remark was true. If in the implication $A \Rightarrow B$ both *A* and *B* are true, we consider the implication to be true. If he graduates and does not work in Yosemite for the summer, his remark was a false statement. When *A* is true and *B* false, $A \Rightarrow B$ is false. Now suppose he doesn't graduate. Whether he works for the summer in Yosemite or not, you could not accuse him of making a false statement. By default, we accept $A \Rightarrow B$ as true if A is false.

1.6 Practice   Summarize this discussion by writing the truth table for $A \Rightarrow B$.

(*Answer, page A-1*)

The **equivalence** connective, $A \Leftrightarrow B$, is shorthand for the statement $(A \Rightarrow B) \wedge (B \Rightarrow A)$. We can write the truth table for equivalence by constructing, one piece at a time, a table for $(A \Rightarrow B) \wedge (B \Rightarrow A)$, as in Figure 1.3.

A	B	$A \Rightarrow B$	$B \Rightarrow A$	$(A \Rightarrow B) \wedge (B \Rightarrow A)$
T	T	T	T	T
T	F	F	T	F
F	T	T	F	F
F	F	T	T	T

Figure 1.3

Notice from this truth table that A and B are *equivalent*; that is, $A \Leftrightarrow B$ is true exactly when A and B have the same truth values.

The connectives we've seen so far are called **binary connectives** because they join two statements together to produce a third statement. Now let's consider a **unary** connective, a connective acting on one statement to produce a second statement. The **negation** connective, A', is a unary connective and is read "not A," "it is false that A," or "it is not true that A".

1.7 Practice   Write the truth table for A'. (It will require only two rows.)   (*Answer, page A-1*)

1.8 Example   Finding the negation of a statement can be tricky. If A is the statement "Peter is tall and fat," then A' is the statement "It is false that Peter is tall and fat," which may be read "Peter is not tall or he is not fat." If A is "Cucumbers are green or seedy," then A' can be read "Cucumbers are neither green nor seedy." Finding negations is even trickier if we allow our statements to include **quantifiers**, words such as "every" or "each" or "some" that tell how many objects have a certain property. For the statement A, "Every book is interesting," A' can be read "There is some book that is not interesting." And if A is "There is an honest pawnbroker," then A' is "Every pawnbroker is dishonest."   □

Using connectives we can string lots of statements together to form a complex statement. In order to reduce the number of parentheses required, we stipulate which connectives should be applied first. The order of precedence is $'$, $\wedge$, $\vee$, $\Rightarrow$, $\Leftrightarrow$. Thus, $A \wedge B \vee C$ means $(A \wedge B) \vee C$; and $A \vee B \Rightarrow C'$ means $(A \vee B) \Rightarrow (C)'$. We write the truth tables for complex statements much as we did for $(A \Rightarrow B) \wedge (B \Rightarrow A)$.

1.9 **Example**   The truth table for the statement $(A \vee B') \wedge (A \vee B)'$ is given in Figure 1.4.

A	B	B'	$A \vee B'$	$A \vee B$	$(A \vee B)'$	$(A \vee B') \wedge (A \vee B)'$
T	T	F	T	T	F	F
T	F	T	T	T	F	F
F	T	F	F	T	F	F
F	F	T	T	F	T	T

Figure 1.4

If we have n statement letters in a statement and we are making a truth table for that statement, how many rows will the truth table have? From truth tables done so far we know that a statement with only one statement letter has two rows in its truth table, and a statement with two statement letters has four rows. The number of rows equals the number of true–false combinations possible among the statement letters. The first statement letter has two possibilities, T and F. For each of these possibilities, the second statement letter has two possible values. Figure 1.5a pictures this two-level

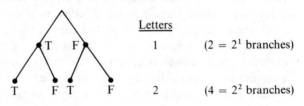

	Letters	
	1	$(2 = 2^1 \text{ branches})$
	2	$(4 = 2^2 \text{ branches})$

Figure 1.5(a)

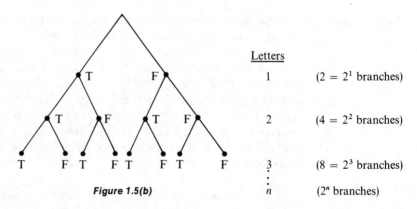

	Letters	
	1	$(2 = 2^1 \text{ branches})$
	2	$(4 = 2^2 \text{ branches})$
	3	$(8 = 2^3 \text{ branches})$
	$\vdots$	
	n	(2^n branches)

Figure 1.5(b)

"tree" with four "branches" showing the four possible combinations of T–F for two statement letters. For n statement letters, we extend the tree to n levels as in Figure 1.5b. The total number of branches then equals 2^n. The total number of rows in a truth table for n statement letters is also 2^n.   □

1.10 Practice Construct truth tables for the following statements:

(a) $(A \Rightarrow B) \Leftrightarrow (B \Rightarrow A)$. (Remember $C \Leftrightarrow D$ is true precisely when C and D have the same truth values.)

(b) $(A \vee A') \Rightarrow (B \wedge B')$

(c) $((A \wedge B') \Rightarrow C')'$

(d) $(A \Rightarrow B) \Leftrightarrow (B' \Rightarrow A')$ (*Answer, page A-1*)

The logical operators AND, OR, and NOT are available in many programming languages. These operators act (in accordance with the truth tables we have defined) upon combinations of true or false expressions to produce an overall truth value. Such truth values provide the decision-making capabilities fundamental to the flow of control in computer programs of any complexity.

TAUTOLOGIES

A statement like item (b) of Practice 1.10 whose truth values are always false is called a **contradiction**. A statement like statement (d) whose truth values are always true is called a **tautology**.

We will list a number of particular tautologies, prove one or two of them, and leave the rest as exercises. In this list, the equality sign stands for equivalence because equivalent statements are equal in truth value. We will use 0 to stand for any contradiction and 1 for any tautology.

Some Tautologies

1a. $A \vee B = B \vee A$ 1b. $A \wedge B = B \wedge A$

2a. $(A \vee B) \vee C = A \vee (B \vee C)$ 2b. $(A \wedge B) \wedge C = A \wedge (B \wedge C)$

3a. $A \vee (B \wedge C)$ 3b. $A \wedge (B \vee C)$
$\quad = (A \vee B) \wedge (A \vee C)$ $\quad = (A \wedge B) \vee (A \wedge C)$

4a. $A \vee 0 = A$ 4b. $A \wedge 1 = A$

5a. $A \vee A' = 1$ 5b. $A \wedge A' = 0$

1.11 Example The truth table in Figure 1.6 verifies tautology 1a and that in Figure 1.7 verifies 4b.

A	B	$A \vee B$	$B \vee A$	$A \vee B \Leftrightarrow B \vee A$
T	T	T	T	T
T	F	T	T	T
F	T	T	T	T
F	F	F	F	T

Figure 1.6

A	1	$A \wedge 1$	$A \wedge 1 \Leftrightarrow A$
T	T	T	T
F	T	F	T

Figure 1.7

Note that only two rows are needed for Figure 1.7 because 1 (a tautology) cannot take on false truth values.   □

1.12 Practice   Verify tautology 5a.   *(Answer, page A-2)*

The tautologies in our list are grouped into five pairs. In each pair, one statement can be obtained from the other by replacing $\wedge$ with $\vee$, $\vee$ with $\wedge$, 0 with 1, and 1 with 0. Each statement in a pair is called the **dual** of the other. This list of tautologies appears in a more general setting in Chapter 3.

If we have a statement of the form $P \Rightarrow Q$ where P and Q are compound statements, we can use a quicker procedure than constructing a truth table to determine whether $P \Rightarrow Q$ is a tautology. We assume that $P \Rightarrow Q$ is *not* a tautology, that P is true and Q false, which allows us to determine truth value possibilities for the component statements making up P and Q. We continue assigning the truth values so determined until every statement letter has a truth value. $P \Rightarrow Q$ will be a tautology *if and only if* some statement letter has been assigned both true and false values by this process. The flowchart in Figure 1.8 summarizes this procedure.

1.13 Example   Consider a statement of the form $(A \Rightarrow B) \Rightarrow (B' \Rightarrow A')$. Here P is $A \Rightarrow B$ and Q is $B' \Rightarrow A'$. We assign truth values

$A \Rightarrow B$ true

$B' \Rightarrow A'$ false

The second assignment implies

B' true

A' false

or

B false

A true

But for A true and $A \Rightarrow B$ true, B must be true. B then has an assignment of both T and F, and therefore $(A \Rightarrow B) \Rightarrow (B' \Rightarrow A')$ is a tautology. We can verify this tautology by constructing its truth table.   □

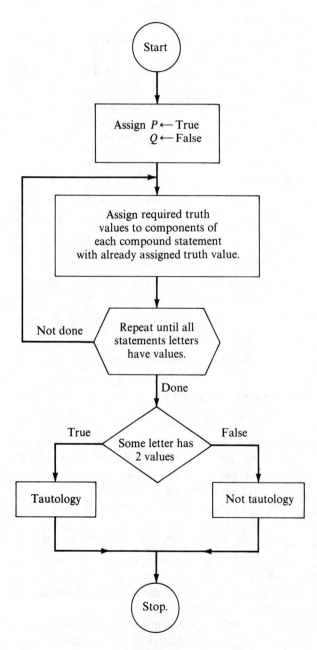

Figure 1.8

√ *CHECKLIST*

Definitions

statement (*p. 2*)	consequent (*p. 4*)
conjunction (*p. 3*)	binary connective (*p. 5*)
disjunction (*p. 4*)	unary connective (*p. 5*)
implication (*p. 4*)	quantifier (*p. 5*)
equivalence (*p. 5*)	contradiction (*p. 7*)
negation (*p. 5*)	tautology (*p. 7*)
antecedent (*p. 4*)	dual of a statement (*p. 8*)

Techniques

Construct truth tables for complex statements.

Recognize tautologies and contradictions.

Main Ideas

Statements and how they can be combined (conjunction, disjunction, etc.).

Truth values for compound statements that depend upon the truth values of their components.

A list of basic tautologies.

EXERCISES SECTION 1.1

Answers to starred items are given at the back of the book.

★ 1. Find the antecedent and consequent in each of the following statements.
  (a) Healthy plant growth follows from sufficient water.
  (b) Increased availability of microcomputers is a necessary condition for further technological advances.
  (c) Errors will be introduced only if there is a modification of the program.
  (d) Fuel savings implies good insulation or all windows are storm windows.

★ 2. Several forms of negation are given for each of the following statements. Which are correct?
  (a) Some people like mathematics.
    (1) Some people dislike mathematics.
    (2) Everybody dislikes mathematics.
    (3) Everybody likes mathematics.

    (b) The answer is either 2 or 3.
        (1) Neither 2 nor 3 is the answer.
        (2) The answer is not 2 or not 3.
        (3) The answer is not 2 and it is not 3.
    (c) All people are tall and thin.
        (1) Someone is short and fat.
        (2) No one is tall and thin.
        (3) Someone is short or fat.

★ 3. Using letters for the component statements, translate the following compound statements into symbolic notation.
    (a) If prices go up, then housing will be plentiful and expensive; but if housing is not expensive, then it will still be plentiful.
    (b) Either going to bed or going swimming is a sufficient condition for changing clothes; however, changing clothes does not mean going swimming.
    (c) Either it will rain or it will snow but not both.
    (d) If Janet wins or if she loses, she will be tired.
    (e) Either Janet will win or, if she loses, she will be tired.

4. Construct truth tables for the following statements, where A, B, and C are statements. Note any tautologies or contradictions.
  ★ (a) $(A \Rightarrow B) \Leftrightarrow A' \vee B$
  ★ (b) $(A \wedge B) \vee C \Rightarrow A \wedge (B \vee C)$
    (c) $A \wedge (A' \vee B')'$
    (d) $A \wedge B \Rightarrow A'$
    (e) $(A \Rightarrow B) \Rightarrow [(A \vee C) \Rightarrow (B \vee C)]$

★ 5. Consider the following Pascal program fragment.

```
FOR COUNT:= 1 TO 5 DO
   BEGIN
      READ(A);
      IF ((A < 5.0) AND (2*A < 10.7)) OR (SQRT(5.0*A) > 5.1)
      THEN
      WRITELN(A)
   END
```

The input values for A are 1.0, 5.1, 2.4, 7.2, 5.3. What are the output values?

★ 6. Suppose that A, B, and C represent conditions that will be true or false when a certain computer program is executed. Suppose further that you want the program to carry out a certain task only when A or B is true (but not both) and C is false. Using A, B, and C and the connectives AND, OR, and NOT, write a statement that will be true only under these conditions.

★ 7. A memory chip from a minicomputer has 2^4 bistable (ON–OFF) memory elements. What is the total number of ON–OFF configurations?

8. Verify that the statements in the list on page 7 are tautologies. (We have already verified 1a, 4b, and 5a.)

9. Use the procedure of Figure 1.8 to prove that $(B' \wedge (A \Rightarrow B)) \Rightarrow A'$ is a tautology.

Section 1.2   *PROOF TECHNIQUES*

THEOREMS

Mathematical results are often expressed as statements of the form "if P, then Q" or $P \Rightarrow Q$, where P and Q may themselves be compound statements. To assert that $P \Rightarrow Q$ is a "theorem" says that $P \Rightarrow Q$ is a tautology, that the implication is always true. According to the truth table for implication, $P \Rightarrow Q$ is true provided we do *not* have the case where P is true and Q false. To prove the theorem, then, we need to show that we cannot have P true and Q false, that is, that whenever P is true, Q is also true. If $P \Rightarrow Q$ is being asserted as a theorem, then P is its **hypothesis** and Q its **conclusion**.

Before we discuss theorem-proving further, let's digress for a few moments. Unfortunately, the reader of a textbook is looking at the static end result of a dynamic process and is not able to share the adventure of developing new ideas. The textbook may say "Prove the following theorem," and the reader knows that the theorem is true, and, furthermore, that it is probably stated in its most polished form. But the researcher did not suddenly acquire a vision of this perfectly worded theorem together with the absolute certainty that it is true and that all he or she must do is work to find a proof. Before a theorem and its proof reach the final, polished form, a combination of inductive and deductive reasoning comes into play. Suppose you're the researcher trying to formulate and then prove a theorem. Say, for example, that you have examined a lot of cases in which whenever P is true, Q is also true. (Thus, you may have looked at 7 or 8 numbers divisible by 6 and observed that all the numbers were divisible by 3 as well.) On the basis of these experiences, you may conjecture: If P, then Q. (If a number is divisible by 6, then it is also divisible by 3.) And the more cases you find where Q follows from P, the more confident you will be in your conjecture. This process illustrates **inductive reasoning**, drawing a conclusion based on experience.

No matter how reasonable the conjecture sounds, however, you will not be satisfied until you have applied **deductive reasoning** to it as well. In this process, you try to verify the truth or falsity of your idea. You produce a logically valid sequence of reasoning that proves the conjecture true (thus making it a theorem), or else you find an example that disproves the conjecture. Often it is very difficult to decide which of these two things you should try to do! Suppose you decide to try to disprove, or refute, the conjecture. You then search for an example in which *P* is true but *Q* is false. You look for a **counterexample** to your conjecture. A single counterexample to a conjecture is sufficient to disprove it. Thus, you could refute our example conjecture if you could find a single number divisible by 6 but not by 3. Since our example is true, such a number does not exist. Of course, hunting for a counterexample and being unsuccessful does *not* constitute a proof that the conjecture is true.

1.14 Example Consider the statement "Every number less than 10 is bigger than 5." Or expressed as an implication, "If a number is less than 10, then it is also bigger than 5." A counterexample to this implication is the number 4. Four is less than 10, but it is not bigger than 5. Of course there are other counter-examples, but one is sufficient to disprove the statement. □

1.15 Practice Provide counterexamples to the following statements.

(a) All animals living in the ocean are fish.
(b) Input to a computer is always given by means of punched paper tape.   (*Answer, page A-2*)

METHODS OF ATTACK

Suppose you decide to try to prove the conjecture. Although a single counter-example is sufficient to refute a conjecture, in general, *many* examples do not prove a conjecture, they only strengthen your inclination to look for a proof. The one exception to this situation occurs when you are making an assertion about a finite collection. In this case, the assertion can be proved true by showing that it is true for each member of the collection. For example, the statement "If a whole number between 1 and 20 is divisible by 6, then it is also divisible by 3" can be proved by simply showing it to be true for all the whole numbers between 1 and 20.

Direct Proof

In general, how can you prove that $P \Rightarrow Q$ is true? In the obvious approach, a **direct proof**, you would assume *P* to be true and then, based on this assump-tion and making use of other facts known to you, you would build a chain of logical reasoning leading to the conclusion that *Q* is true. Two difficulties arise here. Can you recognize the facts that will be helpful, and can you arrange a logical progression of statements that begins with "*P* is true" and ends with "*Q* is true"? Unfortunately, there is no formula for constructing proofs, no algorithm or computer program for general theorem proving. Experience is helpful, not only because you get better with practice, but also because a proof that works for one theorem may be modified to work for a new but similar theorem.

The facts available to use in a proof fall into two categories: purely logical facts and facts related to the particular subject matter with which the conjecture deals (graph theory, probability, number theory, etc.) As you continue to study a subject, your storehouse of usable facts increases. Part of the purpose of this textbook is to help you accumulate "facts" about theoretical computer science.

1.16A Example   Part A. A proof using logic without quantifiers.

In this example, we will prove a theorem making use only of logical facts.

Theorem: The bucket is not rusting, but if the water faucet leaks then the bucket is rusting; therefore, the water faucet is not leaking.

The component statements of the theorem are

A: The water faucet leaks.

B: The bucket is rusting.

The theorem may be symbolized as

$$[B' \wedge (A \Rightarrow B)] \Rightarrow A'$$

This statement is of the form $P \Rightarrow Q$ where P is $B' \wedge (A \Rightarrow B)$ and Q is A'. By constructing a truth table for $P \Rightarrow Q$ or using the procedure described in Figure 1.8, we can verify that it is a tautology. The only logical facts we need to use are the truth value definitions of the connectives. Notice that it is simply the *form* of P and Q that allows us to say that $P \Rightarrow Q$ is a theorem; we do not really know whether the water faucet is leaking or not!   □

1.16B Example   Part B. A proof using logic with quantifiers.

More subtle logical facts involving quantifiers are also available to us. Consider another theorem:

Every sloolihoth is joysome. Percy is a sloolihoth, therefore Percy is joysome.

If we break this down into simple statements, as we did the previous example, we get the components

A: Every sloolihoth is joysome.

B: Percy is a sloolihoth.

C: Percy is joysome.

The theorem is then symbolized as

$$(A \wedge B) \Rightarrow C$$

Clearly, this is not a tautology. Yet by our understanding of quantifiers, we are willing to accept as valid reasoning an argument of the form "if all members of a class of objects have some property, then a specific member of the class has that property." Once again, the *form* of the argument is important rather than the content since the individual phrases do not even make sense.   □

In theorems pertaining to a particular topic, the *content* of the argument becomes important. We make use not only of logical facts, but also of any other information we may have about the topic, such as definitions or previously proved theorems.

1.17   Example   We will give a direct proof of our example theorem, "If a number is divisible by 6, then it is also divisible by 3." The theorem asserts something about an arbitrary number, so we must be careful to be completely general; we will let x represent any number. Thus, we want to assume that the hypothesis, x is divisible by 6, is true and then be able to deduce that the conclusion, x is divisible by 3, is true. We have to make use of the definition of divisibility— a is divisible by b if a equals the product of an integer times b—and also other arithmetic properties.

Hypothesis: x is divisible by 6.

$$x = k \cdot 6 \qquad \text{for some integer } k \text{ (definition of divisibility)}$$

$$6 = 2 \cdot 3 \qquad \text{(known fact about numbers)}$$

$$x = k(2 \cdot 3) \qquad \text{(replacement of a quantity by its equal)}$$

$$x = (k \cdot 2)3 \qquad \text{(known fact about multiplication)}$$

$$k \cdot 2 \text{ is an integer} \qquad \text{(known fact about integers)}$$

Conclusion: x is divisible by 3 (definition of divisibility).   □

1.18   Practice   Give a direct proof of the theorem "If a number is divisible by 6, then twice that number is divisible by 4." Show each step in going from hypothesis to conclusion.   *(Answer, page A-2)*

Contraposition

If you have tried diligently and failed to produce a direct proof for your conjecture, but still feel the conjecture is probably true, there are some variants on the direct proof technique you might try. Practice 1.10(d) showed that $A \Rightarrow B$ and $B' \Rightarrow A'$ are equivalent. $B' \Rightarrow A'$ is the **contrapositive** of $A \Rightarrow B$. Thus, even when P and Q are compound statements, $P \Rightarrow Q$ and its contrapositive, $Q' \Rightarrow P'$, have the same truth values, and $P \Rightarrow Q$ is a tautology exactly when $Q' \Rightarrow P'$ is a tautology. Therefore, to prove the theorem $P \Rightarrow Q$, it would be sufficient to prove the contrapositive, $Q' \Rightarrow P'$ (**proof by contra-position**). You would try to find a direct proof beginning with Q' and ending with P'. We have already discussed direct proofs, so the only new idea here is figuring out what the contrapositive is.

1.19 Example   The contrapositive of the theorem "If a number is divisible by 6, then it is also divisible by 3" is "If a number is not divisible by 3, then it is not divisible by 6." The easiest proof of the theorem is the direct proof given in Example 1.17. A proof by contraposition is more subtle.

Hypothesis: x is not divisible by 3.

$x \neq k \cdot 3$ for *any* integer k (This is the negation of divisibility by 3.)

$x \neq (2k_1)3$ for *any* integer k_1 ($2k_1$ would be an integer k, ruled out above.)

$x \neq k_1(2 \cdot 3)$ for *any* integer k_1 (multiplication facts)

$x \neq k_1 \cdot 6$ for *any* integer k_1 (number fact)

Conclusion: x is not divisible by 6 (negation of divisibility by 6).   □

1.20 Practice   Write the contrapositive of each statement in Practice 1.5.   *(Answer, page A-2)*

Practice 1.10(a) showed that the statements $A \Rightarrow B$ and $B \Rightarrow A$ are not equivalent. $B \Rightarrow A$ is the **converse** of $A \Rightarrow B$. If an implication is true, its converse may be true or false. Therefore, you cannot prove $P \Rightarrow Q$ by looking at $Q \Rightarrow P$.

1.21 Example   The implication "If $a > 5$, then $a > 2$" is true, but its converse "If $a > 2$, then $a > 5$" is false.   □

1.22 Practice   Write the converse of each statement in Practice 1.5.   *(Answer, page A-3)*

Theorems are often stated in the form P if and only if Q, meaning P if Q and P only if Q, or $Q \Rightarrow P$ and $P \Rightarrow Q$. To prove such a theorem, you must prove both an implication and its converse. Remember that any "if and only if" theorem requires you to prove both directions.

Contradiction

In addition to contraposition and direct proof, another proof technique you might use is **proof by contradiction**. Again we will let 0 stand for any contradiction, that is, any statement whose truth value is always false. ($A \wedge A'$ would be such a statement.) Once more, suppose you are trying to prove $P \Rightarrow Q$. By constructing a truth table, we see that $P \Rightarrow Q$ is equivalent to

$P \wedge Q' \Rightarrow 0$, so to prove the theorem $P \Rightarrow Q$, it is sufficient to prove $P \wedge Q' \Rightarrow 0$. Therefore, in a proof by contradiction you assume both the hypothesis and the negation of the conclusion to be true and then try, based on these assumptions, to arrive at some contradiction.

1.23 Example   Let's use proof by contradiction on the statement "If a number added to itself gives itself, then the number is 0." Let x represent any number. The hypothesis is $x + x = x$ and the conclusion is $x = 0$. To do a proof by contradiction, assume $x + x = x$ and $x \neq 0$. Then $2x = x$ and $x \neq 0$. Because $x \neq 0$, we can divide both sides of the equation $2x = x$ by x and arrive at the contradiction $2 = 1$. Hence, $x + x = x \Rightarrow x = 0$.   □

Induction

There is one final proof technique applicable to certain types of situations. To illustrate how the technique works, imagine that you are climbing an infinitely high ladder. How do you know whether you will be able to reach an arbitrarily high rung? Suppose we make the following two assertions about your climbing abilities:

(1) You can reach the first rung.
(2) Once you get to a rung, you can always climb to the next one up. (Notice that this assertion is an implication.)

If both statement (1) and the implication in (2) are true, then by (1) you can get to rung 1 and therefore by (2) you can get to rung 2; by (2) again, you can get to rung 3; by (2) again you can get to rung 4, and so on. You can climb as high as you wish. Both assertions here are necessary. If only (1) is true, you have no guarantee of getting beyond the first rung, and if only (2) is true, you may never be able to get started. Let's assume the rungs of the ladder are numbered by positive integers—1, 2, 3, 4, 5, ... Now think of a specific property a number might have. Instead of "reaching an arbitrarily high rung" we can talk about an arbitrary, positive integer having that property. We will use the shorthand notation $P(n)$ to mean that the positive integer n has the property P. How can we use the ladder-climbing technique to prove that for all positive integers n, we have $P(n)$? The two assertions we need to prove are

(i) $P(1)$ (1 has property P)
(ii) for any positive integer k, $P(k) \Rightarrow P(k + 1)$ (if any number has property P, so does the next number)

If we can prove both (i) and (ii), then $P(n)$ holds for any positive integer n, just as you could climb to an arbitrary rung on the ladder. To prove (i), we need only show the property holds for the number 1, usually a trivial task. To prove (ii), an implication, we assume $P(k)$ is true and show, based on this assumption, that $P(k + 1)$ is true. But assuming that the property holds for the number k is not assuming what we ultimately want to prove. It is merely the way to proceed to prove implication (ii) true. The technique we have described here, outlined in flowchart form in Figure 1.9, is **proof by mathematical induction**. It is a proof technique suggested whenever a theorem begins in the form "For all positive integers n, show that . . ."

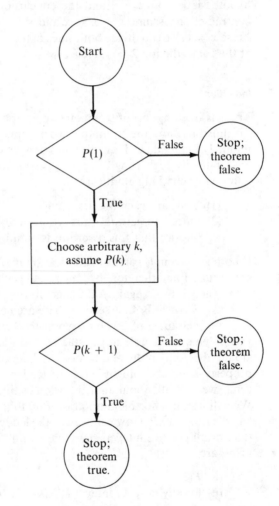

Figure 1.9

1.24 **Example**   Prove that for any positive integer n, $2^n > n$. A proof by mathematical induction is called for. $P(1)$ is the assertion $2^1 > 1$, which is surely true. Now we assume $P(k)$, $2^k > k$, and try to conclude $P(k + 1)$, $2^{k+1} > k + 1$.

Beginning with the left side of $P(k + 1)$, we note that $2^{k+1} = 2^k \cdot 2$. Using the inductive assumption $2^k > k$ and multiplying both sides of this inequality by 2, we get $2^k \cdot 2 > k \cdot 2$. We complete the argument

$$2^{k+1} = 2^k \cdot 2 > k \cdot 2 = k + k \geq k + 1$$

or

$$2^{k+1} > k + 1 \qquad\qquad\qquad \square$$

1.25 Example   Prove that the equation

$$1 + 3 + 5 + \cdots + (2n - 1) = n^2$$

is true for any positive integer n.

Again using mathematical induction, $P(1)$ is $1 = 1^2$. We assume $P(k)$,

$$1 + 3 + 5 + \cdots + (2k - 1) = k^2$$

and try to show $P(k + 1)$,

$$1 + 3 + 5 + \cdots + [2(k + 1) - 1] = (k + 1)^2$$

The left side of $P(k + 1)$ can be rewritten to show the next to last term:

$$1 + 3 + 5 + \cdots + (2k - 1) + [2(k + 1) - 1]$$

This expression contains the left side of $P(k)$, and we can substitute from $P(k)$. Thus,

$$\begin{aligned}
&1 + 3 + 5 + \cdots + [2(k + 1) - 1]\\
&= 1 + 3 + 5 + \cdots + (2k - 1) + [2(k + 1) - 1]\\
&= k^2 + [2(k + 1) - 1]\\
&= k^2 + [2k + 2 - 1]\\
&= k^2 + 2k + 1\\
&= (k + 1)^2 \qquad\qquad\qquad \square
\end{aligned}$$

1.26 Practice   In the Pascal program fragment

        ISUM := 0;
        FOR I := 1 TO N DO ISUM := ISUM + I;

the variable ISUM is calculated to be $1 + 2 + 3 + \cdots + n$. Prove by mathematical induction that for any positive integer n

$$1 + 2 + 3 + \cdots + n = \frac{n(n + 1)}{2}.$$

Thus, the above program fragment could be replaced by

        ISUM := N*(N + 1) DIV 2;   (*Answer, page A-3*)

It may be appropriate for the first step of the induction process to begin at 0 or at 2 or 3, instead of at 1. The same principle applies, no matter where you first hop on the ladder. Also, an induction proof may be called for when it is not quite as obvious as in the above examples.

1.27   Example   A programming language might be designed with the following convention regarding multiplication: a single factor requires no parentheses, but the product "a times b" must be written as $(a)b$. We want to show that any product of factors can be written with an even number of parentheses. The proof is by induction on the number of factors. For a single factor, there are 0 parentheses, an even number. Assume that for any product of k factors there are an even number of parentheses. Now consider a product P of $k + 1$ factors. P can be written in the form r times s where r has k factors and s is a single factor. By the inductive hypothesis, r has an even number of parentheses. Then we write r times s as $(r)s$, adding 2 more parentheses to the even number of parentheses in r, and giving P an even number of parentheses.   □

✔ CHECKLIST

Definitions

hypothesis (*p. 12*)
conclusion (*p. 12*)
inductive reasoning (*p. 12*)
deductive reasoning (*p. 12*)
counterexample (*p. 13*)
contrapositive (*p. 15*)

converse (*p. 16*)
direct proof (*p. 13*)
proof by contraposition (*p. 16*)
proof by contradiction (*p. 16*)
proof by mathematical
   induction (*p. 18*)

Techniques

Look for a counterexample.

Attempt direct proofs, proofs by contraposition, contradiction, and induction.

Main Ideas

Inductive reasoning to formulate a conjecture based upon experience.

Deductive reasoning either to refute a conjecture by finding a counterexample or to prove the conjecture.

In proving a conjecture, logical facts and facts about the particular subject involved can be used.

If we cannot directly prove a conjecture, we can try proof by contra-position or contradiction.

Mathematical induction to prove properties of positive integers.

EXERCISES SECTION 1.2

★ 1. Write the converse and the contrapositive of each statement in Exercise 1, Section 1.1.

2. Provide counterexamples to the following statements:
   (a) Every geometric figure with four right angles is a square.
   (b) If a real number is not positive, then it must be negative.
   (c) All people with red hair have green eyes or are tall.
   (d) All people with red hair have green eyes and are tall.

★ 3. Give a direct proof that the product of odd integers is odd. (Hint: how can you write a typical odd integer?)

4. Prove by contradiction that the product of odd integers is odd.

5. Prove by contraposition that if a number x is positive, so is $x + 1$.

★ 6. Let x and y be positive numbers, and prove that $x < y$ if and only if $x^2 < y^2$.

7. Prove that for any positive integer n,

$$2 + 6 + 10 + \cdots + (4n - 2) = 2n^2$$

★ 8. In the Pascal program fragment

```
ISUM := 0; I := 0;
REPEAT
  I := I + 2;
  ISUM := ISUM + I
UNTIL I = 2*N;
```

the variable ISUM is calculated to be $2 + 4 + 6 + \cdots + 2n$. Prove that for any positive integer n,

$$2 + 4 + 6 + \cdots + 2n = n(n + 1).$$

Thus, the above program segment could be replaced by

```
ISUM := N*(N + 1);
```

★ 9. Prove that for any positive integer n, the number $2^{2n} - 1$ is divisible by 3.

10. Prove that $2^n < n!$ for any positive integer $n \geq 4$.

11. A string of 0's and 1's is to be processed and converted to an even-parity string by adding a parity bit to the end of the string. The parity bit is

initially 0. When a 0 character is processed, the parity bit remains unchanged. When a 1 character is processed, the parity bit is switched from 0 to 1 or from 1 to 0. Prove that the number of 1's in the final string, that is, including the parity bit, is always even. (Hint: consider various cases.)

12. What is wrong with the following "proof" by mathematical induction? We will prove that for any positive integer n, n is equal to 1 more than n. Assume that $P(k)$ is true,

$$k = k + 1$$

Adding 1 to both sides of this equation, we get

$$k + 1 = k + 2$$

thus,

$$P(k + 1) \text{ is true}$$

13. What is wrong with the following "proof" by mathematical induction? We will prove that all computers are built by the same manufacturer. In particular, we will prove that in any collection of n computers where n is a positive integer, all of the computers are built by the same manufacturer. We first prove $P(1)$, a trivial process because in any collection consisting of one computer, there is only one manufacturer. Now we assume $P(k)$; that is, in any collection of k computers, all the computers were built by the same manufacturer. To prove $P(k + 1)$, we consider any collection of $k + 1$ computers. Pull one of these $k + 1$ computers (call it HAL) out of the collection. By our assumption, the remaining k computers all have the same manufacturer. Let HAL change places with one of these k computers. In the new group of k computers, all have the same manufacturer. Thus, HAL's manufacturer is the same one that produced all the other computers, and all $k + 1$ computers have the same manufacturer.

14. The induction principle used in this section can be summarized as follows:

$$\left. \begin{array}{l} \text{(1) } P(1) \text{ true} \\ \text{(2) } P(k) \text{ true} \Rightarrow P(k + 1) \text{ true} \end{array} \right\} \Rightarrow P(n) \text{ true for all } n$$

We call this the principle of **weak induction**, as opposed to the principle of **strong induction**, which is:

$$\left. \begin{array}{l} \text{(i) } P(1) \text{ true} \\ \text{(ii) } P(r) \text{ true for all } r \le k \Rightarrow P(k + 1) \text{ true} \end{array} \right\} \Rightarrow P(n) \text{ true for all } n$$

Notice that in implication (ii) there may be many facts available to help us conclude that $P(k + 1)$ is true, but that in implication (2), the only available fact is that $P(k)$ is true. Implications (ii) and (2) are not directly equivalent. If (2) can be proved, so can (ii), but not conversely. However, the two induction principles themselves are equivalent; that is, the two major implications are equivalent. In other words, if we accept weak induction as a valid principle, then strong induction is valid, and conversely. The converse is easy to see. If we accept strong induction as valid reasoning, then weak induction is surely valid because we can say that we concluded $P(k + 1)$ from $P(r)$ for all $r \leq k$ even though we only used the single condition $P(k)$. However, the real difficulty arises when we accept weak induction as a valid principle and want to show that strong induction is valid. Why may we conclude that $P(n)$ is true for all n when we used so much more evidence just to conclude $P(k + 1)$? We attack this problem by pointing out that, based on the principle of weak induction, we can prove that every nonempty set of positive integers has a smallest member (Well-Ordering Principle). Now we assume that (i) and (ii) have been shown, and want to conclude $P(n)$ for all n. Complete the proof by letting T be the subset of the positive integers defined by

$$T = \{t \mid P(t) \text{ is not true}\},$$

and showing that $T = \varnothing$ (the empty set).

Section 1.3   SETS

As we have said, the purpose of this chapter is to define terminology. However, if we look up a word in a dictionary, its definition is given in terms of other words, which are defined using still other words, and so on. We have to have a starting point for our definitions; ours will be the idea of a **set**. We will not formally define what a set is. Nonetheless, we have an informal understanding of a set as a collection of objects characterized by some defining property, so that any particular object either does or does not have the property and, thus, either does or does not belong to the set.

NOTATION

We use capital letters to denote sets and the symbol $\in$ to denote membership in a set. Thus $a \in A$ means that object a is a member, or element, of set A, and $b \notin A$ means that object b is not an element of set A. Curly braces $\{\ \}$ are used when describing sets.

1.28   Example   If $A = \{$Monday, Wednesday, Friday$\}$, then Monday $\in A$ and Tuesday $\notin A$.
$\square$

Two sets are **equal** if they have the same elements. Order in which the elements are written does not matter, so, for example, $\{$Monday, Wednesday, Friday$\} = \{$Wednesday, Monday, Friday$\}$. Also, each element of a set is listed only once; it would be redundant to list it again.

In describing a particular set, we have to identify the property characterizing its elements. For a finite set, we do this simply by listing the elements, as in set A of Example 1.28. It is impossible to list all of the elements of an infinite set. For some infinite sets, however, we can indicate a pattern for listing elements indefinitely. Thus, we might talk about the set $\{2, 4, 6, 8, 10, \ldots\}$ when we mean the set of all positive even integers. Although this is a common practice, the danger exists that the pattern the reader sees is not the one the writer had in mind. It is better to actually give the defining property in so many words and write $\{2, 4, 6, 8, 10, \ldots\}$ as $\{x \mid x$ is a positive even integer$\}$, read as "the set of all x such that x is a positive even integer". In general, a set whose elements are characterized as having property P is described as $\{x \mid P(x)\}$.

1.29   Practice   List the elements of each of the following sets.

(a) $\{x \mid x$ is an integer and $3 < x \le 7\}$.

(b) $\{x \mid x$ is a month with exactly 30 days$\}$.

(c) $\{x \mid x = y^3$ for $y \in \{0, 1, 2\}\}$.   (*Answer, page A-3*)

1.30   Practice   Give the defining property of each set below.

(a) $\{1, 4, 9, 16\}$

(b) $\{$the butcher, the baker, the candlestick maker$\}$

(c) $\{2, 3, 5, 7, 11, 13, 17, \ldots\}$ (*Answer, page A-3*)

It is convenient to name certain standard sets so that we can refer to them easily. We will use

$\mathbb{N} = $ the set of all nonnegative integers

$\mathbb{Z} = $ the set of all integers

$\mathbb{Q} = $ the set of all rational numbers

$\mathbb{R} = $ the set of all real numbers

$\mathbb{C} = $ the set of all complex numbers

Sometimes we will also want to talk about the set with no elements (the **empty set**, or **null set**), denoted by $\varnothing$ or $\{\ \}$.

RELATIONSHIPS BETWEEN SETS

For $A = \{2, 3, 5, 12\}$ and $B = \{2, 3, 4, 5, 9, 12\}$, every member of A is also a member of B. When this happens, A is said to be a subset of B.

1.31 Practice Complete the definition: A is a **subset** of B if $x \in A \Rightarrow$ _____. *(Answer, page A-3)*

If A is a subset of B, we denote this by $A \subseteq B$. If $A \subseteq B$ but $A \neq B$ (there is at least one element of B that is not an element of A), then A is a **proper subset** of B, denoted by $A \subset B$.

1.32 Example Let

$$A = \{1, 7, 9, 15\}$$
$$B = \{7, 9\}$$
$$C = \{7, 9, 15, 20\}$$

Then the following statements (among others) are all true:

$$B \subseteq C$$
$$B \subseteq A$$
$$B \subset A$$
$$A \nsubseteq C$$
$$15 \in C$$
$$\{7, 9\} \subseteq B$$
$$\{7\} \subset A$$
$$\emptyset \subseteq C \quad (x \in \emptyset \Rightarrow x \in C \text{ is true because } x \in \emptyset \text{ is false.}) \qquad \square$$

1.33 Practice Let

$$A = \{x \mid x \in \mathbb{N} \text{ and } x \geq 5\}$$
$$B = \{10, 12, 16, 20\}$$
$$C = \{x \mid x = 2y \text{ for } y \in \mathbb{N}\}$$

Which of the following statements are true?

(a) $B \subseteq C$
(b) $B \subset A$
(c) $A \subseteq C$
(d) $26 \in C$
(e) $\{11, 12, 13\} \subseteq A$
(f) $\{11, 12, 13\} \subset C$

(g) $\{12\} \in B$
(h) $\{12\} \subseteq B$
(i) $\{x \mid x \in \mathbb{N} \text{ and } x < 20\} \nsubseteq B$
(j) $5 \subseteq A$
(k) $\{\emptyset\} \subseteq B$
(l) $\emptyset \notin A$ *(Answer, page A-3)*

We know that A and B are equal sets if they have the same elements. We can restate this equality in terms of subsets; $A = B$ if and only if $A \subseteq B$ and

$B \subseteq A$. Proving set inclusion in both directions is the usual way to establish equality of two sets.

1.34 Example   We will prove that $\left\{x \mid x \in \mathbb{N} \text{ and } 0 \leq \dfrac{x}{4} < 1\right\} = \{x \mid x \in \mathbb{N} \text{ and } x^2 < 15\}$.

Let $A = \left\{x \mid x \in \mathbb{N} \text{ and } 0 \leq \dfrac{x}{4} < 1\right\}$, and $B = \{x \mid x \in \mathbb{N} \text{ and } x^2 < 15\}$. To show that $A = B$, we show $A \subseteq B$ and $B \subseteq A$. For $A \subseteq B$, we must choose an arbitrary member of A, that is, anything satisfying the defining property of A, and show that it also satisfies the defining property of B. Any member of A is a nonnegative integer x satisfying the inequality $0 \leq \dfrac{x}{4} < 1$ or $0 \leq x < 4$. Thus, the members of A are 0, 1, 2, and 3. The square of each of these numbers is less than 15. Hence, each member of A is a member of B, and $A \subseteq B$. Conversely, the members of B are those nonnegative integers with a square of less than 15. These numbers are 0, 1, 2, and 3, each of which satisfies the inequality $0 \leq \dfrac{x}{4} < 1$, so $B \subseteq A$.   □

1.35 Practice   Let $A = \{x \mid \cos(x/2) = 0\}$ and $B = \{x \mid \sin x = 0\}$. Prove that $A \subseteq B$.
                                                                                    (Answer, page A-3)

SETS OF SETS

For a set S, we can form a new set whose elements are all of the subsets of S. This new set is called the **power set** of S, $\mathscr{P}(S)$. $\mathscr{P}(S)$ will always have at least $\varnothing$ and S itself as members.

1.36 Practice   For $A = \{1, 2, 3\}$ what is $\mathscr{P}(A)$?   *(Answer, page A-3)*

In Practice 1.36, A has 3 elements and $\mathscr{P}(A)$ has 8 elements. Try finding $\mathscr{P}(S)$ for other sets S until you can guess the answer to the following Practice problem.

1.37 Practice   If S has n elements, then $\mathscr{P}(S)$ has _____ elements. (Does your answer work for $n = 0$ too?)   *(Answer, page A-3)*

There are several ways we can show that for a set S with n elements, $\mathscr{P}(S)$ will have 2^n elements. A formal proof would use induction. For the basis step of the induction, we let $n = 0$. The only set with 0 elements is $\varnothing$. The only subset of $\varnothing$ is $\varnothing$, so $\mathscr{P}(\varnothing) = \{\varnothing\}$, a set with $1 = 2^0$ elements. We assume that for any set with k elements, the power set has 2^k elements.

Now let S have $k + 1$ elements, and put one of these elements, call it x, aside. The remaining set has k elements, so by our inductive assumption, its power set has 2^k elements. Each of these elements is also a member of $\mathscr{P}(S)$. The only members of $\mathscr{P}(S)$ not counted by this procedure are those including element x. All the subsets including x can be found by taking all those subsets not including x (of which there are 2^k) and throwing in the x; thus, there will be 2^k subsets including x. Altogether, there are 2^k subsets without x and 2^k subsets with x, or $2^k + 2^k = 2 \cdot 2^k = 2^{k+1}$ subsets. Therefore, $\mathscr{P}(S)$ has 2^{k+1} elements.

Analogy with the truth tables of Section 1.1 is another way to show that $\mathscr{P}(S)$ has 2^n elements for a set S with n elements. There, we had n statement letters and showed that there were 2^n true–false combinations among these letters. But we can also think of each true–false combination as representing a particular subset, with T indicating membership and F indicating nonmembership in that subset. (For example, the row of the truth table with all statement letters F corresponds to the empty set.) Thus, the number of true–false combinations among n statement letters equals the number of subsets of a set with n elements; both are 2^n.

BINARY AND UNARY OPERATIONS

By itself a set is not very interesting until we *do* something with its elements. In the set $\mathbb{Z}$, for example, we can perform several arithmetic operations on the integers. We might subtract two integers, or we might take the negative of an integer. Subtraction is a binary operation on $\mathbb{Z}$; taking the negative of a number is a unary operation on $\mathbb{Z}$. A binary operation acts on two numbers, and a unary operation acts on one number. To see exactly what is involved in a binary operation, let's analyze subtraction more closely. For any two integers x and y, $x - y$ will produce an answer, and only one answer, and that answer will always be an integer. And finally, subtraction is performed on an **ordered pair** of numbers. For example, $7 - 5$ does not produce the same result as $5 - 7$. An ordered pair is denoted by (x, y), where x is the first component of the ordered pair and y is the second component. Order is important in an ordered pair; thus, the sets $\{1, 2\}$ and $\{2, 1\}$ are equal, but the ordered pairs $(1, 2)$ and $(2, 1)$ are not equal. Two ordered pairs (x, y) and (u, v) are equal only when $x = u$ and $y = v$.

1.38 Practice   Given that $(2x - y, x + y) = (7, -1)$, solve for x and y.   (*Answer, page A-3*)

1.39 Practice   Let $S = \{3, 4\}$. List all the ordered pairs (x, y) of elements of S.
(*Answer, page A-3*)

We will generalize these properties of subtraction on the integers to define a binary operation $\circ$ on a set S.

1.40   Definition   ∘ is a **binary operation** on a set S if for every ordered pair (x, y) of elements of S, $x \circ y$ exists, is unique, and is a member of S.   □

The two properties that $x \circ y$ always exists and is unique are described by saying that the binary operation ∘ is **well defined**. The property that $x \circ y$ always belongs to S is described by saying that S is **closed** under the operation ∘.

1.41   Example   Addition, subtraction, and multiplication are all binary operations on $\mathbb{Z}$. For example, when we perform addition on the ordered pair of integers (x, y), $x + y$ exists and is a unique integer.   □

A "candidate" ∘ can fail to be a binary operation on a set S in any of three ways: (1) there may be certain elements $x, y \in S$ for which $x \circ y$ does not exist; (2) there may be certain elements $x, y \in S$ for which $x \circ y$ gives more than one result; or (3) there may be elements $x, y \in S$ for which $x \circ y$ does not belong to S.

1.42   Example   Division is not a binary operation on $\mathbb{Z}$ because $x \div 0$ does not exist.   □

1.43   Example   Subtraction is not a binary operation on $\mathbb{N}$ because $\mathbb{N}$ is not closed under subtraction. $(1 - 10 \notin \mathbb{N}.)$ □

1.44   Example   Define $x \circ y$ on $\mathbb{N}$ by:

$$x \circ y = \begin{cases} 1 & \text{if } x \geq 5 \\ 0 & \text{if } x \leq 5 \end{cases}$$

Then, by the first part of the definition for ∘, $5 \circ 1 = 1$, but by its second part, $5 \circ 1 = 0$. Thus, ∘ is not well defined on $\mathbb{N}$.

For # to be a **unary operation** on a set S means that for any $x \in S$, $x^{\#}$ is well defined, and S is closed under #; in other words, for any $x \in S$, $x^{\#}$ exists, is unique, and is a member of S. We do not have a unary operation if any of these conditions is not met.

1.45   Example   Let $x^{\#}$ be defined by $x^{\#} = -x$, so that $x^{\#}$ is the negative of x. Then # is a unary operation on $\mathbb{Z}$, but not on $\mathbb{N}$ because $\mathbb{N}$ is not closed under #.   □

From these examples it is clear that whether ∘ (or #) is a binary (or unary) operation depends not only on its definition, but also on the set involved.

1.46   Practice   Which of the following are neither binary nor unary operations on the given sets? Why not?

(a) $x \circ y = x \div y$; S = set of all positive integers
(b) $x \circ y = x \div y$; S = set of all positive rational numbers
(c) $x \circ y = x^y$; $S = \mathbb{R}$
(d) $x \circ y$ = maximum of x and y; $S = \mathbb{N}$
(e) $x^{\#} = \sqrt{x}$; S = set of all positive real numbers
(f) $x^{\#}$ = solution to equation $(x^{\#})^2 = x$; $S = \mathbb{C}$ (*Answer, page A-3*)

1.47 Example   Let S be the set of statements as discussed in Section 1.1. Then the connectives $\wedge$, $\vee$, $\Rightarrow$, and $\Leftrightarrow$ are binary operations on S. The connective $'$ is a unary operation on S.   □

So far, all of our binary operations have been defined by means of a description or an equation. Suppose S is a finite set, $S = \{x_1, x_2, \ldots, x_n\}$. Then a binary operation $\circ$ on S can be defined by an array, or table, where element i, j (*i*th row and *j*th column) denotes $x_i \circ x_j$.

1.48 Example   Let $S = \{2, 5, 9\}$, and let $\circ$ be defined by the array

$\circ$	2	5	9
2	2	2	9
5	5	9	2
9	5	5	9

Thus, $2 \circ 5 = 2$ and $9 \circ 2 = 5$. $\circ$ is a binary operation on S.   □

Operations on Sets

We have generally seen operations where the elements operated upon are numbers, but we can also operate on sets. Given an arbitrary set S, we can define some binary and unary operations on the set $\mathscr{P}(S)$. A binary operation on $\mathscr{P}(S)$ must act on any two subsets of S to produce a unique subset of S. There are at least two natural ways this could happen.

1.49 Example   Let S be the set of all students at Scoobydoo U. Then the members of $\mathscr{P}(S)$ are sets of students. Let A be the set of football players at Scoobydoo U, and let B be the set of basketball players. Both A and B belong to $\mathscr{P}(S)$. One new set of students consists of anybody who plays either sport (or both sports) and is called the union of A and B. Another new set consists of anybody who plays both sports. This set (which might be empty) is called the intersection of A and B.   □

1.50 Definition   Let $A, B \in \mathscr{P}(S)$. The **union** of A and B, denoted by $A \cup B$, is $\{x \mid x \in A$ or $x \in B\}$.   □

1.51 Practice   Let $A, B \in \mathscr{P}(S)$. Complete the following definition: The **intersection** of A and B, denoted by $A \cap B$, is $\{x \mid \underline{\hspace{2cm}}\}$.   (*Answer, page A-3*)

1.52 Example   Let $A = \{1, 3, 5, 7, 9\}$ and $B = \{3, 5, 6, 10, 11\}$. Here we may consider A and B as members of $\mathscr{P}(\mathbb{N})$. Then $A \cup B = \{1, 3, 5, 6, 7, 9, 10, 11\}$ and $A \cap B = \{3, 5\}$. Both $A \cup B$ and $A \cap B$ are members of $\mathscr{P}(\mathbb{N})$.   □

We can use *Venn diagrams* to visualize the binary operations of union and intersection. The shaded areas in Figures 1.10 and 1.11 indicate the set

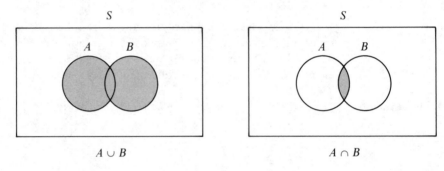

$A \cup B$ $A \cap B$

Figure 1.10   **Figure 1.11**

that results from performing the binary operation on the two given sets. We will define one unary operation on $\mathscr{P}(S)$.

1.53 Definition   For a set $A \in \mathscr{P}(S)$, the **complement** of A, A', is $\{x \mid x \in S \text{ and } x \notin A\}$.   □

1.54 Practice   Illustrate A' in a Venn diagram.   (*Answer, page A-4*)

A final binary operation on sets A and B is **set difference**: $A - B = \{x \mid x \in A \text{ and } x \notin B\}$. This operation can be rewritten as $A - B = \{x \mid x \in A \text{ and } x \in B'\}$ and, finally, as $A - B = A \cap B'$.

1.55 Practice   Illustrate $A - B$ in a Venn diagram.   (*Answer, page A-4*)

Two sets A and B such that $A \cap B = \varnothing$ are said to be **disjoint**. Thus, $A - B$ and $B - A$, for example, are disjoint sets.

1.56 Example   Let

$$A = \{x \mid x \text{ is an even nonnegative integer}\}$$

$$B = \{x \mid x = 2y + 1, y \in \mathbb{N}\}$$

$$C = \{x \mid x = 4y, y \in \mathbb{N}\}$$

be subsets of $\mathbb{N}$. Then A and B are disjoint sets, $A \cup B = \mathbb{N}$, $A \cup C = A$, $A' = B$, $A - C = \{x \mid x = 4y + 2, y \in \mathbb{N}\}$, and $B \subset C'$.   □

1.57 Practice   Let

$$A = \{1, 2, 3, 5, 10\}$$
$$B = \{2, 4, 7, 8, 9\}$$
$$C = \{5, 8, 10\}$$

be subsets of $S = \{1, 2, 3, 4, 5, 6, 7, 8, 9, 10\}$. Find

(a) $A \cup B$
(b) $A - C$
(c) $B' \cap (A \cup C)$ (*Answer, page A-4*)

SET IDENTITIES

There are many set equalities involving the operations of union, intersection, difference, and complementation that are true for all subsets of a given set S. Because they are independent of the particular subsets used, these equalities are called set identities. Some basic set identities follow.

Basic Set Identities

1a. $A \cup B = B \cup A$ 1b. $A \cap B = B \cap A$
2a. $(A \cup B) \cup C = A \cup (B \cup C)$ 2b. $(A \cap B) \cap C = A \cap (B \cap C)$
3a. $A \cup (B \cap C)$ 3b. $A \cap (B \cup C)$
  $= (A \cup B) \cap (A \cup C)$ $= (A \cap B) \cup (A \cap C)$
4a. $A \cup \emptyset = A$ 4b. $A \cap S = A$
5a. $A \cup A' = S$ 5b. $A \cap A' = \emptyset$

(Note that 2a allows us to write $A \cup B \cup C$ with no need for parentheses; 2b allows us to write $A \cap B \cap C$.)

1.58 Example   Let's prove identity 3a. We might draw a Venn diagram for each side of the equation and see that they look the same. However, the identity 3a is supposed to hold for all subsets A, B, and C, and whatever picture we draw cannot be completely general. Thus, if we draw A and B disjoint, that's a special case, but if we draw A and B not disjoint, that doesn't take care of the case where A and B are disjoint. To bypass this difficulty, let's prove set equality by proving set inclusion in each direction. Thus, we want to prove $A \cup (B \cap C) \subseteq (A \cup B) \cap (A \cup C)$ and also $(A \cup B) \cap (A \cup C) \subseteq A \cup (B \cap C)$. To show that $A \cup (B \cap C) \subseteq (A \cup B) \cap (A \cup C)$, we let x be an arbitrary member of $A \cup (B \cap C)$. Then we can proceed as follows:

$$x \in A \cup (B \cap C) \Rightarrow x \in A \text{ or } x \in (B \cap C)$$
$$\Rightarrow x \in A \text{ or } (x \in B \text{ and } x \in C)$$
$$\Rightarrow x \in A \text{ or } x \in B \text{ and } x \in A \text{ or } x \in C$$
$$\Rightarrow x \in A \cup B \text{ and } x \in A \cup C$$
$$\Rightarrow x \in (A \cup B) \cap (A \cup C).$$

To show that $(A \cup B) \cap (A \cup C) \subseteq A \cup (B \cap C)$, we reverse the above argument. □

1.59 Practice   Prove identity 4a.   *(Answer, page A-4)*

These identities may have reminded you of the list of tautologies in Section 1.1, page 7. (If not, check back and compare.) More on this similarity in Chapter 3.

Once we have proved the set identities in this list, we can use them to prove other set identities, much as we use algebraic identities like $(x - y)^2 = x^2 - 2xy + y^2$ to rewrite algebraic expressions.

1.60 Example   We can use the basic set identities to prove

$$(A \cup (B \cap C)) \cap [(A' \cup (B \cap C)) \cap (B \cap C)'] = \varnothing$$

for A, B, and C any subsets of S. In the following proof, the number to the right is that of the identity used to validate each step.

$$(A \cup (B \cap C)) \cap [(A' \cup (B \cap C)) \cap (B \cap C)']$$

$= [(A \cup (B \cap C)) \cap (A' \cup (B \cap C))] \cap (B \cap C)'$	2b
$= [((B \cap C) \cup A) \cap ((B \cap C) \cup A')] \cap (B \cap C)'$	1a (twice)
$= [(B \cap C) \cup (A \cap A')] \cap (B \cap C)'$	3a
$= [(B \cap C) \cup \varnothing] \cap (B \cap C)'$	5b
$= (B \cap C) \cap (B \cap C)'$	4a
$= \varnothing$	5b □

The **dual** for each set identity in our list also appears in the list. The dual is obtained by interchanging $\cup$ and $\cap$, and interchanging S and $\varnothing$. Example 1.60's dual is $(A \cap (B \cup C)) \cup [(A' \cap (B \cup C)) \cup (B \cup C)'] = S$, which we could prove true by replacing each basic set identity used in the proof of Example 1.60 with its dual. Because this method always works, any time we have proved a set identity by using the basic identities, we have also proved its dual.

1.61 Practice   (a) Using the basic set identities, establish the set identity

$$(C \cap (A \cup B)) \cup ((A \cup B) \cap C') = A \cup B.$$

(*A*, *B*, and *C* are any subsets of *S*.)

(b) State the identity you now know as a "bonus."   (*Answer*, *page A-4*)

There are two ways to prove a set identity: (1) establish set inclusion in each direction; or (2) verify the identity (or its dual) by using already established identities.

CARTESIAN PRODUCT

There is one final operation we will define using elements of $\mathscr{P}(S)$.

1.62 Definition   Let *A* and *B* be subsets of *S*. The **Cartesian product (cross product)** of *A* and *B*, denoted by $A \times B$, is defined by: $A \times B = \{(x, y) \mid x \in A \text{ and } y \in B\}$.   □

Thus, the Cartesian product of two sets, *A* and *B*, is the set of all ordered pairs whose first component comes from *A* and whose second comes from *B*. Cross product is not a binary operation on $\mathscr{P}(S)$. Although it acts on an ordered pair of members of $\mathscr{P}(S)$ and gives a unique result, the resulting set is not, in general, a subset of *S*; that is, it is not a member of $\mathscr{P}(S)$. The closure property for a binary operation fails to hold.

We will often be interested in the cross product of a set with itself, and we will abbreviate $A \times A$ by A^2; in general, we use A^n to mean the set of all ordered *n*-tuples $(x_1, \ldots, x_n)$ of elements of *A*.

1.63 Practice   Let $A = \{1, 2\}$ and $B = \{3, 4\}$.

(a) Find $A \times B$.
(b) Find $B \times A$.
(c) Find A^2.
(d) Find A^3.   (*Answer*, *page A-4*)

✓ CHECKLIST

Definitions

equal sets (*p. 24*)	union of sets (*p. 29*)
empty set (*p. 24*)	intersection of sets (*p. 30*)
subset (*p. 25*)	disjoint sets (*p. 30*)
proper subset (*p. 25*)	complement of a set (*p. 30*)
power set (*p. 26*)	set difference (*p. 30*)
ordered pair (*p. 27*)	Cartesian product (cross product)
binary operation (*p. 28*)	of sets (*p. 33*)
unary operation (*p. 28*)	

Techniques

Describe sets by listing elements and by a defining property.

Prove that one set is a subset of another.

Find the power set of a set.

Check whether something is a binary or unary operation on a set.

Form new sets by taking union, intersection, complement, and cross product of sets.

Prove set identities by (1) showing set inclusion in each direction; or (2) using the basic set identities.

Main Ideas

Sets and how they can be related (equal, subset, etc.) and combined (union, intersection, etc.).

Notation for standard sets.

The power set of a set with n elements, which has 2^n elements.

Binary and unary operations on sets.

Basic set identities exist (in dual pairs) and can be used to prove other set identities; once an identity is proved in this manner, its dual is also true.

EXERCISES SECTION 1.3

★ 1. Given the description of a set A as $A = \{2, 4, 8, \ldots\}$, do you think $16 \in A$? (Check the answer given for this problem.)

★ 2. Let

$$R = \{1, 3, \pi, 4.1, 9, 10\}$$

$$S = \{\{1\}, 3, 9, 10\}$$

$$T = \{1, 3, \pi\}$$

$$U = \{\{1, 3, \pi\}, 1\}$$

Which of the following are true? For those that are not, why not?

(a) $S \subseteq R$ (g) $T \subset R$

(b) $1 \in R$ (h) $\{1\} \in S$

(c) $1 \in S$ (i) $\emptyset \subseteq S$

(d) $1 \subseteq U$ (j) $T \subseteq U$

(e) $\{1\} \subseteq T$ (k) $T \in U$

(f) $\{1\} \subseteq S$ (l) $T \notin R$

★ 3. Which of the following are true where A, B, and C represent arbitrary sets?

(a) $A \cap \varnothing = \varnothing$

(b) $\{\varnothing\} = \varnothing$

(c) $\{\varnothing\} = \{0\}$

(d) $\varnothing \in \{\varnothing\}$

(e) $\varnothing \subseteq A$

(f) $\varnothing \in A$

(g) $\{\varnothing\} = \{\{\varnothing\}\}$

(h) If $A \subset B$ and $B \subseteq C$, then $A \subset C$

(i) If $A \neq B$ and $B \neq C$, then $A \neq C$

(j) If $A \in B$ and $B \nsubseteq C$, then $A \notin C$

(k) $\varnothing \times A = \varnothing$

(l) $\varnothing \cap \{\varnothing\} = \varnothing$

4. Program QUAD finds and prints solutions to quadratic equations of the form $ax^2 + bx + c = 0$. Program EVEN lists all the even integers from $-2n$ to $2n$. Let Q denote the set of values output by QUAD and E denote the set of values output by EVEN.

(a) Show that for $a = 1$, $b = -2$, $c = -24$, and $n = 50$, $Q \subseteq E$.

(b) Show that for the same values of a, b, and c, but a value for n of 2, $Q \nsubseteq E$.

★ 5. (a) Find $\mathscr{P}(S)$ for $S = \{\varnothing, \{\varnothing\}, \{\varnothing, \{\varnothing\}\}\}$.

(b) Find $\mathscr{P}(\mathscr{P}(S))$ for $S = \{a, b\}$.

6. (a) Recall that ordered pairs must have the property that $(x, y) = (u, v)$ if and only if $x = u$ and $y = v$. Prove that $\{\{x\}, \{x, y\}\} = \{\{u\}, \{u, v\}\}$ if and only if $x = u$ and $y = v$. Therefore, although we know that $(x, y) \neq \{x, y\}$, we can define the ordered pair (x, y) as the set $\{\{x\}, \{x, y\}\}$.

(b) Show by an example that we cannot define the ordered triple (x, y, z) as the set $\{\{x\}, \{x, y\}, \{x, y, z\}\}$.

★ 7. Which of the following are binary or unary operations on the given sets? For those that are not, why not?

(a) $x \circ y = x + 1$; $S = \mathbb{N}$

(b) $x \circ y = x + y - 1$; $S = \mathbb{N}$

(c) $x \circ y = \begin{cases} x - 1 & \text{if } x \text{ is odd} \\ x & \text{if } x \text{ is even} \end{cases}$ $S = \mathbb{Z}$

(d) $x^{\#} = \ln x$; $S = \mathbb{R}$

(e) $x^{\#} = x^2$; $S = \mathbb{Z}$

(f)

$\circ$	1	2	3
1	1	2	3
2	2	3	4
3	3	4	5

$S = \{1, 2, 3\}$

(g) $x \circ y =$ the taller of x and y; $S =$ set of all people living in Arkansas

(h) $x \circ y =$ that person, x or y, whose name appears first in an alphabetic sort; $S =$ set of 10 people with different names

(i) $x \circ y = \begin{cases} 1/x & \text{if} \quad x \text{ is positive} \\ 1/(-x) & \text{if} \quad x \text{ is negative} \end{cases}$ $S = \mathbb{R}$

(j) $x \circ y = $ matrix addition; $S = $ set of all 2×3 matrices with integer entries

8. How many different binary operations can be defined on a set with n elements? (Hint: think about filling in a table.)

9. We have written binary operations in **infix** notation where the operation symbol appears between the two operands, as in $A + B$. Evaluation of a complicated arithmetic expression is more efficient when the operations are written in **postfix** notation where the operation symbol appears after the two operands, as in $AB+$. Many compilers change expressions in a computer program from infix to postfix form. One way to produce an equivalent postfix expression from an infix expression is to write the infix expression with a full set of parentheses, move each operator to replace its corresponding right parenthesis, and then eliminate all remaining parentheses. Thus,

$$A * B + C$$

becomes, when fully parenthesized,

$$((A * B) + C)$$

and the postfix notation is

$$AB * C+$$

Rewrite each of the following in postfix notation.
(a) $(A + B) * (C - D)$
(b) $A ** B - C * D$
(c) $A * C + B/(C + D * B)$

★ 10. Let

$$A = \{2, 4, 5, 6, 8\}$$

$$B = \{1, 4, 5, 9\}$$

$$C = \{x \mid x \in \mathbb{Z} \text{ and } 2 \leq x < 5\}$$

be subsets of $S = \{0, 1, 2, 3, 4, 5, 6, 7, 8, 9\}$. Find:
(a) $(A \cap B)'$ (d) $(B - A)' \cap (A - B)$
(b) $C - B$ (e) $(C' \cup B)'$
(c) $(C \cap B) \cup A'$ (f) $B \times C$

★ 11. Consider the following subsets of $\mathbb{Z}$:

$$A = \{x \mid x = 3y \text{ for } y \in \mathbb{Z} \text{ and } y \geq 4\}$$

$$B = \{x \mid x = 2y \text{ for } y \in \mathbb{Z}\}$$

$$C = \{x \mid x \in \mathbb{Z} \text{ and } |x| \leq 10\}$$

Describe each of the following sets in terms of A, B, C, and set operations:
(a) set of all odd integers
(b) $\{-10, -8, -6, -4, -2, 0, 2, 4, 6, 8, 10\}$
(c) $\{x \mid x = 6y \text{ for } y \in \mathbb{Z} \text{ and } y \geq 2\}$
(d) $\{-9, -7, -5, -3, -1, 1, 3, 5, 7, 9\}$
(e) $\{x \mid x = 2y + 1 \text{ for } y \in \mathbb{Z} \text{ and } y \geq 5\} \cup \{x \mid x = 2y - 1 \text{ for } y \in \mathbb{Z} \text{ and } y \leq -5\}$

★ 12. For each of the following statements, find conditions on sets A and B to make the statement true.
(a) $A \cup B = A$
(b) $A \cap B = A$
(c) $A \cup \emptyset = \emptyset$
(d) $B - A = \emptyset$
(e) $A \cup B \subseteq A \cap B$

13. (a) If S is any finite set, we denote the number of elements in S by $|S|$. Show that for A and B finite sets, $|A \cup B| = |A| + |B| - |A \cap B|$.
(b) Show that for finite sets A, B, and C,

$$|A \cup B \cup C| = |A| + |B| + |C| - |A \cap B| - |A \cap C| - |B \cap C| + |A \cap B \cap C|$$

(Hint: Write $A \cup B \cup C$ as $A \cup (B \cup C)$.)

(c) Thirty-five students in a programming class did the same programming assignment on the same machine. These students fell into the following categories: 19 wrote programs that ran in ≤ 2 minutes; 19 wrote programs that produced output of ≤ 15 pages; and 15 wrote programs with ≤ 5 variable names. Also, 9 students wrote programs that ran in ≤ 2 minutes and produced output of ≤ 15 pages; 7 students wrote programs that produced output of ≤ 15 pages and used ≤ 5 variable names; 6 students wrote programs that ran in ≤ 2 minutes and used ≤ 5 variable names. How many students wrote programs that ran in ≤ 2 minutes, produced ≤ 15 pages of output, and used ≤ 5 variable names?

14. Verify the basic set identities on page 31 by showing set inclusion in each direction. (We have already done 3a and 4a.)

15. A and B are subsets of S. Prove the following set identities by showing set inclusion in each direction.

(a) $(A \cup B)' = A' \cap B'$
(b) $(A \cap B)' = A' \cup B'$ $\Big\}$ DeMorgan's Laws

(c) $A \cup (B \cap A) = A$
(d) $(A \cap B')' \cup B = A' \cup B$
(e) $(A \cap B) \cup (A \cap B') = A$

16. (a) A, B, and C are subsets of S. Prove the following set identities using the basic set identities listed in this section.

⋆ (1) $(A \cup B) \cap (A \cup B') = A$
(2) $[((A \cap C) \cap B) \cup ((A \cap C) \cap B')] \cup (A \cap C)' = S$
(3) $(A \cup C) \cap [(A \cap B) \cup (C' \cap B)] = A \cap B$

(b) State the dual of each of the above identities.

17. The operations of set union and set intersection can be extended to apply to a family of sets. We may describe the family as the collection of all sets A_i, where i takes on any of the values of a fixed set I. Here, I is called the **index set** for the family. The union of the family, $\bigcup_{i \in I} A_i$, is defined by $\bigcup_{i \in I} A_i = \{x \mid x \text{ is a member of some } A_i\}$. The intersection of the family, $\bigcap_{i \in I} A_i$, is defined by $\bigcap_{i \in I} A_i = \{x \mid x \text{ is a member of each } A_i\}$.

(a) Let $I = \{1, 2, 3, 4, \ldots\}$ and for each $i \in I$, let A_i be the set of real numbers in the interval $(-1/i, 1/i)$. What is $\bigcup_{i \in I} A_i$? What is $\bigcap_{i \in I} A_i$?

(b) Let $I = \{1, 2, 3, 4, \ldots\}$ and for each $i \in I$, let A_i be the set of real numbers in the interval $[-1/i, 1/i]$. What is $\bigcup_{i \in I} A_i$? What is $\bigcap_{i \in I} A_i$?

Section 1.4   COMBINATORICS

Combinatorics is that branch of mathematics that deals with counting. Often we want to know how many members there are in some finite set. This trivial-sounding question can be difficult to answer. We have already answered two "how many" questions; how many rows are there in a truth table with n statement letters, and how many subsets are there for a set with n elements? (Actually, as we've noted, these can be thought of as the same question.) Counting questions are important whenever we have finite resources (how much storage does a particular computer program require? how many users can a given computer configuration support?) or whenever we are interested in efficiency (how many computations does a particular algorithm involve?). Counting is also the basis for probability and statistics.

FUNDAMENTAL COUNTING PRINCIPLE

Many counting problems can be solved by applying the Fundamental Counting Principle. But before stating it, let's reconsider the tree of Figure 1.5b. Remember we used it to count the number of rows in a truth table with n statement letters. There are two possible outcomes when choosing the truth value of the first statement letter, two outcomes for choosing the truth value of the second statement letter, and so on. The tree structure illustrates that the total number of possible outcomes is the *product* of the number of outcomes for each step. This type of reasoning is behind the Fundamental Counting Principle.

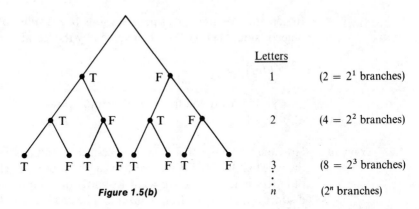

	Letters	
	1	($2 = 2^1$ branches)
	2	($4 = 2^2$ branches)
	3	($8 = 2^3$ branches)
	$\vdots$	
	n	(2^n branches)

Figure 1.5(b)

1.64 Fundamental Counting Principle If there are n_1 possible outcomes for a first event and then n_2 possible outcomes for a second event, there are $n_1 \cdot n_2$ possible outcomes for the sequence of the two events. □

1.65 Example The last part of your telephone number contains 4 digits. How many 4-digit numbers are there? The first digit can be any one of 10 possibilities, as can the second, third, and fourth, so there are $10 \cdot 10 \cdot 10 \cdot 10 = 10{,}000$ different numbers. But suppose the same digit can't be used twice; then how many 4-digit numbers are there? There are 10 choices for the first digit, only 9 choices for the second because we can't use what we used before, and so on. There are $10 \cdot 9 \cdot 8 \cdot 7 = 5{,}040$ different numbers. □

1.66 Practice If a woman has 7 blouses and 5 skirts, how many outfits can she put together? *(Answer, page A-4)*

PERMUTATIONS

Notice that in the 4-digit telephone number problem, 1259 would be distinguished from 2951. An ordered arrangement of objects is called a **permutation**. To count permutations, the factorial function is used. For a positive interger n, **n factorial** is defined as $n(n-1)(n-2)\cdots 1$ and denoted by $n!$; $0!$ is defined to have the value 1.

The problem of 4-digit numbers with no repetitions amounts to looking at a set of 10 objects (the digits) and counting the number of permutations, or arrangements, there are on 4 distinct objects chosen from the 10. The answer, $10 \cdot 9 \cdot 8 \cdot 7$, can be expressed in terms of factorials as follows:

$$10 \cdot 9 \cdot 8 \cdot 7 = \frac{10 \cdot 9 \cdot 8 \cdot 7 \cdot 6 \cdots 2 \cdot 1}{6 \cdots 2 \cdot 1} = \frac{10!}{6!} = \frac{10!}{(10 - 4)!}$$

In general, the number of permutations of r distinct objects chosen from n objects is denoted by $P(n, r)$ and given by the formula

$$P(n, r) = \frac{n!}{(n - r)!} \tag{1}$$

As a special case, the number of permutations of all n objects is $P(n, n) = n!/0! = n!$

1.67 Example   How many 3-letter words (not necessarily meaningful) can be formed from the word COMPILER if no letters can be repeated? Here the arrangement of letters matters, and we want to know the number of permutations of 3 distinct objects taken from 8 objects. The answer is $P(8, 3) = 8!/5! = 336$. □

1.68 Practice   In how many ways can a president and vice-president be selected from a group of 20 people?   *(Answer, page A-4)*

1.69 Practice   In how many ways can six people be seated in a row of six chairs?   *(Answer, page A-4)*

Counting problems are not always as straightforward as the ones we've seen. The next example uses formula (1) as part of the Fundamental Counting Principle.

1.70 Example   The library has 4 books on operating systems, 7 on programming, and 3 on data structures. Let's see how many ways these books can be arranged on a shelf, given that all books on the same subject must be together. If we think of the 3 subjects and how they can be arranged, there are 3! different orderings of subject matter. But for each of these, there are 4! arrangements of the operating systems books, 7! arrangements of the programming books, and 3! arrangements of the data structures books. Thus, by the Fundamental Counting Principle, the final number of arrangements of all the books is $(3!)(4!)(7!)(3!) = 4,354,560$. □

COMBINATIONS

Sometimes we want to select r objects from a set of n objects, but we don't care how they are arranged. Then we are counting the number of **combinations** of r distinct objects chosen from n objects, denoted by $C(n, r)$. For each such combination, there are $r!$ ways to permute the r objects. By the Fundamental

Counting Principle, the number of permutations of r distinct objects chosen from n objects is the product of the number of ways to choose the objects, $C(n, r)$, multiplied by the number of ways to arrange the objects chosen, $r!$. Thus,

$$C(n, r) \cdot r! = P(n, r)$$

or

$$C(n, r) = \frac{P(n, r)}{r!} = \frac{n!}{r!(n - r)!}$$

1.71 Example How many 5-card poker hands can be dealt from a 52-card deck? Here order does not matter; $C(52, 5) = 52!/(5!47!) = 2,598,960$. □

1.72 Practice In how many ways can a committee of 3 be chosen from a group of 12 people? (*Answer, page A-4*)

1.73 Example In how many ways can 3 freshmen and 5 sophomores be selected from among the 19 freshmen and 34 sophomores in a class? Because we have two events, selecting freshmen and selecting sophomores, we again apply the Fundamental Counting Principle. Each event is a combinations problem. There are $C(19, 3)$ ways to choose the freshmen, and $C(34, 5)$ ways to choose the sophomores. Hence, the final answer is

$$C(19, 3) \cdot C(34, 5) = \frac{19!}{3!16!} \cdot \frac{34!}{5!29!} = (1,938)(278,256)$$ □

✔ **CHECKLIST**

Definitions

combinatorics (*p. 38*)
Fundamental Counting Principle (*p. 39*)
factorial (*p. 39*)
permutation (*p. 39*)
combination (*p. 40*)

Techniques

Use the Fundamental Counting Principle.

Find the number of permutations of r objects chosen from n objects.

Find the number of combinations of r objects chosen from n objects.

Main Ideas

There are formulas to help us count the number of ways to arrange, or simply the number of ways to select, objects from a set.

EXERCISES SECTION 1.4

★ 1. Compute the value of the following expressions:
   (a) $P(7, 2)$
   (b) $P(8,5)$
   (c) $C(10, 7)$
   (d) $C(9, 2)$

2. What is the significance of $C(n, n)$; compute its value. What is the significance of $C(n, 1)$; compute its value.

★ 3. How many 3-digit numbers < 600 can be made using the digits 8, 6, 4, and 2?

4. How many 8-bit binary numbers are there?

★ 5. A telephone conference call is being placed from Central City to Booneville by way of Cloverdale. There are 45 trunk lines from Central City to Cloverdale, and 13 from Cloverdale to Booneville. How many different ways could the call be placed?

★ 6. A palindrome is a word that reads the same forwards and backwards. How many 5-letter, one-word, English-language palindromes are possible?

7. How many Social Security numbers are possible?

★ 8. An identifier in BASIC must be either a single letter or a letter followed by a single digit. How many identifiers are possible?

9. How many batting orders are possible for a 9-man baseball team?

10. How many permutations of the letters in COMPUTER are there? How many of these end in a vowel?

★ 11. In how many different ways can you pull 4 cards from a standard 52-card deck? In how many different ways can you pull 4 aces from such a deck?

12. In how many different ways can you seat 11 men and 8 women in a row if the men and women are each to sit together?

13. Quality control wants to test 25 microprocessor chips from the 300 manufactured each day. In how many ways can this be done?

★ 14. In how many ways can a jury of 5 men and 7 women be selected from a panel of 17 men and 23 women?

15. In how many ways can a card hand consisting of 3 spades and 2 hearts be selected from a standard 52-card deck?

16. A box contains 5 dimes and 7 quarters. Find the number of sets of 4 coins. Find the number of sets of 4 coins of which 2 are dimes and 2 are quarters. Find the number of sets of 4 coins in which all are dimes or all are quarters.

★ 17. A committee of 3 is chosen from a set of 5 Democrats, 3 Republicans, and 4 Independents. In how many ways can this be done? In how many ways can this be done if the committee must contain at least 1 Independent?

18. A hostess wishes to invite 6 dinner guests from a list of 14 friends. In how many ways can she choose her guests? In how many ways can she choose her guests if 2 of her friends dislike each other, and neither will come if the other is present? In how many ways can she choose her guests if two of her friends are very fond of each other and one won't come without the other?

FOR FURTHER READING

For further discussion of first order logic, see [1] or [2]. A different formalization of first order logic appears in [3]. Many excellent examples of symbolizing English sentences as statements in a first order logic are given in [4]. Three readable texts on set theory are [5], [6], and [7]. A good book on combinatorics, also containing some graph theory, is [8], and the basics of combinatorics are also covered in Anderson's little volume [9].

1. Elliott Mendelson, *Introduction to Mathematical Logic*, 2nd ed. Princeton, N.J.: Van Nostrand, 1979.

2. Angelo Margaris, *First Order Mathematical Logic*. Waltham, MA: Blaisdell, 1967.

3. Alfred B. Manaster, *Completeness, Compactness and Undecidability*. Englewood Cliffs, N.J.: Prentice-Hall, 1975.

4. Patrick Suppes, *Introduction to Logic*. Princeton, N.J.: Van Nostrand, 1957.

5. J. Donald Monk, *Introduction to Set Theory*. New York: McGraw-Hill, 1969.

6. Robert R. Stoll, *Set Theory and Logic*. San Francisco: W. H. Freeman and Company, 1963.

7. Martin M. Zuckerman, *Sets and Transfinite Numbers*. New York: Macmillan, 1974.

8. C. L. Liu, *Introduction to Combinatorial Mathematics*. New York: McGraw-Hill, 1968.

9. Ian Anderson, *A First Course in Combinatorial Mathematics*. Oxford: Clarendon Press, 1974.

Chapter 2

HOW TO SPEAK MATHEMATICS: BASIC VOCABULARY II

Chapter 2 contains still more vocabulary. If you feel your spirits flagging, reread the introduction to Chapter 1!

Section 2.1 **RELATIONS**

BINARY RELATIONS

If we learn that two people, Henrietta and Horace, are related, we understand that there is some family connection between them, that (Henrietta, Horace) stands out from other ordered pairs of people because there is a relationship (cousins, sister/brother, or whatever) that Henrietta and Horace satisfy. The mathematical analogue is to distinguish certain ordered pairs of objects from other ordered pairs because the components of the distinguished pairs satisfy some relationship that the components of the other pairs do not.

2.1 Example Let $S = \{1, 2\}$ and $T = \{2, 3\}$, then $S \times T = \{(1, 2), (1, 3), (2, 2), (2, 3)\}$. If we are interested in the relationship of equality, then $(2, 2)$ is the only distinguished element of $S \times T$, that is, the only ordered pair whose components are equal. If we are interested in the relationship of one number being less than another, we would choose $(1, 2)$, $(1, 3)$, and $(2, 3)$ as the distinguished ordered pairs of $S \times T$. ☐

In Example 2.1, we could pick out the distinguished ordered pairs (x, y) by saying that $x = y$ or that $x < y$. Similarly, the notation $x \rho y$ indicates the ordered pair (x, y) satisfies a relationship ρ. The relation ρ may be defined by some verbal description or simply by listing the distinguished pairs ρ selects.

2.2 Example Let $S = \{1, 2\}$ and $T = \{2, 3, 4\}$. On the set $S \times T = \{(1, 2), (1, 3), (1, 4), (2, 2), (2, 3), (2, 4)\}$, a relation ρ can be defined by $x \rho y \Leftrightarrow x = \frac{1}{2}y$ Thus, $(1, 2)$ and $(2, 4)$ satisfy ρ. Or the same ρ could be defined by saying that $\{(1, 2), (2, 4)\}$ is the set of ordered pairs satisfying ρ. ☐

We have been talking about binary relations, relations between two objects. In Example 2.2, one way to define the binary relation ρ is to specify a subset of $S \times T$. Formally, a binary relation actually is a subset of $S \times T$.

2.3 Definition   Given sets S and T, a **binary relation** ρ on $S \times T$ is a subset of $S \times T$.   □

Generally, however, there will also be some verbal description of the relation ρ's elements satisfy that makes ρ an interesting subset of $S \times T$.

2.4 Example   Let $S = \{1, 2\}$ and $T = \{2, 3, 4\}$. If ρ is defined on $S \times T$ by $\rho = \{(2, 3), (2, 4)\}$, then $2 \rho 3$ and $2 \rho 4$ but not, for instance, $1 \rho 4$. Here ρ seems to have no obvious verbal interpretation.   □

2.5 Example   Let $S = \{1, 2\}$ and $T = \{2, 3, 4\}$. Let ρ be given by $x \rho y \Leftrightarrow x + y$ is odd. Then $(1, 2) \in \rho$, $(1, 4) \in \rho$, and $(2, 3) \in \rho$.   □

2.6 Practice   For each of the following binary relations ρ on $N \times N$, decide which of the given ordered pairs belong to ρ.

(a) $x \rho y \Leftrightarrow x = y + 1$;   (2, 2), (2, 3), (3, 3), (3, 2)
(b) $x \rho y \Leftrightarrow x$ divides y;   (2, 4), (2, 5), (2, 6)
(c) $x \rho y \Leftrightarrow x$ is odd;   (2, 3), (3, 4), (4, 5), (5, 6)
(d) $x \rho y \Leftrightarrow x > y^2$;   (1, 2), (2, 1), (5, 2), (6, 4), (4, 3)

(*Answer, page A-5*)

We can define *n*-ary relations by generalizing the definition for binary relations.

2.7 Practice   Complete the following definition: Given sets $S_1, S_2, \ldots, S_n$, an ***n*-ary relation** (***n*-ary predicate**) on $S_1 \times S_2 \times \cdots \times S_n$ is _____.   (*Answer, page A-5*)

As a special case, a **unary relation (unary predicate)** ρ on a set S is just a particular subset of S. An element $x \in S$ satisfies or fails to satisfy ρ according to whether x does or does not belong to the subset.

Often we will be interested in a binary relation on a set S, meaning the relation is a subset of S^2. In general, an *n*-ary relation on S is a subset of S^n, a set of ordered *n*-tuples of members of S.

Operations on Relations

Suppose B is the set of all binary relations on a given set S. We can define two binary operations and one unary operation on B. Let ρ and σ belong to B. We define a new binary relation, $\rho + \sigma$, on S (that is, a new member of B) by

$$x(\rho + \sigma)y \Leftrightarrow x \,\rho\, y \quad \text{or} \quad x \,\sigma\, y$$

A second new binary relation, $\rho \cdot \sigma$, on S is defined by

$$x(\rho \cdot \sigma)y \Leftrightarrow x \,\rho\, y \quad \text{and} \quad x \,\sigma\, y$$

And finally a new binary relation, ρ', on S is given by

$$x \,\rho'\, y \Leftrightarrow \text{not } x \,\rho\, y$$

2.8 Practice   Since ρ and σ above are actually subsets of S^2, express the new relations $\rho + \sigma$, $\rho \cdot \sigma$, and ρ' in terms of these subsets.   *(Answer, page A-5)*

2.9 Practice   Let ρ and σ be binary relations on $\mathbb{N}$ defined by $x \,\rho\, y \Leftrightarrow x = y$, $x \,\sigma\, y \Leftrightarrow x < y$.

Give verbal descriptions for (a), (b), and (c).
(a) What is the relation $\rho + \sigma$?
(b) What is the relation ρ'?
(c) What is the relation σ'?
(d) What is the relation $\rho \cdot \sigma$? (Give a set description.)
    (Answer, page A-5)

The following facts about the operations $+$, $\cdot$, and $'$ on relations are immediate consequences of Practice 2.8 and the basic set identities found in Section 1.3.

1a. $\rho + \sigma = \sigma + \rho$	1b. $\rho \cdot \sigma = \sigma \cdot \rho$
2a. $(\rho + \sigma) + \gamma = \rho + (\sigma + \gamma)$	2b. $(\rho \cdot \sigma) \cdot \gamma = \rho \cdot (\sigma \cdot \gamma)$
3a. $\rho + (\sigma \cdot \gamma) = (\rho + \sigma) \cdot (\rho + \gamma)$	3b. $\rho \cdot (\sigma + \gamma) = (\rho \cdot \sigma) + (\rho \cdot \gamma)$
4a. $\rho + \varnothing = \rho$	4b. $\rho \cdot S^2 = \rho$
5a. $\rho + \rho' = S^2$	5b. $\rho \cdot \rho' = \varnothing$

PROPERTIES OF RELATIONS

A binary relation on a set S may have certain properties. For example, the relation ρ of equality on S, $(x, y) \in \rho \Leftrightarrow x = y$, has three properties: (1) for any $x \in S$, $x = x$, or $(x, x) \in \rho$; (2) for any $x, y \in S$, if $x = y$ then $y = x$, or $(x, y) \in \rho \Rightarrow (y, x) \in \rho$; and (3) for any $x, y, z \in S$, if $x = y$ and $y = z$, then

$x = z$, or $((x, y) \in \rho$ and $(y, z) \in \rho) \Rightarrow (x, z) \in \rho$. These three properties make the equality relation reflexive, symmetric, and transitive.

2.10 Definition Let ρ be a binary relation on a set S. Then,

ρ is **reflexive** means: $x \in S \Rightarrow (x, x) \in \rho$

ρ is **symmetric** means: $(x, y \in S \wedge (x, y) \in \rho) \Rightarrow (y, x) \in \rho$

ρ is **transitive** means: $(x, y, z \in S \wedge (x, y) \in \rho \wedge (y, z) \in \rho) \Rightarrow (x, z) \in \rho$ □

2.11 Example Consider the relation $\leq$ on the set $\mathbb{N}$. This relation is reflexive because for any nonnegative integer x, $x \leq x$. It is also a transitive relation because for any nonnegative integers x, y, and z, $x \leq y$ and $y \leq z$ implies $x \leq z$. However, $\leq$ is not symmetric; $3 \leq 4$ does not imply $4 \leq 3$. In fact, for any $x, y \in \mathbb{N}$, if both $x \leq y$ and $y \leq x$, then $y = x$. This characteristic is described by saying that $\leq$ is antisymmetric. □

2.12 Definition Let ρ be a binary relation on a set S. Then ρ is **antisymmetric** means

$(x, y \in S \wedge (x, y) \in \rho \wedge (y, x) \in \rho) \Rightarrow x = y$ □

2.13 Example Let $S = \mathscr{P}(\mathbb{N})$. Define a binary relation ρ on S by $A \rho B \Leftrightarrow A \subseteq B$. Then ρ is reflexive because every set is a subset of itself. ρ is transitive because if A is a subset of B and B is a subset of C, then A is a subset of C. And ρ is antisymmetric because if A is a subset of B and B is a subset of A, then A and B are equal sets. □

The properties of symmetry and antisymmetry for binary relations are not precisely opposites. The equality relation on a set S is both symmetric and antisymmetric. However, the equality relation on S, or a subset of it, is the only relation having both these properties. To illustrate, suppose ρ is a symmetric and antisymmetric relation on S, and let $(x, y) \in \rho$. By symmetry, it follows that $(y, x) \in \rho$. But by antisymmetry, $x = y$. Thus, only equal elements can be related.

2.14 Practice Test the following binary relations on the given sets S for reflexivity, symmetry, antisymmetry, and transitivity:

(a) $S = \mathbb{N}$; $x \rho y \Leftrightarrow x + y$ is even
(b) $S = \mathbb{N}$; $x \rho y \Leftrightarrow x$ divides y
(c) S = the set of all lines in the plane; $x \rho y \Leftrightarrow x$ is parallel to y or x coincides with y
(d) $S = \mathbb{N}$; $x \rho y \Leftrightarrow x = y^2$
(e) $S = \{0, 1\}$; $x \rho y \Leftrightarrow x = y^2$

(f) $S = \{x \mid x \text{ is a person living in Peoria}\}$; $x \rho y \Leftrightarrow x$ is older than y

(g) $S = \{x \mid x \text{ is a student in your class}\}$; $x \rho y \Leftrightarrow x$ sits in the same row as y

(h) $S = \{1, 2, 3\}$; $\rho = \{(1, 1), (2, 2), (3, 3), (1, 2), (2, 1)\}$

(Answer, page A-5)

For the rest of this section we will concentrate on two types of binary relations, characterized by which properties (reflexivity, symmetry, antisymmetry, transitivity) they satisfy.

PARTIAL ORDERINGS

2.15 Definition   A binary relation on a set S that is reflexive, antisymmetric, and transitive is called a **partial ordering** on S.   □

From the previous examples and Practice 2.14, we have the following instances of partial orderings:

on $\mathbb{N}$: $x \rho y \Leftrightarrow x \leq y$

on $\mathscr{P}(\mathbb{N})$: $A \rho B \Leftrightarrow A \subseteq B$

on $\mathbb{N}$: $x \rho y \Leftrightarrow x$ divides y

on $\{0, 1\}$: $x \rho y \Leftrightarrow x = y^2$

If ρ is a partial ordering on S, then the ordered pair (S, ρ) is called a **partially ordered set**. We will denote an arbitrary, partially ordered set by $(S, \preceq)$ (in any *particular* case, $\preceq$ has some definite meaning such as "less than or equal to," "is a subset of," "divides", and so on).

Let $(S, \preceq)$ be a partially ordered set. Then $\preceq$ is a set of ordered pairs of elements of S. Let $A \subseteq S$. If we select from $\preceq$ the ordered pairs of elements of A, this new set is called the *restriction* of $\preceq$ to A and is a partial ordering on A. (Do you see why?) For instance, once we know the relation "x divides y" is a partial ordering on $\mathbb{N}$, we automatically know that "x divides y" is a partial ordering on $\{1, 2, 3, 6, 12, 18\}$.

We want to introduce some terminology about partially ordered sets. Let $(S, \preceq)$ be a partially ordered set. If $x \preceq y$, then either $x = y$ or $x \neq y$. If $x \preceq y$ but $x \neq y$, we write $x \prec y$ and say that x is a **predecessor** of y, or y is a **successor** of x. A given y may have many predecessors, but if $x \prec y$ and there is no z with $x \prec z \prec y$, then x is an **immediate predecessor** of y.

2.16 Practice   Consider the relation "x divides y" on $\{1, 2, 3, 6, 12, 18\}$.

(a) Write the ordered pairs (x, y) of this relation.

(b) Write all the predecessors of 6.

(c) Write all the immediate predecessors of 6.   *(Answer, page A-5)*

We can graph the partially ordered set $(S, \preceq)$ if S is finite. Each of the elements of S is represented by a dot, called a **node**, or **vertex**, of the graph. If x is an immediate predecessor of y, then the node for y is placed above the node for x, and the two nodes are connected by a straight line segment.

2.17 Example   Consider $\mathscr{P}(\{1, 2\})$ under the relation of set inclusion. This is a partially ordered set. (We already know that $(\mathscr{P}(\mathbb{N}), \subseteq)$ is a partially ordered set.) The elements of $\mathscr{P}(\{1, 2\})$ are $\varnothing$, $\{1\}$, $\{2\}$, and $\{1, 2\}$. The binary relation $\subseteq$ consists of the following ordered pairs:

$$(\varnothing, \varnothing) \quad (\{1\}, \{1\}) \quad (\{2\}, \{2\}) \quad (\{1, 2\}, \{1, 2\}) \quad (\varnothing, \{1\}) \quad (\varnothing, \{2\})$$
$$(\varnothing, \{1, 2\}) \quad (\{1\}, \{1, 2\}) \quad (\{2\}, \{1, 2\})$$

The graph of this partially ordered set appears in Figure 2.1. Notice that although $\varnothing$ is not an immediate predecessor of $\{1, 2\}$, it is a predecessor of $\{1, 2\}$ (shown on the graph by the chain of upward line segments connecting $\varnothing$ with $\{1, 2\}$).

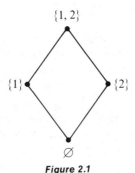

Figure 2.1

2.18 Practice   Draw the graph for the relation "x divides y" on $\{1, 2, 3, 6, 12, 18\}$.

(Answer, page A-5)

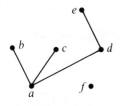

Figure 2.2

The graph of a partially ordered set conveys all the information about the partial ordering. We could reconstruct the set of ordered pairs making up the partial ordering just by looking at the graph. Thus, given the graph in Figure 2.2 of a partial ordering $\preceq$ on a set $\{a, b, c, d, e, f\}$, we could conclude that $\preceq$ is the set

$$\{(a, a), (b, b), (c, c), (d, d), (e, e), (f, f), (a, b), (a, c), (a, d), (a, e), (d, e)\}$$

Two elements of S may be unrelated in a partial ordering of S. In Example 2.17, $\{1\}$ and $\{2\}$ are unrelated; so are 2 and 3, and 12 and 18 in Practice 2.18. And in Figure 2.2, f is not related to any other element. A

Figure 2.3

partial ordering in which every element of the set is related to every other element is called a **total ordering (chain)**. The graph for a total ordering looks like Figure 2.3. The relation $\leq$ on $\mathbb{N}$ is a total ordering.

Again, let $(S, \preceq)$ be a partially ordered set. If there is a $y \in S$ with $y \preceq x$ for all $x \in S$, then y is a **least element** of the partially ordered set. A least element, if it exists, is unique. If y and z are both least elements, then $y \preceq z$ because y is least and $z \preceq y$ because z is least; by antisymmetry, $y = z$. An element $y \in S$ is **minimal** if there is no $x \in S$ with $x \prec y$. Similar definitions apply for greatest element and maximal elements.

2.19 Practice   Define **greatest element** and **maximal element** in a partially ordered set $(S, \preceq)$.

(*Answer, page A-5*)

2.20 Example   In the partially ordered set of Practice 2.18, 1 is both least and minimal. Twelve and 18 are both maximal, but there is no greatest element.   □

A least element is always minimal, and a greatest element is always maximal. In a totally ordered set, a minimal element is a least element, and a maximal element is a greatest element.

2.21 Practice   Draw the graph for a partially ordered set with four elements having two minimal elements but no least element, two maximal elements but no greatest element, and in which each element is related to two other elements.

(*Answer, page A-5*)

Partial orderings satisfy the properties of reflexivity, antisymmetry, and transitivity. Another type of binary relation, which we'll study next, satisfies a different set of properties.

EQUIVALENCE RELATIONS

2.22 Definition   A binary relation on a set S that is reflexive, symmetric, and transitive is called an **equivalence relation** on S.   □

We have already come upon the following examples of equivalence relations.

On any set S: $x \rho y \Leftrightarrow x = y$

On $\mathbb{N}$: $x \rho y \Leftrightarrow x + y$ is even

On the set of all lines in the plane: $x \rho y \Leftrightarrow x$ is parallel to y or coincides with y

On $\{0, 1\}$: $x \rho y \Leftrightarrow x = y^2$

On $\{x \mid x$ is a student in your class$\}$; $x \rho y \Leftrightarrow x$ sits in the same row as y

On $\{1, 2, 3\}$: $\rho = \{(1, 1), (2, 2), (3, 3), (1, 2), (2, 1)\}$

We can illustrate an important feature of an equivalence relation on a set by looking at $S = \{x \mid x$ is a student in your class$\}$, $x \rho y \Leftrightarrow x$ sits in the same row as y. Let's group all those students in set S who are related to one another. We come up with Figure 2.4. We have "partitioned" the set S into subsets such that everyone in the class belongs to one and only one subset.

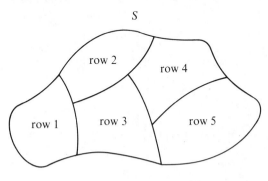

Figure 2.4

2.23 Definition   A **partition** of a set S is a collection of disjoint subsets of S whose union equals S.   □

Any equivalence relation, as we will see, partitions the set on which it is defined. The subsets making up the partition, often called the **blocks** of the partition, are formed as in the above case by grouping together related elements.

For ρ an equivalence relation on a set S and $x \in S$, we let $[x]$ denote the set of all members of S related to x, called the **equivalence class** of x. Thus,

$$[x] = \{y \mid y \in S \land x \rho y\}$$

2.24 Example   In the case where $x \rho y \Leftrightarrow x$ sits in the same row as y, suppose that John, Chuck, Jose, Judy, and Ted all sit in row 3. Then $[\text{John}] = \{$John, Chuck, Jose, Judy, Ted$\}$. Also $[\text{John}] = [\text{Ted}] = [\text{Judy}]$, and so on. There can be more than one name for the same equivalence class.   □

For ρ an equivalence relation on S, the distinct equivalence classes of members of S form a partition of S. To satisfy the definition of a partition, we must show (1) that the union of these distinct classes equals S, and (2) that the distinct classes are disjoint. To show that the union of the classes equals

S is easy since it is essentially a set equality, and we prove it by showing set inclusion in each direction. Every equivalence class is a subset of S, so the union of classes is also contained in S. For the opposite direction, let $x \in S$. Then $x \rho x$ (reflexivity of ρ); thus, $x \in [x]$, and every member of S belongs to some equivalence class, hence to the union of classes.

Now let $[x]$ and $[z]$ be two distinct classes; that is, $[x] \neq [z]$. We need to show that $[x] \cap [z] = \varnothing$. We will do a proof by contradiction. Therefore, we assume that $[x] \cap [z] \neq \varnothing$, and, thus, that there is a $y \in S$ such that $y \in [x] \cap [z]$.

$y \in [x] \cap [z]$	(assumption)
$y \in [x], y \in [z]$	(definition of $\cap$)
$x \rho y, z \rho y$	(definition of $[x]$ and of $[z]$)
$x \rho y, y \rho z$	(symmetry of ρ)
$x \rho z$	(transitivity of ρ)

Now we can show that $[x] = [z]$; we prove set inclusion in each direction. Let

$q \in [z]$	($[z] \neq \varnothing$)

Then

$z \rho q$	(definition of $[z]$)
$x \rho z$	(from above)
$x \rho q$	(transitivity of ρ)
$q \in [x]$	(definition of $[x]$)
$[z] \subseteq [x]$	(definition of $\subseteq$)
$[x] \subseteq [z]$	(Practice 2.25 below)
$[x] = [z]$	($[z] \subseteq [x]$ and $[x] \subseteq [z]$)

Therefore, $[x] = [z]$, which is a contradiction because $[x]$ and $[z]$ are distinct. Thus, our assumption was wrong; $[x] \cap [z] = \varnothing$, and distinct equivalence classes are disjoint.

2.25 Practice   Show that $[x] \subseteq [z]$.

(*Answer, page A-6*)

We have shown that an equivalence relation on a set determines a partition. The converse is also true. Given a partition of a set S, we define a relation ρ by $x \rho y \Leftrightarrow x$ is in the same subset of the partition as y.

2.26 Practice   Show that ρ, as defined above, is an equivalence relation on S, that is, show that ρ is reflexive, symmetric, and transitive.

(*Answer, page A-6*)

We have proved the following about equivalence relations.

2.27 Theorem   An equivalence relation on a set S determines a partition of S, and a partition of a set S determines an equivalence relation on S.   □

2.28 Example   The equivalence relation on $\mathbb{N}$ given by $x\,\rho\,y \Leftrightarrow (x + y$ is even$)$ determines a partition of $\mathbb{N}$ into two equivalence classes. If x is an even number, then for any even number y, $x + y$ is even, and $y \in [x]$. All even numbers form one class. If x is an odd number and y is any odd number, $x + y$ is even and $y \in [x]$. All odd numbers form the second class. The partition can be pictured as in Figure 2.5. Notice again that an equivalence class may have more than one name, or representative. In this example, $[2] = [8] = [1048]$, and so on, $[1] = [17] = [947]$, and so on.   □

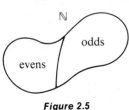

Figure 2.5

2.29 Practice   For each of the following equivalence relations describe the corresponding equivalence classes.

  (a) On the set of all lines in the plane, $x\,\rho\,y \Leftrightarrow x$ is parallel to y or x coincides with y.
  (b) On the set $\mathbb{N}$, $x\,\rho\,y \Leftrightarrow x = y$
  (c) On $\{1, 2, 3\}$, $\rho = \{(1, 1), (2, 2), (3, 3), (1, 2), (2, 1)\}$.   (*Answer, page A-6*)

Partitioning a set into equivalence classes is helpful because it is often convenient to treat the classes themselves as entities. We will conclude this section with two examples where this is the case (other examples occur throughout much of the rest of the text).

2.30 Example   Let $S = \{a/b \mid a, b \in \mathbb{Z}, b \neq 0\}$. S is therefore the set of all fractions. Two fractions such as $1/2$ and $2/4$ are said to be equivalent. Formally, a/b is equivalent to c/d, denoted by $a/b \sim c/d$, if and only if $ad = bc$. We will show that the binary relation $\sim$ on S is an equivalence relation. First, $a/b \sim a/b$ because $ab = ba$. Also if $a/b \sim c/d$, then $ad = bc$, or $cb = da$ and $c/d \sim a/b$. Hence, $\sim$ is reflexive and symmetric. To show that $\sim$ is transitive, let $a/b \sim c/d$ and $c/d \sim e/f$. Then $ad = bc$ and $cf = de$. Multiplying the first equation by f and the second by b, we get $adf = bcf$ and $bcf = bde$. Therefore, $adf = bde$, or $af = be$. Thus, $a/b \sim e/f$, and $\sim$ is transitive. Some sample equivalence classes of S formed by this equivalence relation are

$$[1/2] = \{1/2, 2/4, 3/6, 4/8, \ldots, -1/-2, -2/-4, -3/-6,$$
$$-4/-8, \ldots\}$$

$$[3/10] = \{3/10, 6/20, 9/30, 12/40, \ldots, -3/-10, -6/-20, -9/-30,$$
$$-12/-40, \ldots\}$$

The set $\mathbb{Q}$ of rational numbers can be regarded as the set of equivalence classes of S. A single rational number, such as $[1/2]$, has many fractions representing it although we customarily use the reduced fractional representation. When we add two rational numbers, such as $[1/2] + [3/10]$, we look for representatives from the classes having the same denominator and add those representatives. Our answer is the class to which the resulting sum belongs, and we usually name the class by using a reduced fraction. Thus, to add $[1/2] + [3/10]$, we represent $[1/2]$ by $5/10$ and $[3/10]$ by $3/10$. The sum of $5/10$ and $3/10$ is $8/10$, and $[8/10]$ is customarily named $[4/5]$. This procedure is so familiar that it is generally written as $1/2 + 3/10 = 4/5$; nonetheless, classes of fractions are being manipulated by means of representatives. □

2.31 Example   We will define a binary relation of **congruence modulo 4** on the set $\mathbb{Z}$ of integers. An integer x is congruent modulo 4 to y, symbolized by $x \equiv_4 y$, or $x \equiv y \pmod 4$, if $x - y$ is an integral multiple of 4. Congruence modulo 4 is an equivalence relation on $\mathbb{Z}$. (Can you prove this?) To construct the equivalence classes, note that $[0]$, for example, will contain all integers differing from 0 by a multiple of 4, such as 4, 8, -12, and so on. The distinct equivalence classes are

$$[0] = \{0, 4, 8, 12, 16, \ldots, -4, -8, -12, -16, \ldots\}$$
$$[1] = \{1, 5, 9, 13, \ldots, -3, -7, -11, -15, \ldots\}$$
$$[2] = \{2, 6, 10, 14, \ldots, -2, -6, -10, -14, \ldots\}$$
$$[3] = \{3, 7, 11, 15, \ldots, -1, -5, -9, -13, \ldots\}$$

There is nothing special about the choice of 4 here, and we can give a definition for congruence modulo n for any positive integer n. It is always an equivalence relation. Later we will look at situations in which the resulting classes themselves are the entities. And, as in Example 2.30, the classes will be manipulated by means of representatives. We will also see how manipulating these classes relates to the arithmetic performed by a computer. □

2.32 Practice   State the definition of **congruence modulo n** for an arbitrary positive integer n.
(Answer, page A-6)

2.33 Practice   What are the equivalence classes corresponding to the relation of congruence modulo 5 on $\mathbb{Z}$?
(Answer, page A-6)

✔ **CHECKLIST**

Definitions

binary relation (*p. 47*)	total ordering (*p. 52*)
n-ary relation (*p. 47*)	least element (*p. 52*)
reflexive relation (*p. 49*)	minimal element (*p. 52*)

symmetric relation (*p. 49*)
antisymmetric relation (*p. 49*)
transitive relation (*p. 49*)
partial ordering (*p. 50*)
predecessor in a partial
  ordering (*p. 50*)
immediate predecessor in a
  partial ordering (*p. 50*)

greatest element (*p. 52*)
maximal element (*p. 52*)
equivalence relation (*p. 52*)
partition (*p. 53*)
equivalence class (*p. 53*)
congruence modulo n (*p. 56*)

Techniques

Test an ordered pair for membership in a binary relation.

Test a binary relation for reflexivity, symmetry, antisymmetry, and transitivity.

Graph a partially ordered set.

Find least, minimal, greatest, or maximal elements in a partially ordered set.

Find the equivalence classes associated with an equivalence relation.

Main Ideas

A binary relation on a set S; formally a subset of $S \times S$, the distinctive relationship satisfied by its members often has a verbal description as well.

Operations on binary relations on a set $(+, \cdot, ')$.

Partially ordered sets and their graphs.

An equivalence relation on a set and the associated equivalence classes, which may themselves be treated as entities. An equivalence relation on a set S always determines a partition of S, and conversely a partition of a set S always defines an equivalence relation on S.

EXERCISES SECTION 2.1

★ 1. Decide which of the given items satisfy the relations.
    (a) ρ a binary relation on $\mathbb{Z}$, $x \, \rho \, y \Leftrightarrow x = -y$;
        $(1, -1), (2, 2), (-3, 3), (-4, -4)$
    (b) ρ a unary relation on $\mathbb{N}$, $x \in \rho \Leftrightarrow x$ is prime;
        $19, 21, 33, 41$
    (c) ρ a ternary relation on $\mathbb{N}$, $(x, y, z) \in \rho \Leftrightarrow x^2 + y^2 = z^2$;
        $(1, 1, 2), (3, 4, 5), (0, 5, 5), (8, 6, 10)$
    (d) ρ a binary relation on $\mathbb{Q}$, $x \, \rho \, y \Leftrightarrow x \leq 1/y$;
        $(1, 2), (-3, -5), (-4, 1/2), (1/2, 1/3)$

★ 2. Let ρ and σ be binary relations on $\mathbb{N}$ defined by $x \rho y \Leftrightarrow x \,|\, y$, $x \sigma y \Leftrightarrow 5x \leq y$. Decide which of the given ordered pairs satisfy the relations.
   (a) $\rho + \sigma$;   (2, 6)   (3, 17)   (2, 1)   (0, 0)
   (b) $\rho \cdot \sigma$;   (3, 6)   (1, 2)   (2, 12)
   (c) ρ';   (1, 5)   (2, 8)   (3, 15)
   (d) σ';   (1, 1)   (2, 10)   (4, 8)

★ 3. Test the following binary relations on the given sets S for reflexivity, symmetry, antisymmetry, and transitivity.
   (a) $S = \mathbb{Q}$
       $x \rho y \Leftrightarrow |x| \leq |y|$
   (b) $S = \mathbb{Z}$
       $x \rho y \Leftrightarrow x - y$ is an integral multiple of 3
   (c) $S = \mathbb{N}$
       $x \rho y \Leftrightarrow x \cdot y$ is even
   (d) $S = \mathbb{N}$
       $x \rho y \Leftrightarrow x$ is odd
   (e) $S =$ set of all squares in the plane
       $S_1 \rho S_2 \Leftrightarrow$ length of side of $S_1 =$ length of side of S_2
   (f) $S =$ set of all singly-dimensioned arrays
       $x \rho y \Leftrightarrow$ number of elements in $x =$ number of elements in y
   (g) $S =$ set of all people in the United States
       $x \rho y \Leftrightarrow x$ is the brother of y

   4. Which of the binary relations of Exercise 3 are equivalence relations? For each equivalence relation, describe the associated equivalence classes.

   5. For each case below, think of a set S and a binary relation ρ on S (different from any in the examples or problems) satisfying the given conditions.
   (a) ρ is reflexive and symmetric but not transitive
   (b) ρ is reflexive and transitive but not symmetric
   (c) ρ is not reflexive, not symmetric, but is transitive
   (d) ρ is reflexive but neither symmetric nor transitive.

★ 6. Let ρ and σ be binary relations on a set S.
   (a) If ρ and σ are reflexive, is $\rho + \sigma$ reflexive? Is $\rho \cdot \sigma$ reflexive?
   (b) If ρ and σ are symmetric, is $\rho + \sigma$ symmetric? Is $\rho \cdot \sigma$ symmetric?
   (c) If ρ and σ are antisymmetric, is $\rho + \sigma$ antisymmetric? Is $\rho \cdot \sigma$ antisymmetric?
   (d) If ρ and σ are transitive, is $\rho + \sigma$ transitive? Is $\rho \cdot \sigma$ transitive?

   7. Graph the partial ordering "x divides y" on the set $\{2, 3, 5, 7, 21, 42, 105, 210\}$. Name any least elements, minimal elements, greatest elements, and maximal elements. Name a totally ordered subset with four elements.

★ 8. Graph each of the two partially ordered sets
   (a) $S = \{1, 2, 3, 5, 6, 10, 15, 30\}$

$$x \rho y \Leftrightarrow x \text{ divides } y$$

(b) $S = \mathscr{P}(\{1, 2, 3\})$

$$A \rho B \Leftrightarrow A \subseteq B$$

What do you notice about the structure of these two graphs?

9. Let (S, ρ) and (T, σ) be two partially ordered sets. Define a relation μ on $S \times T$ by $(s_1, t_1) \mu (s_2, t_2) \Leftrightarrow s_1 \rho s_2$ and $t_1 \sigma t_2$. Show that μ is a partial ordering on $S \times T$.

10. (a) Let $(S, \preceq)$ be a partially ordered set and define a new binary relation $\succeq$ on S by $x \succeq y \Leftrightarrow y \preceq x$. Show that $(S, \succeq)$ is a partially ordered set, called the *dual* of $(S, \preceq)$.

(b) If $(S, \preceq)$ is a finite, partially ordered set with the graph shown in Figure 2.6, draw the graph of $(S, \succeq)$.

(c) Let $(S, \preceq)$ be a totally ordered set and let $X = \{(x, x) | x \in S\}$. Show that the set difference $\succeq - X$ equals the set $\preceq'$.

Figure 2.6

11. A computer program is to be written that will generate a dictionary or the index for a book. We will assume a maximum length of n characters per word. Thus, we are given a set S of words of length $\leq n$, and we want to produce a linear list of these words arranged in alphabetical (lexicographical) order. There is a natural total ordering $\preceq$ on alphabetic characters ($a \prec b, b \prec c$, etc.), and we will assume our words contain only alphabetic characters. We want to define a total ordering $\preceq$ on S that will arrange the members of S lexicographically.

Let $X = (x_1, \ldots, x_j)$ and $Y = (y_1, \ldots, y_k)$ be members of S with $j \leq k$. Let β (for blank) be a new symbol and fill out X with $k-j$ blanks on the right. X can now be written $(x_1, \ldots, x_k)$. Let β precede any alphabetic character. Then $X \preceq Y$ if

(1) $x_1 \neq y_1$ and $x_1 \preceq y_1$

or

(2) $x_1 = y_1, x_2 = y_2, \ldots, x_m = y_m$ $(m \leq k)$

$x_{m+1} \neq y_{m+1}$ and $x_{m+1} \preceq y_{m+1}$

Otherwise $Y \preceq X$.

Note that because the ordering $\preceq$ on alphabetic characters is a total ordering, if $Y \preceq X$ by "otherwise," then there exists $m \leq k$ such that $x_1 = y_1, x_2 = y_2, \ldots, x_m = y_m, x_{m+1} \neq y_{m+1}$ and $y_{m+1} \preceq x_{m+1}$.

(a) Show that $\preceq$ on S as defined above is a total ordering.

(b) Apply the total ordering described to the words boo, bug, be, bah, bugg. Note why each word precedes the next.

★ 12. Exercise 11 discusses a total ordering on a set of words of length $\leq n$ that will produce a linear list in alphabetical order. Suppose we want to generate a list of all the distinct words in a text (for example, a compiler must create a symbol table of variable names). As in Exercise 11, we will assume that the words contain only alphabetic characters because there is a natural precedence relation already existing ($a \prec b$, $b \prec c$, etc.). If numeric or special characters are involved, they must be assigned a precedence relation with alphabetic characters (the collating sequence must be determined). If we list words alphabetically, it is a fairly quick procedure to decide when the word currently being processed is new, but to fit the new word into place all successive words must be moved one unit down the line. If the words are listed in order of processing, new words are simply tacked on the end and no rearranging is necessary, but each word being processed has to be compared with each member of the list to determine if it is new. Thus, both logical linear lists have disadvantages.

We will describe a listing process called a **binary tree search** that usually allows a quick determination of whether a word is new, and no juggling to fit it into place if it is, thus combining the advantages of both methods above. Suppose we want to process the phrase "when in the course of human events." The first word in the text is used to label the first node of a graph.

when
•

Once a node is labeled, it drops down a left and right arc, putting two unlabeled nodes below the one just labeled.

Figure 2.7

When the next word in the text is processed, comparison begins with the first node. When the word being processed alphabetically precedes the label of a node, the left arc is taken, and when the label precedes the word, the right arc is taken until the word becomes the label of the first unlabeled node it reaches. (If the word equals a node label, the next word in the text is processed.) This procedure continues for the entire text. Thus,

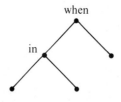

Figure 2.8

then

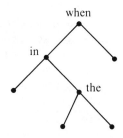

Figure 2.9

then

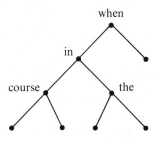

Figure 2.10

until finally

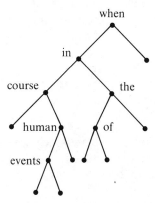

Figure 2.11

By traversing the nodes of this graph in the proper order (described by always processing the left nodes below a node first, then the node, then the right nodes below it), an alphabetical listing "course, events, human, in, of, the, when" is produced.

(a) The type of graph shown in Figure 2.11 is called a **tree**. Turned upside down it could be viewed as the graph of a partial ordering $\leq$. What would be the least element? Is there a greatest element? Which of the following would belong to $\leq$:(in, of), (the, of), (in, events), (course, of)?

Here the tree structure contains more information than the partial ordering because we are interested not only in whether a word w_1 precedes a word w_2 but also whether w_2 is to the left or right of w_1.

(b) Use the binary tree search to graph "Old King Cole was a merry old soul." Considering the graph (upside down) as a partial ordering, name the maximal elements.

13. (a) Compute the total number of possible partitions of a three-element set.

(b) Compute the total number of possible partitions of a four-element set.

14. Given two partitions π_1 and π_2 of a set S, π_1 is a **refinement** of π_2 if each block of π_1 is a subset of a block of π_2. Show that refinement is a partial ordering on the set of all partitions of S.

15. Let S be the set of all statements with n statement letters as defined in Section 1.1. Let ρ be a binary relation on S defined by $P \rho Q \Leftrightarrow (P \Leftrightarrow Q$ is a tautology). Show that ρ is an equivalence relation on S and describe the resulting equivalence classes.

16. (a) Show that the EQUIVALENCE statement in the FORTRAN language defines an equivalence relation on variable names. Describe the resulting equivalence classes.

(b) Show that the DEFINED attribute of the DECLARE statement in PL/I defines an equivalence relation on variable names. Describe the resulting equivalence classes.

Section 2.2   FUNCTIONS

In this section we will discuss functions, which are essentially special cases of binary relations. This view of a function is a rather sophisticated one, however, and we will work up to it gradually.

THE DEFINITION

Function is a common enough word even in a nontechnical context. A newsmagazine may have an article on national unemployment that says something like "unemployment varies depending upon the region of the country" or "unemployment is a function of the region of the country." It may illustrate this functional relationship by a graph such as Figure 2.12. From the graph we see that each region of the country has some unique unemployment figure associated with it but that no region has more than one number associated with it and that both the Mid-Atlantic and the South have the same figure, 5%.

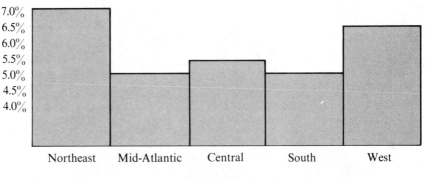

Figure 2.12

Of course we use mathematical functions in algebra and calculus. The equation $g(x) = x^3$ expresses a functional relationship between values for x and corresponding values that result when x is replaced in the equation by its values. Thus an x value of 2 has the number $2^3 = 8$ associated with it. (This number is expressed as $g(2) = 8$.) Similarly, $g(1) = 1^3 = 1$, $g(-1) = (-1)^3 = -1$, and so on. For each x value, the corresponding $g(x)$ value is unique. If we were to graph this function on a rectangular coordinate system, the points $(2, 8)$, $(1, 1)$ and $(-1, -1)$ would be points on the graph. If we allow x to take on any real number value, the resulting graph is the continuous curve shown in Figure 2.13.

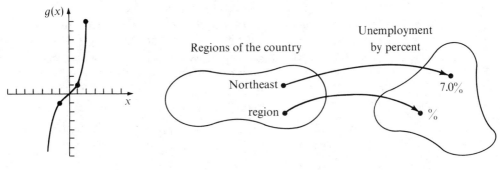

Figure 2.13   Figure 2.14

The function in the unemployment example above could be described as follows. We could set the stage by the diagram in Figure 2.14, indicating that the function always starts with a given region of the country and that a particular unemployment percentage is associated with that region. The association itself is described by the set of ordered pairs

{(Northeast, 7.0%), (Mid-Atlantic, 5.0%), (Central, 5.5%),
(South, 5.0%), (West, 6.5%)}

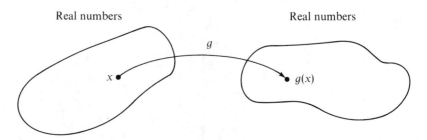

Figure 2.15

For the algebraic example $g(x) = x^3$, Figure 2.15 shows that the function always starts with a given real number and that a second real number is associated with it. The association itself is described by $\{(x, g(x)) | g(x) = x^3\}$, or, simply, $g(x) = x^3$. This set includes $(2, 8) (1, 1) (-1, -1)$, but because it is an infinite set, we cannot list all its members, we have to describe them.

From the above examples, we can conclude that there are three parts to a function: (1) a set of starting values; (2) a set from which associated values come, and (3) the association itself. The set of starting values is called the **domain** of the function, and the set from which associated values come is called the **codomain** of the function. Thus, the picture for an arbitrary function f is shown in Figure 2.16. Here, f is a function from S to T, symbolized $f : S \to T$. The association itself is a set of ordered pairs, each of the form (s, t) where $s \in S$, $t \in T$, and t is the value from T the function associates with the value s from S; $t = f(s)$. Hence, the association is a subset of $S \times T$ (a binary relation on $S \times T$). But the important property of this relation is that every member of S must have one and only one T value associated with it, so every $s \in S$ will appear only once as the first component of an (s, t) pair. (This property does not prevent a given T value from appearing more than once.)

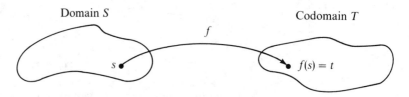

Figure 2.16

We are now ready for a formal definition of a function.

2.34 Definition   Let S and T be sets. A **function (mapping)** $f:S \to T$ is a subset of $S \times T$ where each member of S appears exactly once as the first component of an ordered pair. S is the **domain** and T the **codomain** of the function. If (s, t) belongs to the function, then t is denoted by $f(s)$; t is the **image** of s under f, s is a **preimage** of t under f, and f is said to **map** s to t. For $A \subseteq S$, $f(A)$ denotes $\{f(a) \mid a \in A\}$.
□

We have talked a lot about values from the sets S and T, but as our unemployment example shows, these values are not necessarily numbers, nor is the association itself necessarily described by an equation.

2.35 Practice   Which of the following are functions from the domain to the codomain indicated? For those that are not, why not?

(a) $f:S \to T$ where $S = T = \{1, 2, 3\}$, $f = \{(1, 1), (2, 3), (3, 1), (2, 1)\}$
(b) $g:\mathbb{Z} \to \mathbb{N}$ where g is defined by $g(x) = |x|$ (the absolute value of x)
(c) $h:\mathbb{N} \to \mathbb{N}$ where h is defined by $h(x) = x - 4$
(d) $f:S \to T$ where S is the set of all people in your hometown, T is the set of all Social Security numbers, and f associates his or her Social Security number with each person
(e) $g:S \to T$ where $S = \{1972, 1973, 1974, 1975\}$,
   $T = \{\$20,000, \$30,000, \$40,000, \$50,000, \$60,000\}$, g is defined by the graph

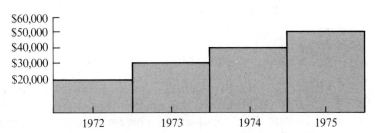

Profits of the American Earthworm Corp.

(f) $h:S \to T$ where S is the set of all 2×2 arrays $\begin{bmatrix} a & b \\ c & d \end{bmatrix}$ of positive integers, $T = N$, and h is defined by $h\left(\begin{bmatrix} a & b \\ c & d \end{bmatrix}\right) = a$

(g) $f:\mathbb{R} \to \mathbb{R}$ where f is defined by $f(x) = 4x - 1$
(h) $g:\mathbb{N} \to \mathbb{N}$ where g is defined by

$$g(x) = \begin{cases} x + 3 & \text{if } x \geq 5 \\ x & \text{if } x \leq 5 \end{cases}$$

(*Answer, page A-6*)

2.36 Practice   (a) For $f: \mathbb{Z} \rightarrow \mathbb{Z}$ given by $f(x) = x^2$, what is the image of -4?

(b) What are the preimages of 9?   *(Answer, page A-6)*

2.37 Example   In Section 1.3 we defined the idea of a binary operation ∘ on a set S. ∘ associates a unique member of S, $x \circ y$, with every (x, y) pair of elements of S. Thus, any binary operation on S is a function: $S \times S \rightarrow S$.   □

Let's return to our earlier example of $g: \mathbb{R} \rightarrow \mathbb{R}$ where g is defined by $g(x) = x^3$. It is common in algebra and calculus to say "consider the function $g(x) = x^3$," implying that the equation *is* the function. Technically, the equation only describes how to compute associated values. The function $h: \mathbb{R} \rightarrow \mathbb{R}$ given by $h(x) = x^3 - 3x + 3(x + 5) - 15$ is the same function as g since it contains the same ordered pairs. However, the equation is different in that it says to do different things to any given x value. On the other hand, the function $f: \mathbb{Z} \rightarrow \mathbb{R}$ given by $f(x) = x^3$ is not the same function as g. The domain has been changed, which changes the set of ordered pairs. The graph of $f(x)$ would consist of discrete (separated) points (see Figure 2.17).

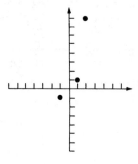

Figure 2.17

Most of the functions we are interested in will have this latter feature. In a digital computer, information is processed in a series of distinct (discrete) steps. Even in situations where one quantity varies continuously with another, we approximate by taking data at discrete, small intervals, much as the graph of $g(x)$ is approximated by the graph of $f(x)$.

Finally, let's look at the function $k: \mathbb{R} \rightarrow \mathbb{C}$ given by $k(x) = x^3$. The equation and domain is the same as for $g(x)$; the codomain has been enlarged, but this does not affect the ordered pairs. Is this considered the same function as $g(x)$? It is not, but to see why we need a further definition.

SPECIAL FUNCTIONS

Onto Functions

Let $f: S \rightarrow T$ be an arbitrary function with domain S and codomain T (see Figure 2.18). Part of the definition of a function is that every member of S

has an image under f and that all the images are members of T; the set R of all such images is called **range** of the function f. Thus, $R = \{f(s)\,|\,s \in S\}$, or $R = f(S)$. Clearly, $R \subseteq T$; the range R is shaded in Figure 2.19. If it should happen that $R = T$, that is, that the range coincides with the co-domain, then the function is called an *onto* function.

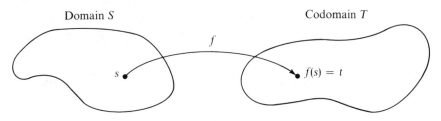

Figure 2.18

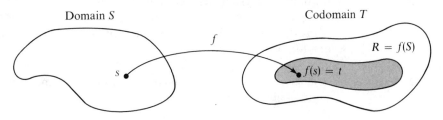

Figure 2.19

2.38 Definition   A function $f: S \to T$ is an **onto** or **surjective** function if the range of f equals the codomain of f. □

In every function with range R and codomain T, $R \subseteq T$. To prove a given function is onto, we must show that $T \subseteq R$; then $R = T$. We must therefore show that an arbitrary member of the codomain is a member of the range, that is, is the image of some member of the domain. On the other hand, if we can produce one member of the codomain that is not the image of any member of the domain, then we have proved that the function is not onto.

2.39 Example   The function $g: \mathbb{R} \to \mathbb{R}$ defined by $g(x) = x^3$ is an onto function. To prove that $g(x)$ is onto, let r be an arbitrary real number, and let $x = \sqrt[3]{r}$. Then x is a real number, so x belongs to the domain of g and $g(x) = (\sqrt[3]{r})^3 = r$. Hence, any member of the codomain is the image under g of a member of the domain. The function $k: \mathbb{R} \to \mathbb{C}$ given by $k(x) = x^3$ is not surjective. There are many complex numbers (i, for example) that cannot be obtained by cubing a real number. Thus, g and k are not equal functions. *For two functions to be equal,* *they must have the same domain, the same codomain, and consist of the same ordered pairs.*   □

2.40   Example   Let $f: \mathbb{Q} \to \mathbb{Q}$ be defined by $f(x) = 3x + 2$. To test whether f is onto, let $q \in \mathbb{Q}$. We want an $x \in \mathbb{Q}$ such that $f(x) = 3x + 2 = q$. When we solve this equation for x, we find that $x = (q - 2)/3$ is the only possible value and is indeed a member of $\mathbb{Q}$. Thus, q is the image of a member of $\mathbb{Q}$ under f, and f is onto. However, the function $h: \mathbb{Z} \to \mathbb{Q}$ defined by $h(x) = 3x + 2$ is not onto because there are many values $q \in \mathbb{Q}$, for example, 0, for which the equation $3x + 2 = q$ has no integer solution.   □

2.41   Practice   Which of the functions found in Practice 2.35 are onto functions?

(*Answer, page A-6*)

One-to-one Functions

The definition of a function guarantees a unique image for every member of the domain. A given member of the range may have more than one preimage, however. In our very first example of a function (unemployment), both Mid-Atlantic and South were preimages of 5%. This function was not one-to-one.

2.42   Definition   A function $f: S \to T$ is **one-to-one**, or **injective** if no member of T is the image under f of two distinct elements of S.   □

To prove that a function is one-to-one, we often assume that there are elements s_1 and s_2 of S with $f(s_1) = f(s_2)$, and then show that $s_1 = s_2$. To prove a function is not injective, we produce an element in the range with two preimages in the domain.

2.43   Example   The function $g: \mathbb{R} \to \mathbb{R}$ defined by $g(x) = x^3$ is one-to-one because if x and y are real numbers with $g(x) = g(y)$, then $x^3 = y^3$ and $x = y$. The function $f: \mathbb{R} \to \mathbb{R}$ given by $f(x) = x^2$ is not injective because, for example, $f(2) = f(-2) = 4$. However, the function $h: \mathbb{N} \to \mathbb{N}$ given by $h(x) = x^2$ is injective since if x and y are nonnegative integers with $h(x) = h(y)$, then $x^2 = y^2$ and because x and y are both nonnegative, $x = y$.   □

2.44   Practice   Which of the functions found in Practice 2.35 are one-to-one functions?

(*Answer, page A-6*)

Bijections

2.45   Definition   A function $f: S \to T$ is **bijective** if it is both one-to-one and onto.   □

2.46   Example   The function $g: \mathbb{R} \to \mathbb{R}$ given by $g(x) = x^3$ is a bijection. The function in part (g) of Practice 2.35 is a bijection. The function $f: \mathbb{R} \to \mathbb{R}$ given by $f(x) = x^2$

is not a bijection (not one-to-one) and neither is the function $k : \mathbb{R} \to \mathbb{C}$ given by $k(x) = x^3$ (not onto).   □

2.47 Definition   A set S is **equivalent** to a set T if there exists a bijection $f : S \to T$.   □

If S is equivalent to T, then all the members of S and of T are paired off by f in a one-to-one correspondence. If S and T are finite sets, this pairing off can only happen when S and T are the same size. With infinite sets, the idea of size gets a bit fuzzy, and we can sometimes prove that a given set is equivalent to what *seems* to be a smaller set.

2.48 Practice   Describe a bijection $f : \mathbb{Z} \to \mathbb{N}$, thus showing that $\mathbb{Z}$ is equivalent to $\mathbb{N}$ even through $\mathbb{N} \subset \mathbb{Z}$.   *(Answer, page A-6)*

COMPOSITION OF FUNCTIONS

Suppose that f and g are functions with $f : S \to T$ and $g : T \to U$. Then for any $s \in S$, $f(s)$ is a member of T, which is also the domain of g. Thus, the function g can be applied to $f(s)$. The result is $g(f(s))$, a member of U; see Figure 2.20. The total effect of taking an arbitrary member s of S, applying the function f and then applying the function g to $f(s)$ is to associate a unique member of U with s. In short, we have created a function: $S \to U$, called the composition function of f and g and denoted by $g \circ f$.

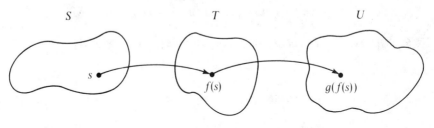

Figure 2.20

2.49 Definition   Let $f : S \to T$, and $g : T \to U$. Then the **composition function**, $g \circ f$, is a function from S to U defined by $(g \circ f)(s) = g(f(s))$.   □

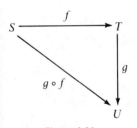

Figure 2.21

We reiterate that the function $g \circ f$ is applied right to left; function f is applied first and then function g.

The diagram in Figure 2.21 also expresses this definition. The corners indicate the domains and codomains of the three functions. The diagram says that, starting with an element of S, if we follow path $g \circ f$ or path f and then path g, we get to the same element in U. Diagrams illustrating that alternate paths produce the same effect are called **commutative diagrams**.

2.50 Practice   Let $f:\mathbb{R} \to \mathbb{R}$ be defined by $f(x) = x^2$. Let $g:\mathbb{R} \to \mathbb{R}$ be the truncation function; that is, $g(x)$ equals the integral part of x, with any digits behind the decimal point dropped.

(a) For $x = 2.3$, what is the value of $(g \circ f)(x)$?
(b) What is the value of $(f \circ g)(x)$?   *(Answer, page A-6)*

From Practice 2.50, we see that order is important in function composition, which should not be surprising. If we were to write a computer program to carry out function composition, we would generally require an assignment statement to compute each function. Reversing the order of composition would then require essentially reversing the order of two program statements, which will almost always change the program results.

Function composition preserves the property of being onto or of being one-to-one. Again, let $f:S \to T$ and $g:T \to U$, but also suppose that both f and g are onto functions. Then the composition function $g \circ f$ is also onto. Recall that $g \circ f:S \to U$, so we must pick an arbitrary $u \in U$ and show that it has a preimage under $g \circ f$ in S. Because g is surjective, there exists $t \in T$ such that $g(t) = u$. And because f is surjective, there exists $s \in S$ such that $f(s) = t$. Then $(g \circ f)(s) = g(f(s)) = g(t) = u$, and $g \circ f$ is an onto function.

2.51 Practice   Let $f:S \to T$, and $g:T \to U$, and assume that both f and g are one-to-one functions. Prove that $g \circ f$ is a one-to-one function. (Hint: assume that $(g \circ f)(s_1) = (g \circ f)(s_2)$.)   *(Answer, page A-6)*

We have now proved the following theorem.

2.52 Theorem   The composition of two bijections is a bijection.   □

PERMUTATION FUNCTIONS

In Theorem 2.52 we are talking about the composition function *when it exists*, that is, when the domain and codomain match up in such a way as to create a composition function. Of course, this is no problem if only one set is involved, that is, if we are only discussing functions that map a set A into A.

2.53 Definition   For a given set A, $S_A = \{f \mid f:A \to A$ and f is a bijection$\}$. S_A is thus the set of all bijections of set A into (and therefore onto) itself; such functions are called **permutations** of A.   □

Our theorem says that if we perform function composition on two members of S_A we get a (unique) member of S_A; thus, function composition is a binary operation on the set S_A.

In Section 1.4 we talked about a permutation of objects in a set as an ordered arrangement of those objects. Similarly, permutation functions represent ordered arrangements of the objects in the domain. If $A = \{1, 2, 3, 4\}$, one permutation function of A, call it f, is given by $f = \{(1, 2)(2, 3)(3, 1)(4, 4)\}$. We can also describe function f in array form by listing the elements of the domain in a row and, directly beneath, the images of these elements under f. Thus,

$$f = \begin{pmatrix} 1 & 2 & 3 & 4 \\ 2 & 3 & 1 & 4 \end{pmatrix}$$

The bottom row is an ordered arrangement of the objects in the top row. Still a shorter way to describe the permutation f is to use cycle notation, and write $f = (1, 2, 3)$—understood to mean that f maps each element listed to the one on its right, the last element listed to the first, and any element of the domain not listed to itself. Therefore, the cycles $(1, 2, 3)$, $(2, 3, 1)$, and $(3, 1, 2)$ all represent f.

2.54 Practice

(a) Let $A = \{1, 2, 3, 4, 5\}$, and let $f \in S_A$ be given in array form by

$$f = \begin{pmatrix} 1 & 2 & 3 & 4 & 5 \\ 4 & 2 & 3 & 5 & 1 \end{pmatrix}$$

Write f in cycle form.
(b) Let $A = \{1, 2, 3, 4, 5\}$, and let $g \in S_A$ be given in cycle form by $g = (2, 4, 5, 3)$. Write g in array form.

(Answer, page A-6)

If f and g are members of S_A for some set A, then $g \circ f \in S_A$, and the action of $g \circ f$ on any member of A is determined right to left by applying function f and then function g. If f and g are cycles, $g \circ f$ is still computed the same way. If $A = \{1, 2, 3, 4\}$ and $f, g \in S_A$ are given by $f = (1, 2, 3)$, and $g = (2, 3)$, then $g \circ f = (2, 3) \circ (1, 2, 3)$. For element 1 of A, $1 \to 2$ under f and $2 \to 3$ under g, so $1 \to 3$ under $g \circ f$. Similarly, $2 \to 3$ under f and $3 \to 2$ under g, so $2 \to 2$ under $g \circ f$. Testing what happens to 3 and 4, we conclude that $g \circ f = (1, 3)$. But if we compute $f \circ g$, we get $(1, 2)$. (Function composition on S_A is, like subtraction, a binary operation where order of the elements matters.)

2.55 Practice

Let $A = \{1, 2, 3, 4, 5\}$, and let f and g belong to S_A. Compute $g \circ f$ for

(a) $f = (5, 2, 3)$; $g = (3, 4, 1)$. Write the answer in cycle form.
(b) $f = (1, 2, 3, 4)$; $g = (3, 2, 4, 5)$. Write the answer in array form.

(Answer, page A-6)

If A is an infinite set, not every permutation of A can be written as a cycle. But even when A is a finite set, not every permutation of A can be written as a cycle. The permutation $g \circ f$ of Practice 2.55 (b) is not a cycle. However, a permutation on a finite set can always be written as the composition of two or more cycles with no common elements. The permutation

$$\begin{pmatrix} 1 & 2 & 3 & 4 & 5 \\ 4 & 2 & 5 & 1 & 3 \end{pmatrix} \text{ is } (1, 4) \circ (3, 5) \text{ or } (3, 5) \circ (1, 4).$$

(When there are no common elements, order of the cycles does not matter.)

2.56 Practice   Write $\begin{pmatrix} 1 & 2 & 3 & 4 & 5 & 6 \\ 2 & 4 & 5 & 1 & 3 & 6 \end{pmatrix}$ as a composition of cycles.   (*Answer, page A-6*)

INVERSE FUNCTIONS

There is one more important property of bijective functions. Let $f : S \to T$ be a bijection. Because f is onto, every $t \in T$ has a preimage in S. And because f is one-to-one, that preimage is unique. We could associate with each element t of T a unique member of S, namely, that $s \in S$ such that $f(s) = t$. This association describes a function g, $g : T \to S$. The picture for f and g is given in Figure 2.22. The domains and codomains of g and f are such that we can form both $g \circ f : S \to S$ and $f \circ g : T \to T$. If $s \in S$, then $(g \circ f)(s) = g(f(s)) = g(t) = s$. Thus, $g \circ f$ maps each element of S to itself. A function on a set S that maps each element to itself is called the **identity function** on S, denoted by i_S. Hence, $g \circ f = i_S$.

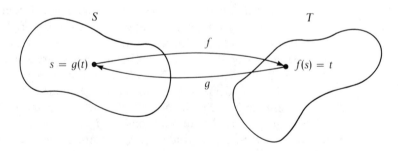

Figure 2.22

2.57 Practice   Show that $f \circ g = i_T$.   (*Answer, page A-6*)

We have now seen that if f is a bijection, $f : S \to T$, then there is a function $g : T \to S$ with $g \circ f = i_S$ and $f \circ g = i_T$. The converse is also true. To prove the converse suppose $f : S \to T$ and there exists $g : T \to S$ with $g \circ f = i_S$ and $f \circ g = i_T$. We can prove that f is a bijection. To show that f is onto, let $t \in T$. Then $t = i_T(t) = (f \circ g)(t) = f(g(t))$. Because $g : T \to S$, $g(t) \in S$, and $g(t)$ is the preimage under f of t. To show that f is one-to-one, suppose $f(s_1) = f(s_2)$. Then $g(f(s_1)) = g(f(s_2))$ and $(g \circ f)(s_1) = (g \circ f)(s_2)$, implying $i_S(s_1) = i_S(s_2)$, or $s_1 = s_2$. Thus, f is a bijection.

2.58 Definition   Let f be a function, $f : S \to T$. If there exists a function $g : T \to S$ such that $g \circ f = i_S$ and $f \circ g = i_T$, then g is called the **inverse function** of f, denoted by f^{-1}.   □

We have proved the following theorem.

2.59 Theorem   Let $f : S \to T$. Then f is a bijection if and only if f^{-1} exists.   □

Actually, we have been a bit sneaky in talking about "the" inverse function of f. What we have shown is that if f is a bijection, this is equivalent to the existence of "an" inverse function. But it is easy to see that there is only one such inverse function. *When you want to prove that something is unique, the standard technique is to assume that there are two such things, and obtain a contradiction.* Thus, suppose f has two inverse functions, f_1^{-1} and f_2^{-1} (existence of either means that f is a bijection). Both f_1^{-1} and f_2^{-1} are functions from T to S; if they are not the same function, then they must act differently somewhere. Assume that there is a $t \in T$ such that $f_1^{-1}(t) \neq f_2^{-1}(t)$. Because f is one-to-one, it follows that $f(f_1^{-1}(t)) \neq f(f_2^{-1}(t))$, or $(f \circ f_1^{-1})(t) \neq (f \circ f_2^{-1})(t)$. But both $f \circ f_1^{-1}$ and $f \circ f_2^{-1}$ are i_T, so $t \neq t$, a contradiction. We are therefore justified in speaking of f^{-1} as "the" inverse function of f. If f is a bijection so that f^{-1} exists, then f is the inverse function for f^{-1}, therefore, f^{-1} is also a bijection.

2.60 Practice $f : \mathbb{R} \to \mathbb{R}$ given by $f(x) = 3x + 4$ is a bijection. Describe f^{-1}.

(*Answer, page A-6*)

A FINAL REMARK

The definition of a function includes functions of more than one variable. We can have a function $f : S_1 \times S_2 \times \cdots \times S_n \to T$ that associates with each ordered n-tuple of elements $(s_1, s_2, \ldots, s_n)$, $s_i \in S_i$, a unique element of T. Recall from Section 1.3 that S^n is the notation for the set of all ordered n-tuples of elements of S.

2.61 Example $f : \mathbb{Z}^2 \to \mathbb{N}$ given by $f(x, y) = x^2 + 2y^2$ maps the ordered pair of integers $(-2, 3)$ to the nonnegative integer 22. This function is not one-to-one because $(2, 3)$ also maps to 22.   □

2.62 Example   Let P be a complex statement (as discussed in Section 1.1) with n statement letters. The truth table for P defines a function mapping $\{T, F\}^n \to \{T, F\}$. This function will be onto unless P is a tautology or a contradiction.   □

✔ *CHECKLIST*

Definitions

function (mapping) (*p. 65*)   bijection (*p. 68*)
domain (*p. 65*)   equivalent sets (*p. 69*)
codomain (*p. 65*)   composition function (*p. 69*)
image (*p. 65*)   commutative diagram (*p. 69*)
preimage (*p. 65*)   permutation function (*p. 70*)
range (*p. 67*)   identity function (*p. 72*)
onto (surjective) function (*p. 67*) inverse function (*p. 73*)
one-to-one (injective)
    function (*p. 68*)

Techniques

Test whether a given relation is a function.

Test a function for being one-to-one or onto.

Find the image of an element under function composition.

Write permutations of a set in array or cycle form.

Main Ideas

Functions, especially bijections.

Composition of functions, which preserves bijectiveness.

Permutations on a set and their composition.

The inverse function of a bijection, itself a bijection.

EXERCISES, SECTION 2.2

★ 1. Which of the following are functions from the domain to the codomain given? Which functions are one-to-one? Which functions are onto? Describe the inverse function for any bijective function.

(a) $f : \mathbb{Z} \to \mathbb{N}$ where f is defined by $f(x) = x^2 + 1$.
(b) $g : \mathbb{N} \to \mathbb{Q}$ where g is defined by $g(x) = 1/x$.
(c) $h : \mathbb{Z} \times \mathbb{N} \to \mathbb{Q}$ where h is defined by $h(z, n) = z/(n + 1)$.
(d) $f : \{1, 2, 3\} \to \{p, q, r\}$ where $f = \{(1, q)(2, r)(3, p)\}$.
(e) $g : \mathbb{N} \to \mathbb{N}$ where g is given by $f(x) = 2^x$.
(f) $h : \mathbb{R}^2 \to \mathbb{R}^2$ where h is defined by $h(x, y) = (y + 1, x + 1)$.

2. Show that each of the following is a function from S to T. Which are surjective (onto)? Which are injective (one-to-one)?
(a) $K : S \to T$ where S is a set of 16 data cards, T is the set of integers

from 1 to 16 (representing sixteen 5-position data fields), and K is given by the following FORTRAN program:

```
DIMENSION NUM(16)
K1 = 16
K2 = 1
DO 100 I = 1, K1
K = (I − 1)/K2 + K2
READ  1000,   (NUM(J), J = 1, K)
100 PRINT 2000, K
    ⋮
1000 FORMAT (16I5)
2000 FORMAT (I6)
END
```

(b) The program is the same as part (a) except S contains 15 cards and line 2 of the program is replaced by K1 = 15.
(c) The program is the same as part (a) except line 3 of the program is replaced by K2 = 2

3. Let f be a function, $f : S \to T$.
   (a) Show that for all subsets A and B of S, $f(A \cap B) \subseteq f(A) \cap f(B)$.
   (b) Show that $f(A \cap B) = f(A) \cap f(B)$ for all subsets A and B of S if and only if f is one-to-one.

★ 4. (a) Let $S = \{a, b, c\}$ and $T = \{p, q\}$. How many functions f are there such that $f : S \to T$? How many of these are onto functions?
   (b) Let S be a set with m elements and T be a set with n elements. How many functions f are there such that $f : S \to T$?
   (c) Let S and T be sets with m and n elements respectively, and assume $m \leq n$. How many injective functions are there from S to T?

5. (a) Let f be a function, $f : S \to S$ where S has n elements. Show that f is one-to-one if and only if f is onto.
   (b) Find an infinite set S and a function $f : S \to S$ such that f is one-to-one but not onto.
   (c) Find an infinite set S and a function $f : S \to S$ such that f is onto but not one-to-one.

★ 6. By the definition of a function f from S to T, f is a subset of $S \times T$ where the image of every $s \in S$ under f is uniquely determined as the second component of the ordered pair (s, t) in f. Now consider any binary relation ρ on $S \times T$. ρ is a subset of $S \times T$ in which some elements of S may not appear at all as first components of an ordered pair and some may appear more than once. We can view ρ as a "nondeterministic function" from a subset of S to T. An $s \in S$ not appearing as the first component of an ordered pair represents an element outside the domain of ρ. For an $s \in S$ appearing once or more as a first component, ρ can select for the image of s any one of the corresponding second components.

Let $S = \{1. 2, 3\}$, $T = \{a, b, c\}$, $U = \{m, n, o, p\}$. Let ρ be a binary relation on $S \times T$ and σ be a binary relation on $T \times U$ defined by

$$\rho = \{(1, a), (1, b), (2, b), (2, c), (3, c)\}$$
$$\sigma = \{(a, m), (a, o), (a, p), (b, n), (b, p), (c, o)\}$$

Thinking of ρ and σ as nondeterministic functions from S to T and T to U, respectively, we can form the composition $\sigma \circ \rho$, a nondeterministic function from S to U.

(a) What is the set of possible images of 1 under $\sigma \circ \rho$?

(b) What is the set of possible images of 2 under $\sigma \circ \rho$? Of 3?

★ 7. (a) Let $f:\mathbb{R} \to \mathbb{Z}$ be defined by $f(x) =$ the greatest integer $\leq x$. Let $g:\mathbb{Z} \to \mathbb{N}$ be defined by $g(x) = x^2$. What is $(g \circ f)(x)$ for $x = -4.7$?

(b) Let f map the set of 2-dimensional, rectangular arrays (matrices) of integers into the integers where f assigns to an array the number that is the sum of all of its elements. Let $g:\mathbb{Z} \to \mathbb{Z}$ be given by $g(x) = 2x$. What is

$$(g \circ f)\begin{bmatrix} 2 & 1 & -4 \\ 6 & 0 & -2 \end{bmatrix}?$$

(c) Let f map strings of alphabetic characters and blank spaces into strings of alphabetic consonants where f takes any string, removes all vowels, and eliminates all blanks. Let g map strings of alphabetic consonants into integers where g maps a string into the number of characters it contains. What is $(g \circ f)$ (abraham lincoln)?

★ 8. (a) Let $A = \{1, 2, 3\}$. Write the elements of S_A, the set of all permutations on A.

(b) Let $A = \{1, 2, \ldots, n\}$. How many elements are in the set S_A?

9. Let A be any set and let S_A be the set of all permutations of A. Let f, g, $h \in S_A$. Prove that the functions $h \circ (g \circ f)$ and $(h \circ g) \circ f$ are equal, showing that we can write $h \circ g \circ f$ with no need for parentheses to indicate grouping.

★ 10. Find the composition of the following cycles representing permutations on $A = \{1, 2, 3, 4, 5, 6, 7, 8\}$. Write your answer as a single cycle or as a composition of cycles with no common elements.

(a) $(1, 3, 4) \circ (5, 1, 2)$

(b) $(2, 7, 8) \circ (1, 2, 4, 6, 8)$

(c) $(1, 3, 4) \circ (5, 6) \circ (2, 3, 5) \circ (6, 1)$ (By Exercise 9, we can omit parentheses indicating grouping.)

(d) $(2, 7, 1, 3) \circ (2, 8, 7, 5) \circ (4, 2, 1, 8)$

★ 11. The "pushdown store," or "stack," is a storage device operating much like a set of plates stacked on a spring in a cafeteria. All storage locations are initially empty. An item of data is added to the top of the stack by a "push" instruction that pushes previously stored items further down in the stack. Only the topmost item on the stack is accessible at any moment, and it is fetched and removed from the stack by a "pop" instruction.

Let's consider strings of integers that are an even number of characters in length; half of the characters are positive integers and the other half are zeros. We process these strings through a pushdown store as follows: as we read from left to right, the push instruction is applied to any nonzero integer, and a zero causes the pop instruction to be applied to the stack and the integer "popped" to be printed. Thus, processing the string 12030040 results in an output of 2314, and processing 12304000 results in an output of 3421. (A string such as 10020340 cannot be handled by this procedure because we cannot pop two integers from a stack containing only one integer.) Both 2314 and 3421 could be thought of as permutations,

$$\begin{pmatrix} 1 & 2 & 3 & 4 \\ 2 & 3 & 1 & 4 \end{pmatrix} \quad \text{and} \quad \begin{pmatrix} 1 & 2 & 3 & 4 \\ 3 & 4 & 2 & 1 \end{pmatrix}$$

respectively, on the set $A = \{1, 2, 3, 4\}$.

(a) What permutation of $A = \{1, 2, 3, 4\}$ is generated by applying this procedure to the string 12003400?

(b) Name a permutation of $A = \{1, 2, 3, 4\}$ that cannot be generated from any string where the digits 1, 2, 3, and 4 appear in order, no matter where the zeros are placed.

12. (a) Let f be a function, $f : S \to T$. If there exists a function $g : T \to S$ such that $g \circ f = i_S$, then g is called a **left inverse** of f. Show that f has a left inverse if and only if f is one-to-one.

(b) Let f be a function, $f : S \to T$. If there exists a function $g : T \to S$ such that $f \circ g = i_T$, then g is called a **right inverse** of f. Show that f has a right inverse if and only if f is onto.

(c) Let $f : \mathbb{N} \to \mathbb{N}$ be given by $f(x) = 3x$. Then f is one-to-one. Find two different left inverse functions for f.

13. Let f and g be bijections, $f : S \to T$, $g : T \to U$. Then f^{-1} and g^{-1} exist. Also, $g \circ f$ is a bijection from S to U. Show that $(g \circ f)^{-1} = f^{-1} \circ g^{-1}$.

14. Let $\mathscr{C}$ be a collection of sets, and define a binary relation ρ on $\mathscr{C}$ as follows: for $S, T \in \mathscr{C}$, $S \rho T \Leftrightarrow S$ is equivalent to T. Show that ρ is an equivalence relation on $\mathscr{C}$.

Section 2.3   GRAPHS

DEFINITIONS

One way to while away the hours on an airplane trip is to look at the material in the seat pockets. This material almost always includes a map showing the routes of the airline you are flying, such as in Figure 2.23. All of the route information could be expressed in paragraph form; for example, there is a direct route between Chicago and Nashville but not between St. Louis and Nashville. The paragraph would be rather long and involved, and we would not be able to assimilate it as quickly and clearly as we can the map. There are many cases where "a picture is worth a thousand words."

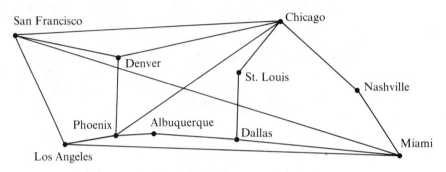

Figure 2.23

Figure 2.23 is a graph. We have also talked about graphs of functions on rectangular coordinate systems. There are many types of graphs—bar graphs, picture graphs, and pie, or circle, graphs, for instance (see Figure 2.24). Each type is a visual representation of data. But our definition of a graph is more restrictive; according to it, only the airline map is a graph.

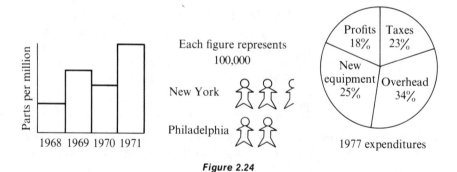

Figure 2.24

Graphs

2.63 Definition   A **graph** is an ordered triple $\langle N, A, f \rangle$ where

N = a nonempty set of **nodes** or **vertices**

A = a set of **arcs** or **edges**

f is a function associating with each arc a an unordered pair x–y of nodes called the **endpoints** of a.   □

Our graphs will always have a finite number of nodes and arcs.

2.64 Example   The set of nodes in the airline map is {Chicago, Nashville, Miami, Dallas, St. Louis, Albuquerque, Phoenix, Denver, San Francisco, Los Angeles}. There are 16 arcs; for example, Phoenix–Albuquerque is an arc (here we are naming an arc by its endpoints), Albuquerque–Dallas is an arc, and so on.   □

2.65 Example   There are four more graphs in Figure 2.25.   □

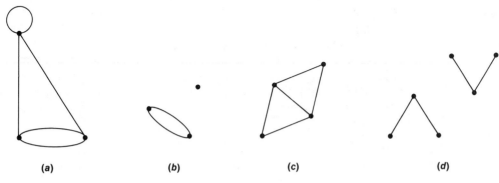

(a)   (b)   (c)   (d)

Figure 2.25

Graphs that at first glance appear different may really be the same according to the definition of a graph. In particular, the shape of an arc is not specified, so that •——• and •〜〰•• are the same graph. Notice that arcs can intersect at points that are not nodes of the graph.

2.66 Practice   According to the definition of a graph, which of the graphs in Figure 2.26 is not the same graph as the one shown in Figure 2.25b?   *(Answer, page A-6)*

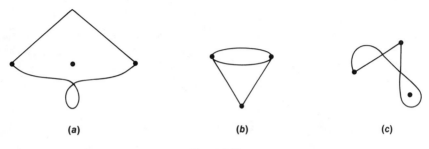

(a)   (b)   (c)

Figure 2.26

2.67 Practice   Sketch a graph having nodes {1, 2, 3, 4, 5}; arcs {$a_1, a_2, a_3, a_4, a_5, a_6$}; and function $f(a_1) = 1\text{--}2$, $f(a_2) = 1\text{--}3$, $f(a_3) = 3\text{--}4$, $f(a_4) = 3\text{--}4$, $f(a_5) = 4\text{--}5$, $f(a_6) = 5\text{--}5$.   *(Answer, page A-6)*

We will need some graph theory terminology. Two nodes in a graph are **adjacent** if they are the endpoints of an arc. In the graph of Practice 2.67, 4 and 5 are adjacent nodes, 3 and 5 are not, and 5 is adjacent to itself. A **loop** in a graph is an arc with endpoints n–n for some node n as in arc a_6 with endpoints 5–5 in Practice 2.67. A graph with no loops is **loop-free**. An **isolated node** is adjacent to no other node, as in Figure 2.25b. The **degree** of a node is the number of arc-ends at that node. In the graph of Practice 2.67 node 2 has degree 1, and nodes, 3, 4, and 5 all have degree 3.

The next two definitions concern the nature of the function f relating arcs to endpoints. Because f is a function, each arc has a unique pair of endpoints. If f is a one-to-one function, then there is at most one arc associated with a pair of endpoints; in this case, the graph is called **simple**. Simple graphs have no parallel arcs. The graph of Practice 2.67 is not simple because of the two arcs, a_3 and a_4, with endpoints 3–4. In Figure 2.25, c and d are simple graphs, a and b are not. A **complete** graph is one in which any two distinct nodes are adjacent. In this case, f is almost an onto function (there does not have to be a loop at every node). Figure 2.25a shows a complete graph.

A **path** from node n_0 to node n_k is a sequence

$$n_0, a_0, n_1, a_1, \ldots, n_{k-1}, a_{k-1}, n_k$$

of nodes and arcs where for each i, the endpoints of arc a_i are n_i–n_{i+1}. In the graph of Practice 2.67, a path from node 2 to node 4 is 2, a_1, 1, a_2, 3, a_4, 4. The **length** of a path is the number of arcs it contains. A graph is **connected** if there is a path from any node to any other node. In Figure 2.25, a and c are connected, b and d are not. A **cycle** in a graph is a path from some node n_0 back to n_0 where no arc appears more than once, n_0 is the only node appearing more than once, and n_0 occurs only at the ends. (Nodes and arcs may be repeated in a path but not, except for node n_0, in a cycle.) In the graph of Practice 2.67, 3, a_3, 4, a_4, 3 is a cycle. A graph with no cycles is **acyclic** —it is loop-free and simple.

Trees

A **tree** is an acyclic, connected graph. Figure 2.27 pictures some trees. Perversely, computer scientists like to draw trees with the "root" at the top.

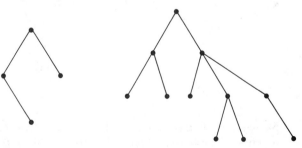

Figure 2.27

2.68 Practice   Answer the following questions for the graph shown in Figure 2.28.

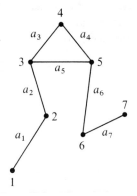

Figure 2.28

(a) Is this graph simple?
(b) Complete?
(c) Connected?
(d) Find two paths from 3 to 6.
(e) Find a cycle.
(f) Name one arc whose removal will make the graph a tree.
(g) Name one arc whose removal will make the graph not connected.

(*Answer, page A-7*)

Directed Graphs

We might want the arcs of a graph to begin at one node and end at another, in which case we would use a directed graph.

2.69 Definition   A **directed graph** is an ordered triple $\langle N, A, f \rangle$ where

N = a set of nodes

A = a set of arcs

f is a function associating with each arc a an ordered pair (x, y) of nodes where x is the **initial point** and y is the **terminal point** of a □

In a directed graph, then, there is a direction associated with each arc. The concept of a path extends as we might expect: a **path** from node n_0 to node n_k is a sequence $n_0, a_0, n_1, a_1, \ldots, n_{k-1}, a_{k-1}, n_k$ where for each i, n_i is the initial point of arc a_i, and n_{i+1} is the terminal point of a_i. If a path exists from node n_0 to node n_k, then n_k is **reachable** from n_0. The definition of a cycle also carries over to directed graphs. A directed graph is **simple** if there is at most one arc from any one node to another.

2.70 Example   In the directed graph of Figure 2.29, there are many paths from node 1 to node 3: 1, a_4, 3, or 1, a_1, 2, a_2, 2, a_2, 2, a_3, 3 are two examples. Node 3 is certainly reachable from 1. Node 1, however, is not reachable from any other node. The cycles in this graph are the loop a_2 and the path 3, a_5, 4, a_6, 3. Figure 2.29 is a simple directed graph.   □

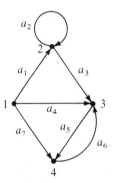

Figure 2.29

Besides imposing direction on the arcs of a graph, we may want other modifications of the basic definition. We would often want the nodes of a graph to carry identifying information, like the names of the cities in the airline route map. We may want the arcs to be "weighted;" for example, we might want to indicate the distances of the various routes in the airline map.

APPLICATIONS

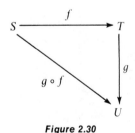

Figure 2.30

Although the idea of a graph is basically very simple, there are an amazing number of situations where relationships between items lend themselves to graphical representation. Not surprisingly, graphs occur frequently in this book. We have used them to visualize partially ordered finite sets. The commutative diagram illustrating composition of functions, Figure 2.30, is a directed graph. Chapter 4 will introduce logic networks and represent them essentially as directed graphs. Directed graphs will also be used to describe finite-state machines in Chapter 7, and the derivations of words in certain formal languages will be shown as trees in Chapter 10 (these are the parse trees generated by a compiler in the course of analyzing a computer program written in a programming language).

Exercise 12 of Section 2.1 describes the organization of data into a tree structure for an efficient search procedure. These trees allow for efficient search through a collection of records to locate a particular record or to determine that it is not in the collection. Such a search would occur when checking for a volume in a library, a patient's medical record in a hospital, an individual's credit record for a loan application, and so on.

We saw that the airline route map was a graph. A representation of any network of transportation routes (a road map, for example), communication lines, or product or service distribution routes such as natural gas pipelines or water mains is a graph. The chemical structure of a molecule is graphically represented. A "family tree" is a graph although the complexities of intermarriage arrangements may not make it a tree in the technical sense. (Information obtained from a family tree is not only interesting, it is useful for research in medical genetics.) The organization chart indicating who reports to whom in a large company is usually a tree. The underlying structure in a flow chart is a directed graph; so is that of an electrical circuit. Directed graphs are used to optimize completion of a complex task, perhaps the manufacture of a product, through procedures called PERT (Program Evaluation and Review Technique) or critical path analysis.

2.71 Practice   Draw the underlying graph in each of the following cases:

(a) Figure 2.31 is a section of road map for part of the state of Arizona.
(b) Figure 2.32 is a representation of an ozone molecule with three oxygen atoms.

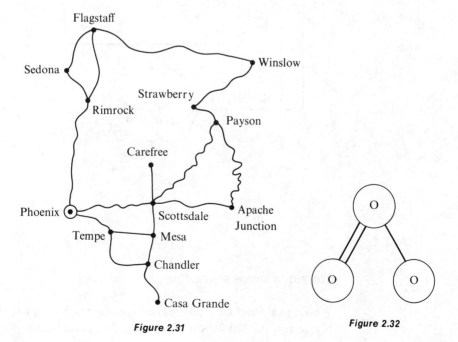

Figure 2.31   Figure 2.32

(c) Figure 2.33 is the flowchart for a computer program to read a sequence of nonnegative integers, print those greater than seven, and stop when a zero input is encountered. (This graph will be directed.)   (*Answer, page A-7*)

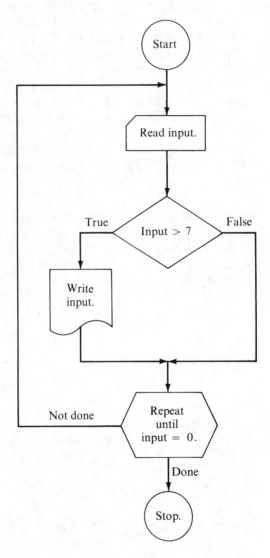

Figure 2.33

COMPUTER REPRESENTATION

We have said that the major advantage of a graph is its visual representation of information. But for storage and manipulation within a computer, this information must be represented in another way—matrix representation, for example.

Suppose a graph has m nodes, numbered as $n_1, n_2, \ldots, n_m$. We can form an $m \times m$ matrix where entry i,j is the number of arcs between nodes n_i and n_j. This matrix is called the **adjacency matrix** A of the graph. Thus,

$$a_{ij} = p \text{ where there are } p \text{ arcs between } n_i \text{ and } n_j$$

2.72 Example

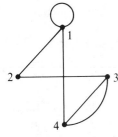

Figure 2.34

The adjacency matrix for the graph in Figure 2.34 is a 4×4 matrix. Entry 1,1 is a 1 due to the loop at node 1. All other elements on the main diagonal are 0. Entry 2,1 (second row, first column) is a 1 because there is one arc between node 2 and node 1, which also means that entry 1,2 is a 1. So far we have

$$A = \begin{bmatrix} 1 & 1 & - & - \\ 1 & 0 & - & - \\ - & - & 0 & - \\ - & - & - & 0 \end{bmatrix}$$

□

2.73 Practice   Complete the adjacency matrix for Figure 2.34.   (*Answer, page A-8*)

The adjacency matrix in Practice 2.73 is symmetric, which will be true for the adjacency matrix of any undirected graph—for if there are p arcs between n_i and n_j, there are certainly p arcs between n_j and n_i. The symmetry of the matrix means that only elements on or below the main diagonal need to be stored. Therefore, for the graph of Figure 2.34, all the information contained in the graph itself is contained in the array below, and the graph could be reconstructed from this array.

$$\begin{bmatrix} 1 & & & \\ 1 & 0 & & \\ 0 & 1 & 0 & \\ 1 & 0 & 2 & 0 \end{bmatrix}$$

An arc between nodes n_i and n_j is a path of length 1. Let's compute the product of an adjacency matrix A with itself. Recalling the definition of matrix multiplication, entry i,j in A^2, $a_{ij}^{(2)}$, is given by

$$a_{ij}^{(2)} = \sum_{k=1}^{m} a_{ik}a_{kj} = a_{i1}a_{1j} + a_{i2}a_{2j} + \cdots + a_{im}a_{mj}$$

If a term in this sum, such as $a_{i2}a_{2j}$, is 0, then either $a_{i2} = 0$ or $a_{2j} = 0$, and there is no path of length 1 from n_i to n_2 or none from n_2 to n_j. Thus, there are no paths of length 2 from n_i to n_j through n_2. If $a_{i2}a_{2j}$ is not 0, it is a product pq of positive integers. Then there are p paths of length 1 from n_i

to n_2, and q paths of length 1 from n_2 to n_j. This result gives pq possible paths of length 2 from n_i to n_j through n_2. The sum of all such terms gives all possible paths of length 2 from n_i to n_j. Thus, A^2 gives the number of all possible paths of length 2 between nodes of the graph.

2.74 Practice   Compute A^2 for the graph of Figure 2.34. Can you find 5 distinct paths of length 2 from node 4 to node 4 on the graph? *(Answer, page A-8)*

2.75 Practice   (a) What information would be contained in entry i, j of matrix A^n?
(b) What type of proof is called for to prove this formally?

(Answer, page A-8)

In a directed graph, the adjacency matrix A reflects the direction of the arcs. For a directed matrix,

$$a_{ij} = p \text{ where there are } p \text{ arcs from } n_i \text{ to } n_j$$

An adjacency matrix for a directed graph will not necessarily be symmetric since an arc from n_i to n_j does not imply an arc from n_j to n_i. Again, by the same argument as before, the entries of A^n give the number of paths of length n from one node to another.

We may want to know whether one node in a directed graph can be reached from another. In a flowchart, for example, the statement(s) corresponding to a node that is not reachable from the start node could be eliminated from the program. For node n_j to be reached from node n_i, there must be a path of some length from n_i to n_j. If we form the adjacency matrix A for the graph and compute $A^2, A^3, A^4, \ldots$, any path would eventually show up as a positive i,j entry in one of these matrices. But we cannot compute an infinite string of matrices. Fortunately, the length of path we should look for is limited. If there are m nodes in the graph, we can write a path with, at most, $m - 1$ arcs (and m nodes) before a node repeats itself. In a longer path, any section of path between repeating nodes is a cycle that can be eliminated, thereby shortening the path.

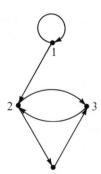

Thus, if a path exists from n_i to n_j, there will be a path from n_i to n_j of length at most $m - 1$. We need only compute $A, A^2, \ldots, A^{m-1}$ to locate such a path. Or we can define a new matrix R where

$$R = A + A^2 + \cdots + A^{m-1}$$

Figure 2.35   Then n_j is reachable from n_i if and only if entry i,j in R is positive.

2.76 Example   Consider the directed graph of Figure 2.35. The adjacency matrix is

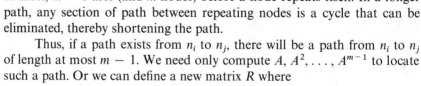

$$A = \begin{bmatrix} 1 & 1 & 0 & 0 \\ 0 & 0 & 1 & 1 \\ 0 & 1 & 0 & 0 \\ 0 & 0 & 1 & 0 \end{bmatrix}$$

□

2.77 Practice   Compute the matrix R for the directed graph of Figure 2.35.

(*Answer, page A-8*)

Matrix representation of a tree is unsatisfactory if we need to know whether one node is to the right or left of another (see Exercise 12, Section 2.1). A more sophisticated representation is required in this event. Also, graphs relating to practical problems can lead to sparse adjacency matrices, that is, matrices with many zeros. This representation wastes computer storage space, and a more efficient representation can be used (see Exercise 10 at the end of this section).

DIRECTED GRAPHS AND BINARY RELATIONS

A simple directed graph defines a binary relation ρ on the set of nodes. Thus, $n_i \rho n_j$ if there is an arc from n_i to n_j. Conversely, if ρ is a binary relation on a set N, we define a corresponding simple directed graph with N the set of nodes and an arc from n_i to n_j whenever $n_i \rho n_j$. Thus, there is a one-to-one correspondence between binary relations and simple directed graphs. If ρ is a reflexive relation, then $n_i \rho n_i$ for every node n_i, and the corresponding graph will have loops at every node, with 1's along the main diagonal of its adjacency matrix. If ρ is symmetric, then $n_i \rho n_j$ implies $n_j \rho n_i$, and the adjacency matrix of the graph is symmetric. When we used graphs to represent partial orderings in Section 2.1, we assumed an upward direction for each arc and eliminated the loops at the nodes. We also eliminated an arc from node n_i to n_j if there was a longer path from n_i to n_j.

2.78 Practice   Explain why the associated binary relation ρ is not antisymmetric for the simple directed graph of Figure 2.29.   (*Answer, page A-8*)

TWO CLASSIC GRAPH THEORY PROBLEMS

Aside from its many applications, graph theory as a separate study has produced a number of interesting problems. We will discuss two here. The first is really a whole class of problems called "path problems."

Path Problems

Swiss mathematician Leonhard Euler (pronounced "oiler") (1707–1783) was intrigued by a puzzle popular among the townfolk of Königsberg, a city in East Prussia, now part of the Kaliningrad Oblast district of the Soviet Union. The river flowing through the city branched around an island. Various bridges over the river created a map as shown in Figure 2.36. The puzzle was to decide whether a person could walk through the city crossing each bridge only once. It is possible, in theory, to answer the question by listing all possible routes, so some dedicated Königsberger could have solved this particular puzzle. Euler's idea was to represent the situation as a graph (see Figure 2.37) where the bridges are arcs and the land masses are nodes. He then solved the general question of when an Euler path exists in any graph.

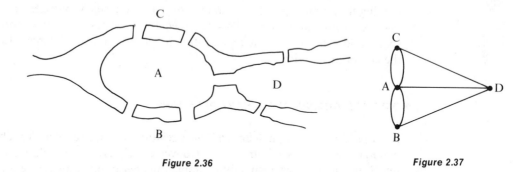

Figure 2.36   Figure 2.37

2.79 Definition   An **Euler path** in a graph is a path that uses each arc only once.   □

2.80 Practice   Do Euler paths exist for either of the graphs of Figure 2.38? (Use trial and error to answer.)   *(Answer, page A-8)*

(a)

Assume all graphs we discuss are connected since otherwise an Euler path generally cannot exist. Whether or not an Euler path exists in a given graph hinges on the degrees of the nodes. A node is **even** if its degree is even and **odd** if its degree is odd. It turns out that in any graph there are always an even number of odd nodes. To see this, choose any graph and let N be the number of odd nodes in it and $N(1)$ the number of nodes of degree 1, $N(2)$ the number of nodes of degree 2, and so on. Then the sum S of the degrees of all the nodes of the graph is

$$S = 1 \cdot N(1) + 2 \cdot N(2) + 3 \cdot N(3) + \cdots + k \cdot N(k) \tag{1}$$

(b)

Figure 2.38

for some k. This sum is, in fact, a count of the total number of arc-ends in the graph. Because the number of arc-ends is twice the number of arcs, S is an even number. We will reorganize equation (1) to group together terms representing odd nodes, and those representing even nodes.

$$S = \underbrace{2 \cdot N(2) + 4 \cdot N(4) + \cdots}_{\text{even nodes}}$$
$$+ \underbrace{1 \cdot N(1) + 3 \cdot N(3) + \cdots + (2n + 1)N(2n + 1)}_{\text{odd nodes}}$$

The sum of the terms representing even nodes is an even number. If we subtract it from both sides of the equation, we get a new equation

$$S' = 1 \cdot N(1) + 3 \cdot N(3) + \cdots + (2n + 1)N(2n + 1) \tag{2}$$

where S' is an even number. Now if we rewrite equation (2) as

$$S' = \underbrace{1 + 1 + \cdots + 1}_{N(1) \text{ terms}} + \underbrace{3 + 3 + \cdots + 3}_{N(3) \text{ terms}} + \cdots$$
$$+ \underbrace{(2n + 1) + (2n + 1) + \cdots + (2n + 1)}_{N(2n + 1) \text{ terms}}$$

we see that there are N terms in the sum, and that each term is an odd number. For the sum of N odd numbers to be even, N must be even. (Can you prove this?) We have proved Theorem 2.81.

2.81   Theorem   The number of odd nodes in any graph is even.   □

Now suppose a graph has an odd node n of degree $2k + 1$, and that an Euler path exists in the graph but that the path does not start at n. Then for each arc we use entering n, there is a new arc to use leaving n until we have used k pairs of arcs. The next time we enter n, there is no new arc on which to leave. Thus, even if our path does not begin at n, it must end at n. Or equivalently, the path either begins or ends at n. If there are more than two odd nodes in the graph, there can be no path, leaving two possible cases where an Euler path might exist—on a graph with no odd nodes or on one with two odd nodes.

Consider the graph with no odd nodes. Pick any node n and begin an Euler path. Whenever you enter a node, you will always have another arc on which to exit until you get back to n. If you have used up every arc of the graph, you are done. If not, there is some node n' of your path with unused arcs. Construct an Euler path beginning and ending at n', much as you did the previous section of path, using all new arcs. Attach this cycle as a side trip on the original path. If you have now used up every arc of the graph, you are done. If not, continue this process until every arc has been covered.

If there are exactly two odd nodes, an Euler path can be started beginning at one odd node and ending at the other. If it has not covered all of the arcs, extra circuits can be patched in as in the previous case.

We now have a complete solution to the Euler path problem.

2.82   Theorem   An Euler path exists in a connected graph if and only if there are no odd nodes or two odd nodes. For the case of no odd nodes, the path can begin at any node and will end there; for the case of two odd nodes, the path can begin at one odd node and end at the other.   □

2.83   Practice   This time using Theorem 2.82, work Practice 2.80.   *(Answer, page A-8)*

2.84   Practice   Is the Königsberg walk possible?   *(Answer, page A-8)*

Another famous mathematician, William Rowan Hamilton (1805–1865), posed a graph theory problem very much like Euler's. He asked how to tell whether a graph has a **Hamiltonian circuit**, a cycle using every node of the graph.

2.85   *Practice*   Do Hamiltonian circuits exist for the graphs of Practice 2.80? (Use trial and error to answer.)   *(Answer, page A-8)*

Theorem 2.82 gives us a simple, efficient procedure, or algorithm, to determine if an Euler path exists on an arbitrary graph. No efficient procedure (trial and error is not efficient!) has been found for determining if a Hamiltonian circuit exists in an arbitrary graph. We will see in Chapter 9 why it is unlikely that such a procedure will ever be found.

However, if we consider only certain types of graphs, we may easily determine if a Hamiltonian circuit exists. In particular, a complete graph has a Hamiltonian circuit because no matter what node you are in, there is always an arc to travel to any unused node, and finally, an arc to return to the starting point.

Suppose we label each arc of a graph with some weighting factor (representing distance, perhaps, or cost). If a Hamiltonian circuit exists for the graph, can we find one with minimum weight? This problem is even more difficult than the Hamiltonian circuit problem. It is sometimes called the "traveling salesman" problem, in honor of the salesman who must visit every city once, return home, and minimize the distance traveled. (The Euler path problem is sometimes called the "highway inspector" problem where the arcs are thought of as roads to be traveled.)

Still assuming the arcs of the graph to be weighted, there are some distance problems with efficient solution procedures. If node n_j is reachable from n_i, there is a procedure to find the path of minimum weight from n_i to n_j. (This problem would be an important one to solve in product distribution where stock in one city must be transported to another city.) A **spanning tree** for a graph is a tree whose set of nodes coincides with the set of nodes for the graph, and whose arcs are (some of) the arcs of the graph. There is a procedure to construct a spanning tree of minimum weight for a graph. (This problem is important in setting up a communications network linking a group of cities.) Adding new nodes and arcs to the graph may result in trees with even less weight (like adding a switchboard in a location other than one of the cities in a communications network).

The Four-Color Problem

The second major graph theory problem we will mention is surely one of the most irritating in the history of mathematics because of its simplicity of statement and difficulty of solution. If a map of various countries, drawn on a sheet of paper, is to be colored so that no two countries with a common border are the same color, how many different colors are required? (We need not worry about countries having only isolated points in common, and we will assume that each country is "connected".) Figure 2.39 shows a map requiring three colors. It is easy to construct a map requiring four colors (see Exercise 17, Section 2.3). No one has ever constructed a map requiring five colors. In fact, a computerized analysis of all possible cases for up to 96 countries has shown that none requires five colors, so if such a map existed, it would have to be very complicated indeed.

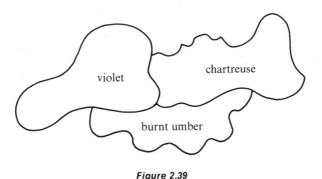

violet

chartreuse

burnt umber

Figure 2.39

The conjecture that four colors are sufficient (the Four-Color Conjecture, or Four-Color Problem) was first proposed to the mathematician Augustus DeMorgan by one of his students in 1852. DeMorgan wrote to Hamilton about the problem. It was not widely discussed, however, until 1878 when Arthur Cayley, one of the most famous mathematicians of the time, announced at a meeting of the London Mathematical Society that he had been unable to solve the problem. A year later, A. B. Kempe, a lawyer, published a "proof" of the four-color conjecture that stood for ten years until it was noted that the proof really only showed that five colors are sufficient.

It is not too difficult to show that five colors are sufficient to color any map. First, we create a dual graph to represent the map. We put one node in each region of the map and an arc between two nodes representing adjacent countries. Then the map-coloring problem becomes one of coloring the nodes of the dual graph so that no two adjacent nodes are the same color. Figure 2.40 shows the dual graph for the map of Figure 2.39. The dual graph will always be simple, loop-free, and **planar**; that is, it can be drawn so that arcs intersect only at nodes. Because we are looking at a single map, the dual graph will also be connected. Now we can prove a relationship that holds between nodes, arcs, and enclosed regions of a connected planar graph.

Figure 2.40

2.86 Theorem

In a connected planar graph with N nodes, A arcs, and R enclosed regions, **Euler's formula**,

$$N - A + R = 1$$

holds.

Here is a good place for an inductive proof. Use induction on the number A of arcs in the graph. Initially, let $A = 0$, which gives us a single node, the simplest graph form. In this case, $N = 1$, $A = 0$, $R = 0$, and the formula holds. Now assume that the formula holds for any connected planar graph with k arcs, and consider the case $A = k + 1$. We can add a new arc to a graph with k arcs in one of two ways: (1) add a new node and connect it by an arc to an old node (see Figure 2.41 a); or (2) connect two old nodes by an arc (see Figure 2.41 b). When procedure (1) is performed, there is one new node and one new arc, so $N - A + R$ stays the same. When procedure (2) is performed, there is one new arc and one new region added, so $N - A + R$ stays the same. Thus, the value of $N - A + R$ has remained constant at 1. □

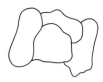

Figure 2.41

2.87 Practice

Draw the dual graph for the map of Figure 2.42 and verify Euler's formula.

(Answer, page A-8)

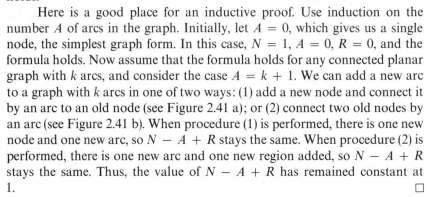

Figure 2.42

Now consider the dual graph of a map and let E be a count of all the edges of enclosed regions. Every enclosed region has at least three edges (the graph is simple and loop-free). Therefore, $E \geq 3R$. At most, each arc in a graph can be counted as the edge of two enclosed regions. However, the outside arcs edge only one region, and arcs extending into a region's interior are not edges at all. Thus, $E \leq 2A$. (The equality holds only in the trivial case of a 1-node graph where both E and A are 0.) Finally, combining these inequalities, $3R \leq 2A$. We can use this inequality to show that there is at least one node of the graph with degree no more than 5 (see Exercise 18, Section 2.3).

We are ready to prove the Five-Color Theorem. Remember that we have translated the map-coloring problem into one of coloring the nodes of a simple, loop-free, connected planar graph so that no two adjacent nodes have the same color. We will use mathematical induction on the number of nodes in such a graph. For the first step of the induction process, it is clear that five colors are sufficient if the number of nodes is less than or equal to 5. Now assume that any such graph with k nodes uses five colors, and then consider such a graph with $k + 1$ nodes. At least one node n of the graph has degree less than or equal to 5; temporarily removing n (and its adjoining arcs) from the graph, we assign a satisfactory coloring (guaranteed by the inductive assumption) to the remaining graph G of k nodes. Now look at the whole graph again. If n has degree less than 5, or if the five nodes adjacent to n do not use five different colors, there is a fifth color left to use for n. Thus, we assume that n is adjacent to five nodes, n_1, n_2, n_3, n_4, n_5, arranged clockwise around n and colored, respectively, colors 1, 2, 3, 4, 5. Pick out all the nodes in the graph colored 1 or 3. Suppose there is no path, using just these nodes, between n_1 and n_3. Then, as far as nodes 1 and 3 are concerned, there are two separate sections of graph, one section containing n_1 and one containing n_3. In the section containing n_1, interchange colors 1 and 3 on all the nodes. Doing this does not disturb the proper coloring of G, it colors n_1 with 3, and leaves color 1 for n. Now suppose there is a path between n_1 and n_3 using only nodes colored 1 or 3. In this case we pick out all nodes in the original graph colored 2 or 4. Is there a path, using just these nodes, between n_2 and n_4? No, there is not. Because of the arrangement of nodes n_1, n_2, n_3, n_4, n_5, such a path would have to cross the path connecting n_1 and n_3. Because the graph is planar, these two paths would have to meet at a node, which would then be colored 1 or 3 from the n_1–n_3 path and 2 or 4 from the n_2–n_4 path— an impossibility. Thus, there is no path using only nodes colored 2 or 4 between n_2 and n_4, and we can rearrange colors as in the previous case.

2.88 Theorem   Five colors are enough to color any map drawn on the plane (Five-Color Theorem).   □

2.89 Practice   Figure 2.43 shows a partially colored graph. A color needs to be assigned to node n. Use the proof of the Five-Color Theorem to color the graph.

(*Answer, page A-8*)

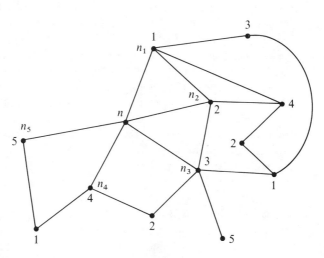

Figure 2.43

After the Five-Color Theorem was established, "proofs" of the Four-Color Conjecture appeared with regularity, subsequently to be shown faulty. Great energy and great minds were applied to the problem, to no avail. Finally, in 1976, two mathematicians at the University of Illinois, Wolfgang Haken and Kenneth Appel, announced a proof of the Four-Color Conjecture.

2.90 *Theorem*   Four colors are enough to color any map drawn on the plane (Four-Color Theorem).   □

To prove the Four-Color Theorem, Haken and Appel used ideas introduced by Kempe in 1879 and patched the flaw in his argument. Kempe defined a normal map as one where no more than three regions meet at a node, and no region encloses another. He then showed that if there is a map requiring five colors, one must exist that is normal and has the least number of regions, called a minimal five-chromatic map. He went on to prove that any normal map must have at least one country with five or fewer adjacent countries, which provided a small set of unavoidable configurations, arrangements of regions that had to occur in every normal map. Kempe then assumed the existence of a normal minimal five-chromatic map, and tried, but failed, to prove that any unavoidable configuration appearing in this map led to a contradiction. Haken and Appel, using a computer, finally developed a set of 1,482 unavoidable configurations, and showed that any of them appearing in a normal minimal five-chromatic map leads to a contradiction.

Until early in the twentieth century, it was thought that for any well-formulated conjecture, such as the Four-Color Problem, there were only two eventual outcomes. If the conjecture were true, someone would eventually produce a proof; it was only a question of accumulating enough theoretical machinery and being sufficiently clever. If the conjecture were false, a counter-example would eventually come to light. This faith in the power of mathematics was shaken by the logician Kurt Gödel who proved in 1931 the existence of statements that, within a very reasonable mathematical framework, were true but could not be proved. (Notice that this is very different from simply saying that no proof has yet been found.) Mathematicians sometimes wondered whether the Four-Color Conjecture could be something of this type. And Haken's and Appel's solution is not the short, neat, elegant type of proof that mathematicians have classically sought. Instead, their proof involves complex computer programs, used not simply to provide numerical answers for a complicated calculation, but to forge the links in a lengthy proof of a "simple" mathematical result. Evidence suggests, furthermore, that it is unlikely that an appreciably shorter proof can be done.

The value of Haken's and Appel's solution to the Four-Color Problem is twofold. Not only does it answer a question that was open for 125 years, it does so by a technique, or at least a technology, undreamed of by the formulators of the problem or by most of those people who have worked on the problem through the years.

✔ CHECKLIST

Definitions

graph (*p. 78*)   path (*p. 80*)
node (*p. 78*)   connected graph (*p. 80*)

Techniques

Find the adjacency matrix for a graph and for a directed graph.

Find which nodes can be reached from another node in a directed graph.

Determine whether an Euler path exists in a graph.

Main Ideas

Graphic representation of many diverse situations.

Matrix representation of graphs.

One-to-one correspondence between binary relations and simple directed graphs.

Simple criterion to determine existence of Euler paths in a graph as opposed to Hamiltonian circuits.

Proof of Five-Color Theorem.

Four-Color Theorem proven by unique use of a computer.

EXERCISES, SECTION 2.3

1. According to the definition of a graph, which of the graphs of Figure 2.44 is not the same as the others?

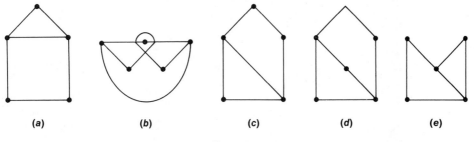

(a)   (b)   (c)   (d)   (e)

Figure 2.44

★ 2. Sketch a picture of each of the following graphs.
   (a) Five nodes, one adjacent to all the others, no others adjacent. Is this a tree?
   (b) A simple graph with three nodes, each of degree 4.
   (c) Four nodes, with cycles of length 1, 2, 3, and 4.

★ 3. (a) Draw a simple, complete, loop-free graph with four nodes.
   (b) Draw a simple, complete, loop-free graph with five nodes.
   (c) Draw a simple, complete, loop-free graph with six nodes.
   (d) Conjecture as to the number of arcs in a simple, complete loop-free graph with n nodes.
   (e) Prove your conjecture.

4. Prove that any tree with n nodes has $n - 1$ arcs.

★ 5. Use the directed graph in Figure 2.45 to answer these:

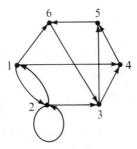

Figure 2.45

   (a) Which nodes are reachable from 3?
   (b) What is the length of the shortest path from 3 to 6?
   (c) Give a path from 1 to 6 of length 8.

6. (a) Write the adjacency matrix A for the graph in Figure 2.46.

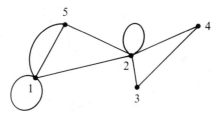

Figure 2.46

(b) The adjacency matrix for an undirected graph is given in "lower triangular form" by

$$
\begin{array}{c}
\quad\ 1\ \ 2\ \ 3\ \ 4 \\
\begin{array}{c} 1 \\ 2 \\ 3 \\ 4 \end{array}
\left[\begin{array}{cccc}
2 & & & \\
1 & 0 & & \\
0 & 1 & 1 & \\
0 & 1 & 2 & 0
\end{array}\right]
\end{array}
$$

Draw the graph.

7. Look at the graph of Figure 2.46. Count the number of paths of length 3 from node 5 to node 2. Then check by computing A^3.

8. (a) Write the adjacency matrix for the directed graph of Figure 2.45.
   (b) The adjacency matrix for a directed graph is given by

$$
\begin{array}{c}
\quad\ \ 1\ \ 2\ \ 3\ \ 4\ \ 5 \\
\begin{array}{c} 1 \\ 2 \\ 3 \\ 4 \\ 5 \end{array}
\left[\begin{array}{ccccc}
0 & 1 & 1 & 0 & 0 \\
0 & 0 & 0 & 0 & 0 \\
0 & 0 & 1 & 1 & 0 \\
0 & 0 & 1 & 0 & 2 \\
1 & 0 & 0 & 0 & 0
\end{array}\right]
\end{array}
$$

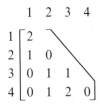

Draw the graph.

9. For the directed graph of Figure 2.45, compute the matrix R to show that nodes 1 and 2 are not reachable from node 3.

★10. A convenient way to represent a graph with a small number of arcs is to use **adjacency lists**. For each node n_i, we list nodes adjacent to n_i. We use "pointers" to get us from one item in the list to the next. Our final representation consists of a two-column array, one column containing the list item and the second containing a pointer to the next item in the list.

Consider the directed graph of Figure 2.47. There is an arc from node 1 to node 2, there are arcs from node 2 to nodes 3 and 4, and there is an arc from node 3 to node 4. In the array for this graph, node 1 has a pointer to row 5, which contains item 2 and indicates an arc from node 1 to node 2. The pointer 0 in row 5 indicates that this is the end of the list for node 1. Node 2 has a pointer to row 6, which contains item 3 and a pointer to row 7; row 7 contains item 4 and a pointer of 0 (end of list for node 2). Thus, there are arcs from node 2 to nodes 3 and 4. Node 3 has a pointer to row 8; row 8 contains item 4 and a pointer of 0, so there is an arc from node 3 to node 4. And finally, node 4 has a pointer of 0,

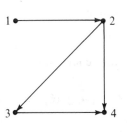

Figure 2.47

	Item	Pointer
nodes 1		5
2		6
3		8
4		0
5	2	0
6	3	7
7	4	0
8	4	0

so there are no arcs emanating from node 4. The total number of entries in the array is 16, exactly the number required for the 4 × 4 adjacency matrix of this graph.

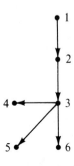

Figure 2.48

(a) Given the graph of Figure 2.48, construct the two-column array representing the graph by adjacency lists.

(b) What is the total number of entries in the array?

(c) What would be the total number of entries in an adjacency matrix for this graph?

11. Let ρ be a binary relation defined on the set $\{0, \pm 1, \pm 2, \pm 4, \pm 16\}$ by $x \rho y \Leftrightarrow y = x^2$. Draw the associated directed graph.

★ 12. (a) Which of the graphs of Figure 2.49 have Euler paths?

(b) Which have Hamiltonian circuits?

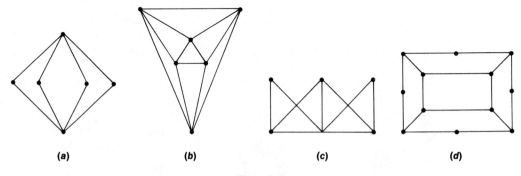

(a)   (b)   (c)   (d)

Figure 2.49

13. Prove that a Hamiltonian circuit always exists in a connected graph where every node has degree 2.

14. Consider a connected graph with $2n$ odd vertices, $n \geq 2$. By Theorem 2.82, an Euler path does not exist for this graph.
    (a) What is the minimum number of disjoint Euler paths, each traveling some of the arcs of the graph, necessary to travel each arc exactly once?
    (b) Show that the minimum number is sufficient.

15. (a) Is it possible to walk in and out of the house of Figure 2.50 so that each door of the house is used exactly once?
    (b) Why or why not?

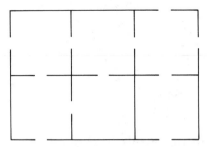

Figure 2.50

16. (a) Find a spanning tree of minimum weight for the labeled graph of Figure 2.51. What is its weight?

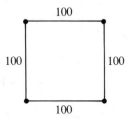

Figure 2.51

    (b) Put a node in the center of the square. Add new arcs from the center to the corners. Find a spanning tree for the new graph, and compute its (approximate) weight.

★17. Draw a map requiring four colors. Draw its dual graph and verify Euler's formula.

18. (a) Consider a simple, loop-free, connected, planar graph. We showed earlier that $3R \leq 2A$. Use this inequality in Euler's formula to show that $A \leq 3N - 3$.

(b) Now use the inequality of part (a) to show that such a graph has at least one node of degree no more than 5. (Hint—use proof by contradiction.)

19. Color the map in Figure 2.52 with four colors. If you have misplaced your crayons, use numbers for colors. Can it be colored with three colors?

Figure 2.52

20. We will prove the Six-Color Theorem without using the dual graph of a map. Instead, we will merely straighten the boundaries of regions, so that the problem of coloring the map shown in Figure 2.53a is represented by the problem of coloring the regions of the graph in Figure 2.53b. First, we assume that no country has a hole in it. Then the graph will always be loop-free, planar, and connected. Also, every node will have degree at least 3.

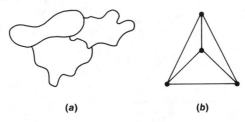

(a)   (b)

Figure 2.53

(a) Show that we can assume our graph to be simple by proving that if six colors are sufficient for coloring a simple graph, they are sufficient for a nonsimple graph as well. (Hint: use temporary small countries at nodes.)

(b) Consider a simple, loop-free, connected, planar graph and assume that every region has at least six sides. Show that $R \leq \frac{1}{3}A$.

(c) Use the result of part (b) to show that in such a graph, $2A \leq 3N - 3$.

(d) Now consider a simple, loop-free, connected, planar graph where every node has degree at least 3. Show that such a graph has at least one region with no more than five sides.

(e) Prove that six colors are sufficient to color any planar map where no country has a hole in it.

(f) Prove that six colors are sufficient to color any planar map. (Hint: cut some temporary slits in the map.)

For Further Reading

Relations and functions are standard fare and without going far afield there is not a great deal more to say about them than what we've already said; however, [1] and [2] parallel what has been done in this chapter. A classical development of graph theory is given in [3]; [4] presents not only the basic ideas of graph theory but also many of its applications. A combinatorial look at graph theory is the theme of [5]. Applications of graphs to computer sicence and algorithms for solution of many graph problems are presented in [6] and [7]. The solution to the Four-Color Problem is discussed by one of its solvers in [8].

1. Charles C. Pinter, *Set Theory*. Reading, Mass.: Addison-Wesley, 1971.

2. Robert R. Stoll, *Sets, Logic, and Axiomatic Theories*, 2nd ed. San Francisco: W. H. Freeman and Company, 1974.

3. Oystein Ore, *Theory of Graphs*. Providence, R.I.: American Mathematical Society Colloquium Publications, vol. 38, 1962.

4. Robert G. Busacker and Thomas L. Saaty, *Finite Graphs and Networks*. New York: McGraw-Hill, 1965.

5. Frank Harary and Edgar M. Palmer, *Graphical Enumeration*. New York: Academic Press, 1973.

6. Donald E. Knuth, *The Art of Computer Programming, Vol. 1*, 2nd ed. Reading, Mass.: Addison-Wesley, 1973.

7. Alfred V. Aho, John E. Hopcroft, and Jeffrey D. Ullman, *The Design and Analysis of Computer Algorithms*. Reading, Mass.: Addison-Wesley, 1974.

8. Wolfgang Haken, "An attempt to understand the Four-Color Problem," *Journal of Graph Theory*, vol. 1, no. 3, 1977.

Chapter 3

STRUCTURES AND SIMULATIONS

You may wonder what the title of this book means (does anyone ever read the title of a textbook?) What is a mathematical structure? This phrase might conjure up images of vastly complicated architecture, or maybe the building that houses the computer center!

In this chapter we will explore what a mathematical structure is. An analogy, if not carried too far, may be helpful. We can think of the skeleton as the basic structure of the human body. People may be thin, fat, short, tall, black, white, and so on, but stripped down to skeletons they all look pretty much alike. The outward physical forms differ, but the inward structure, the shape and arrangement of bones, is the same. There are certain essentials, an underlying sameness, linking all the different specimens. Mathematical structures attempt to get at the underlying similarities in situations that outwardly may appear different. The structures are defined to model, or simulate, the common characteristics. This simulation of common characteristics encountered in diverse situations we will call Simulation I.

Once we have some structures defined, we can talk about another kind of simulation. Perhaps it is possible for one occurrence of a particular structure to do the job of (simulate) another instance of that structure. This sort of simulation, Simulation II, will help us classify occurrences of structures and recognize those that are really the same.

These ideas are necessarily vague at present but will become more concrete as you read this chapter.

Section 3.1   **STRUCTURES—SIMULATION I**

THE BOOLEAN ALGEBRA STRUCTURE

Development

We have just said that mathematical structures are defined to simulate common characteristics encountered in diverse situations. Let's review three situations we saw in Chapters 1 and 2.

In Section 1.1 we studied statement logic. We used the symbols $\vee$ and $\wedge$ to stand for disjunction and conjunction, respectively, and A' to denote the negation of a statement A. Equality between statements meant that the statements were equivalent (they had the same truth values). Zero stood for any contradiction, a statement with truth values always false, and 1 stood for any tautology, a statement with truth values always true. Recall the list of tautologies.

1a. $A \vee B = B \vee A$ 1b. $A \wedge B = B \wedge A$
2a. $(A \vee B) \vee C = A \vee (B \vee C)$ 2b. $(A \wedge B) \wedge C = A \wedge (B \wedge C)$
3a. $A \vee (B \wedge C) = (A \vee B) \wedge (A \vee C)$ 3b. $A \wedge (B \vee C) = (A \wedge B) \vee (A \wedge C)$
4a. $A \vee 0 = A$ 4b. $A \wedge 1 = A$
5a. $A \vee A' = 1$ 5b. $A \wedge A' = 0$

In Section 1.3, we studied set identities among the subsets of a set S. Here $\cup$ and $\cap$ denoted union and intersection of sets, A' was the complement of a set A, and $\varnothing$ was the empty set. Recall the list of set identities.

1a. $A \cup B = B \cup A$ 1b. $A \cap B = B \cap A$
2a. $(A \cup B) \cup C = A \cup (B \cup C)$ 2b. $(A \cap B) \cap C = A \cap (B \cap C)$
3a. $A \cup (B \cap C) = (A \cup B) \cap (A \cup C)$ 3b. $A \cap (B \cup C) = (A \cap B) \cup (A \cap C)$
4a. $A \cup \varnothing = A$ 4b. $A \cap S = A$
5a. $A \cup A' = S$ 5b. $A \cap A' = \varnothing$

And in Section 2.1, for ρ and σ members of the set of all binary relations on a given set S, we defined binary relations $\rho + \sigma$, $\rho \cdot \sigma$, and ρ'. Then

1a. $\rho + \sigma = \sigma + \rho$ 1b. $\rho \cdot \sigma = \sigma \cdot \rho$
2a. $(\rho + \sigma) + \gamma = \rho + (\sigma + \gamma)$ 2b. $(\rho \cdot \sigma) \cdot \gamma = \rho \cdot (\sigma \cdot \gamma)$
3a. $\rho + (\sigma \cdot \gamma) = (\rho + \sigma) \cdot (\rho + \gamma)$ 3b. $\rho \cdot (\sigma + \gamma) = (\rho \cdot \sigma) + (\rho \cdot \gamma)$
4a. $\rho + \varnothing = \rho$ 4b. $\rho \cdot S^2 = \rho$
5a. $\rho + \rho' = S^2$ 5b. $\rho \cdot \rho' = \varnothing$

These three lists of properties are similar. Disjunction of statements, union of sets, sum of binary relations—all seem to play the same roles in their respective environments, as do conjunction of statements, intersection of sets, and product of binary relations. A contradiction seems to correspond to the empty set, and a tautology to S for sets, or S^2 for binary relations.

Now suppose we try to formally characterize the similarities in these three cases. In each case we are talking about items from a set: a set of statement letters, a set of subsets of a set S, or a set of binary relations on a set S. In each case we have two binary operations and one unary operation on the members of the set: disjunction/conjunction/negation, union/intersection/complementation, or sum/product/complementation. In each case there are two distinguished elements of the set: 0/1, $\varnothing/S$, or $\varnothing/S^2$. And, finally, there are the 10 properties that hold in each case. Whenever all these features occur, we will say that we have a Boolean algebra.

Definition

3.1   Practice   Complete Definition 3.2.   (*Answer, page A-9*)

3.2   Definition   A **Boolean algebra** is a set B on which is defined two binary operations, $+$ and $\cdot$; one unary operation, $'$; and in which there are two distinct elements, 0 and 1, such that the following properties hold for all $x, y, z \in B$:

1a. $x + y = y + x$	1b. —————— (commutative properties)
2a. ——————	2b. —————— (associative properties)
3a. ——————	3b. —————— (distributive properties)
4a. ——————	4b. $x \cdot 1 = x$ (identities)
5a. ——————	5b. —————— (complements) □

Your definition should now look like this:

3.2   Definition   A **Boolean algebra** is a set B on which is defined two binary operations, $+$ and $\cdot$; one unary operation, $'$; and in which there are two distinct elements, 0 and 1, such that the following properties hold for all $x, y, z \in B$:

1a. $x + y = y + x$	1b. $x \cdot y = y \cdot x$ (commutative properties)
2a. $(x + y) + z$ $= x + (y + z)$	2b. $(x \cdot y) \cdot z$ $= x \cdot (y \cdot z)$ (associative properties)
3a. $x + (y \cdot z)$ $= (x + y) \cdot (x + z)$	3b. $x \cdot (y + z)$ $= x \cdot y + x \cdot z$ (distributive properties)
4a. $x + 0 = x$	4b. $x \cdot 1 = x$ (identities)
5a. $x + x' = 1$	5b. $x \cdot x' = 0$ (complements) □

What, then, is the Boolean algebra structure? It is a formalization that simulates, or models, the three cases we have considered (and perhaps others as well). There is a subtle philosophical distinction between the formalization itself, the *idea* of the Boolean algebra structure, and any instance of the formalization, such as the three cases with which we started. Nevertheless, we will often use the term "Boolean algebra" to describe both the idea and its occurrences. Doing this should not be confusing. We often have a mental idea ("chair," for example,), and whenever we encounter a concrete example of the idea, we also call it by our word for the idea.

The formalization helps us focus on the essential features common to all examples of Boolean algebras, and we can use these features, these facts available from the definition of a Boolean algebra, to prove other facts about Boolean algebras. Then these new facts hold in any particular instance of a Boolean algebra. (If we ascertain that in a typical human skeleton "the thighbone is connected to the kneebone," then we don't need to reconfirm this in every person we meet.) This type of generalization is typical of modern, or abstract, mathematics.

We denote a Boolean algebra by $[B, +, \cdot, ', 0, 1]$.

3.3 Example   Let $B = \{0, 1\}$ and define binary operations $+$ and $\cdot$ on B by $x + y = \max(x, y)$, $x \cdot y = \min(x, y)$. Thus,

+	0	1		$\cdot$	0	1
0	0	1		0	0	0
1	1	1		1	0	1

Define the unary operation $'$ by the table

$'$	
0	1
1	0

that says $0' = 1$ and $1' = 0$. Then $[B, +, \cdot, ', 0, 1]$ is a Boolean algebra. We can verify the 10 properties by checking all possible cases. Thus, for property 2b, associativity of $\cdot$, we show that

$$(0 \cdot 0) \cdot 0 = 0 \cdot (0 \cdot 0) = 0$$

$$(0 \cdot 0) \cdot 1 = 0 \cdot (0 \cdot 1) = 0$$

$$(0 \cdot 1) \cdot 0 = 0 \cdot (1 \cdot 0) = 0$$

$$(0 \cdot 1) \cdot 1 = 0 \cdot (1 \cdot 1) = 0$$

$$(1 \cdot 0) \cdot 0 = 1 \cdot (0 \cdot 0) = 0$$

$$(1 \cdot 0) \cdot 1 = 1 \cdot (0 \cdot 1) = 0$$

$$(1 \cdot 1) \cdot 0 = 1 \cdot (1 \cdot 0) = 0$$

$$(1 \cdot 1) \cdot 1 = 1 \cdot (1 \cdot 1) = 1$$

For property 4a, we show that

$$0 + 0 = 0$$

$$1 + 0 = 1$$

□

Properties

We can prove other properties holding in any Boolean algebra by using the properties in the definition.

3.4 Example   The **idempotent property**, $x + x = x$, holds in any Boolean algebra. Thus,

$$
\begin{aligned}
x + x &= (x + x) \cdot 1 & \text{(4b)} \\
&= (x + x) \cdot (x + x') & \text{(5a)} \\
&= x + (x \cdot x') & \text{(3a)} \\
&= x + 0 & \text{(5b)} \\
&= x & \text{(4a)} \qquad \square
\end{aligned}
$$

As with set identities, each property in the definition of a Boolean algebra has its **dual** as part of the definition, where the dual is obtained by interchanging $+$ and $\cdot$, and 1 and 0. Therefore, every time a new property about Boolean algebras is proved, its dual also holds.

3.5 Example   The dual of the property in Example 3.4, $x \cdot x = x$, is true in any Boolean algebra. $\square$

3.6 Practice   (a) What does the idempotent property of Example 3.4 say in the context of statement logic?
(b) What does it say in the context of sets?   *(Answer, page A-9)*

Once a property about Boolean algebras is proved, we can use it to prove new properties.

3.7 Practice   (a) Prove that the property $x + 1 = 1$ holds in any Boolean algebra. Give a reason for each step.
(b) What is the dual property?   *(Answer, page A-9)*

More properties of Boolean algebras appear in Exercise 4 at the end of this section.

For x an element of a Boolean algebra B, the element x' is called the **complement** of x. We could also define the complement of x as that element x' such that

$$x + x' = 1 \text{ and } x \cdot x' = 0$$

because the element with these properties is unique. To prove this, suppose x_1 and x_2 are two elements of B with

$$x + x_1 = 1, \qquad x \cdot x_1 = 0$$

and

$$x + x_2 = 1, \qquad x \cdot x_2 = 0$$

Then

$$x_1 = x_1 + 0 \qquad \text{(4a)}$$
$$= x_1 + (x \cdot x_2) \qquad (x \cdot x_2 = 0)$$
$$= (x_1 + x) \cdot (x_1 + x_2) \qquad \text{(3a)}$$
$$= (x + x_1) \cdot (x_1 + x_2) \qquad \text{(1a)}$$
$$= 1 \cdot (x_1 + x_2) \qquad (x + x_1 = 1)$$
$$= (x_1 + x_2) \cdot 1 \qquad \text{(1b)}$$
$$= x_1 + x_2 \qquad \text{(4b)}$$

A similar argument shows that

$$x_2 = x_2 + x_1$$

Then

$$x_1 = x_1 + x_2 = x_2 + x_1 = x_2$$

and $x_1 = x_2$.

We need to examine what *uniqueness* means in the context of statement logic. We just concluded $x_1 = x_2$. If the Boolean algebra being discussed is statement logic, how do we interpret this conclusion? Recall that we used an equal sign to denote that two statements are equivalent, that is, have the same truth values. Thus, for statement letters A and B, the statements $A \vee (B \vee A)$ and $B \vee A$ are equal. (Write the truth tables for each.) Statements $A \wedge A'$ and $B \wedge (A \wedge A')$ are also equal even though $A \wedge A'$ does not involve B. The truth table for $A \wedge A'$ can be written with a B column and four rows by simply ignoring the truth value of B. Then the statements $A \wedge A'$ and $B \wedge (A \wedge A')$ clearly have the same truth values.

3.8 Practice   Prove that $0' = 1$ and $1' = 0$. (Hint: $1' = 0$ will follow by duality from $0' = 1$. To show $0' = 1$, show that 1 satisfies the two properties of the complement of 0.)   *(Answer, page A-9)*

There are many ways to define a Boolean algebra. Indeed, in Definition 3.2, we could have omitted the associative properties since these can be derived from the remaining properties of the definition. In Chapter 4 we will see an application of Boolean algebra to computer logic, and Boolean algebra will also indicate how to physically realize the finite-state machines of Chapter 7.

DEVELOPMENT OF OTHER STRUCTURES

A Boolean algebra is our first example of a mathematical structure. We will generalize from it to see what might be involved in the definition of any structure. First, there will be one or more sets of objects, with perhaps some elements of the set or sets singled out as special in some way. There may be binary or unary operations defined on the sets as in a Boolean algebra, or, in other cases, functions may be defined using the sets as domains and codomains. Finally, there will be a list of properties that the elements of the set(s) obey under the operations or functions.

Our definition of the Boolean algebra structure came about as a simulation of three particular mathematical situations—statement logic, subsets of a set, and binary relations on a set. For the remainder of this section, we will consider four more mathematical situations, explore their properties, and then define another structure simulating these four cases.

For the first mathematical situation, let's look at 2×2 matrices where the matrix entries are integers. We will denote the set of such matrices by $M_2(\mathbb{Z})$. Now what happens when we multiply such matrices? Notice by the definition of matrix multiplication that for $\begin{bmatrix} a_{11} & a_{12} \\ a_{21} & a_{22} \end{bmatrix}$ and $\begin{bmatrix} b_{11} & b_{12} \\ b_{21} & b_{22} \end{bmatrix}$ in $M_2(\mathbb{Z})$,

$$\begin{bmatrix} a_{11} & a_{12} \\ a_{21} & a_{22} \end{bmatrix} \begin{bmatrix} b_{11} & b_{12} \\ b_{21} & b_{22} \end{bmatrix} \begin{bmatrix} a_{11}b_{11} + a_{12}b_{21} & a_{11}b_{12} + a_{12}b_{22} \\ a_{21}b_{11} + a_{22}b_{21} & a_{21}b_{12} + a_{22}b_{22} \end{bmatrix}$$

The resulting product exists and is a unique member of $M_2(\mathbb{Z})$ (note that it has integer entries), so matrix multiplication is a binary operation on $M_2(\mathbb{Z})$. If X, Y, and Z belong to $M_2(\mathbb{Z})$, then a simple, if tedious, calculation will confirm that

$$(X \cdot Y) \cdot Z = X \cdot (Y \cdot Z)$$

This fact is called the **associative property** of multiplication on $M_2(\mathbb{Z})$. (In general, associativity of a binary operation means that grouping of elements does not affect the answer.) Also, the 2×2 matrix I, where $I = \begin{bmatrix} 1 & 0 \\ 0 & 1 \end{bmatrix}$ belongs to $M_2(\mathbb{Z})$ and has the property that for any $X \in M_2(\mathbb{Z})$,

$$X \cdot I = I \cdot X = X$$

I is called the **identity matrix**.

3.9 Practice   Show that for any $X \in M_2(\mathbb{Z})$, $X \cdot I = I \cdot X = X$. (Hint: if X is a typical member of $M_2(\mathbb{Z})$, then X can be written as $\begin{bmatrix} a_{11} & a_{12} \\ a_{21} & a_{22} \end{bmatrix}$.)   (*Answer, page A-9*)

To summarize, matrix multiplication is a binary operation on $M_2(\mathbb{Z})$, the multiplication is associative, and there is an identity matrix.

In the second mathematical situation we want to consider, we will limit ourselves to 2×2 matrices where the elements on the main diagonal are integers and those off the main diagonal are zero. We will denote this set of matrices by $M_2^D(\mathbb{Z})$. Thus a typical element of $M_2^D(\mathbb{Z})$ is a matrix of the form

$$\begin{bmatrix} a_{11} & 0 \\ 0 & a_{22} \end{bmatrix}$$

where a_{11} and a_{22} are integers.

3.10 Practice   Prove that matrix multiplication is a binary operation on $M_2^D(\mathbb{Z})$.

(*Answer, page A-9*)

Notice that $M_2^D(\mathbb{Z}) \subseteq M_2(\mathbb{Z})$, which allows us to conclude immediately that multiplication on $M_2^D(\mathbb{Z})$ is associative. If X, Y, Z belong to $M_2^D(\mathbb{Z})$, they also belong to $M_2(\mathbb{Z})$, and we already know that the equation $(X \cdot Y) \cdot Z = X \cdot (Y \cdot Z)$ is true in $M_2(\mathbb{Z})$. Also, the identity matrix I belongs to $M_2^D(\mathbb{Z})$ and for any X in $M_2^D(\mathbb{Z})$, $X \cdot I = I \cdot X = X$ because this equation is true for any X in $M_2(\mathbb{Z})$. Therefore, $M_2^D(\mathbb{Z})$ under matrix multiplication has all of the properties that $M_2(\mathbb{Z})$ has under multiplication. But there is another property as well. For X and Y any members of $M_2^D(\mathbb{Z})$,

$$X \cdot Y = Y \cdot X$$

This fact is called the **commutative property** of multiplication on $M_2^D(\mathbb{Z})$. (In general, commutativity of a binary operation means that the order of the elements does not affect the answer.)

3.11 Practice   (a) Show that matrix multiplication on $M_2^D(\mathbb{Z})$ is commutative.
(b) Is matrix multiplication on $M_2(\mathbb{Z})$ commutative? Prove or find a counterexample.   (*Answer, page A-9*)

Our third mathematical case is a little less exotic. Here we will consider the set $\mathbb{R}$ of real numbers under ordinary addition. Certainly addition is a binary operation on $\mathbb{R}$. For any real numbers x, y, and z,

$$(x + y) + z = x + (y + z)$$

so that addition on $\mathbb{R}$ is associative. Next we look for an element of $\mathbb{R}$ that will have the same effect in the real numbers under addition as did I in $M_2(\mathbb{Z})$ under matrix multiplication. Recall that for all $X \in M_2(\mathbb{Z})$,

$$X \cdot I = I \cdot X = X$$

Now consider any $x \in \mathbb{R}$, with the operation this time being addition. An identity element i must be a real number such that for any $x \in \mathbb{R}$,

$$x + i = i + x = x$$

3.12 Practice   What is the identity element in the set $\mathbb{R}$ under addition?   (*Answer, page A-9*)

3.13 Practice   Is addition on $\mathbb{R}$ commutative?   (*Answer, page A-9*)

To summarize this case, addition is a binary operation on $\mathbb{R}$ that is associative and commutative, and an identity element exists. But again there is a new property. For every $x \in \mathbb{R}$, there is an element of $\mathbb{R}$ called the **inverse** of x, namely $-x$, such that

$$x + (-x) = (-x) + x = 0$$

3.14 Practice   Translate the "inverse element" property into a corresponding statement about $M_2^D(\mathbb{Z})$ under multiplication by completing the following: for each $X \in M_2^D(\mathbb{Z})$, there is a matrix in $M_2^D(\mathbb{Z})$ called the inverse of X, denoted by X^{-1}, such that

$$\underline{\hspace{1cm}} = \underline{\hspace{1cm}} = \underline{\hspace{1cm}}$$

(Remember that in $M_2^D(\mathbb{Z})$, the operation is multiplication and the identity is I.)   (*Answer, page A-9*)

3.15 Example   Consider the matrix

$$\begin{bmatrix} 2 & 0 \\ 0 & 2 \end{bmatrix}$$

of $M_2^D(\mathbb{Z})$. If this element were to have an inverse matrix in $M_2^D(\mathbb{Z})$, it would be of the form

$$\begin{bmatrix} a_{11} & 0 \\ 0 & a_{22} \end{bmatrix}$$

and have the property that

$$\begin{bmatrix} 2 & 0 \\ 0 & 2 \end{bmatrix}\begin{bmatrix} a_{11} & 0 \\ 0 & a_{22} \end{bmatrix} = \begin{bmatrix} a_{11} & 0 \\ 0 & a_{22} \end{bmatrix}\begin{bmatrix} 2 & 0 \\ 0 & 2 \end{bmatrix} = \begin{bmatrix} 1 & 0 \\ 0 & 1 \end{bmatrix}$$

By the definition of matrix multiplication, the only matrix satisfying this equation has $a_{11} = a_{22} = 1/2$, and it is not a member of $M_2^D(\mathbb{Z})$. Thus, unlike real numbers under addition, not every member of $M_2^D(\mathbb{Z})$ has an inverse element in $M_2^D(\mathbb{Z})$. This example can also be used to show that not every element of $M_2(\mathbb{Z})$ has an inverse element in $M_2(\mathbb{Z})$. $\square$

Our final example looks at the set $\mathbb{R}^+$ of positive real numbers under multiplication.

3.16 Practice   Consider $\mathbb{R}^+$ under multiplication.

(a) Is multiplication a binary operation on $\mathbb{R}^+$?
(b) Is multiplication on $\mathbb{R}^+$ associative? What equation must be satisfied?
(c) Is multiplication on $\mathbb{R}^+$ commutative? What equation must be satisfied?

(d) What equation would an identity element have to satisfy? Is there an identity element?

(e) What equation expresses the inverse element property? Does every member of $\mathbb{R}^+$ have an inverse?

(Answer, page A-10)

3.17 Practice   If we change to $\mathbb{R}$ under multiplication, will your answers to the questions of Practice 3.16 change?   *(Answer, page A-10)*

Our final case, $\mathbb{R}^+$ under multiplication, has all of the properties of the previous case, $\mathbb{R}$ under addition.

Now, what kind of structure can we define that will simulate all four of these cases? All four cases involve one set and a binary operation on that set. Properties common to all four cases are associativity and the existence of an identity element.

3.18 Definition $[S, \cdot]$ is a **monoid** if S is a nonempty set, $\cdot$ is a binary operation on S, and

(a) $\cdot$ is associative

(b) there is an element $i \in S$ (identity element) such that for any $s \in S$, $s \cdot i = i \cdot s = s$ □

The monoid structure, like the Boolean algebra structure, is an abstraction of properties found in a number of different situations. In the definition of a monoid we denoted the binary operation by a multiplication symbol, but in any particular monoid, this symbol has to be properly interpreted. In our four examples of monoids, the interpretation would be matrix multiplication, matrix multiplication, real number addition, and real number multiplication, respectively.

Only our first example, $[M_2(\mathbb{Z}), \cdot]$, is a monoid and nothing more. $[M_2^D(\mathbb{Z}), \cdot]$ is not only a monoid but a commutative monoid. The last two cases, $[\mathbb{R}, +]$ and $[\mathbb{R}^+, \cdot]$, are not only monoids but also examples of a structure called a commutative group.

We defined the monoid structure in an attempt to model some arithmetic properties we had observed in various contexts. In Chapter 5 we will look at several other structures that simulate arithmetic properties. Then, in Chapter 7, we will design a structure to simulate a computer, and in Chapter 9 consider a structure to simulate the most general sort of computation process.

✔ **CHECKLIST**

Definitions

Boolean algebra (*p. 104*)
dual (of a Boolean algebra property) (*p. 106*)

complement (of a Boolean algebra element) (*p. 106*)

monoid (*p. 111*)

Techniques

Prove properties about Boolean algebras.

Decide whether something is a Boolean algebra.

Decide whether something is a monoid.

Main Ideas

Mathematical structures as abstractions of common properties found in diverse situations (Simulation I).

EXERCISES, SECTION 3.1

★ 1. Let $B = \{0, 1, a, a'\}$. $+$ and $\cdot$ are binary operations on B. $'$ is a unary operation on B defined by the table

0	1
1	0
a	a'
a'	a

Suppose you know that $[B, +, \cdot, ', 0, 1]$ is a Boolean algebra. Making use of the properties that must hold in any Boolean algebra, fill in the tables defining the binary operations $+$ and $\cdot$.

$+$	0	1	a	a'
0				
1				
a				
a'				

$\cdot$	0	1	a	a'
0				
1				
a				
a'				

2. Define two binary operations $+$ and $\cdot$ on the set $\mathbb{Z}$ of integers by $x + y = \max(x, y)$ and $x \cdot y = \min(x, y)$.

   (a) Show that the commutative, associative, and distributive properties of a Boolean algebra hold for these two operations on $\mathbb{Z}$.

   (b) Show that no matter what element of $\mathbb{Z}$ is chosen to be 0, the property $x + 0 = x$ of a Boolean algebra fails to hold.

3. Let S be the set $\{0, 1\}$. Then S^2 is the set of all ordered pairs of 0's and 1's; $S^2 = \{(0, 0), (0, 1), (1, 0), (1, 1)\}$. Consider the set B of all functions

mapping S^2 to S. For example, one such function, $f(x, y)$, is given by

$$f(0, 0) = 0$$
$$f(0, 1) = 1$$
$$f(1, 0) = 1$$
$$f(1, 1) = 1$$

(a) How many elements are in B?

(b) For f_1 and f_2 members of B, and $(x, y) \in S^2$, define

$$(f_1 + f_2)(x, y) = \max(f_1(x, y), f_2(x, y))$$
$$(f_1 \times f_2)(x, y) = \min(f_1(x, y), f_2(x, y))$$
$$f'_1(x, y) = \begin{cases} 1 & \text{if } f_1(x, y) = 0 \\ 0 & \text{if } f_1(x, y) = 1 \end{cases}$$

Suppose

$$f_1(0, 0) = 1 \qquad f_2(0, 0) = 1$$
$$f_1(0, 1) = 0 \qquad f_2(0, 1) = 1$$
$$f_1(1, 0) = 1 \qquad f_2(1, 0) = 0$$
$$f_1(1, 1) = 0 \qquad f_2(1, 1) = 0$$

What are the functions $f_1 + f_2, f_1 \cdot f_2$, and f'_1?

(c) Prove that $[B, +, \cdot, ', 0, 1]$ is a Boolean algebra where 0 and 1 are defined by

$$0(0, 0) = 0 \qquad 1(0, 0) = 1$$
$$0(0, 1) = 0 \qquad 1(0, 1) = 1$$
$$0(1, 0) = 0 \qquad 1(1, 0) = 1$$
$$0(1, 1) = 0 \qquad 1(1, 1) = 1$$

4. Prove the following properties of Boolean algebras. Give a reason for each step.

★ (a) $(x')' = x$

★ (b) $x + (x \cdot y) = x$ $x \cdot (x + y) = x$ (absorption properties)

(c) $(x + y)' = x' \cdot y'$ $(x \cdot y)' = x' + y'$ (DeMorgan's laws)

★ (d) $x \cdot [y + (x \cdot z)] = (x \cdot y) + (x \cdot z)$

$x + [y \cdot (x + z)] = (x + y) \cdot (x + z)$ (modular properties)

(e) $(x + y) \cdot (x' + y) = y$ $(x \cdot y) + (x' \cdot y) = y$

(f) $(x + y) + (y \cdot x') = x + y$ $(x \cdot y) \cdot (y + x') = x \cdot y$

5. Prove that the 0 element in any Boolean algebra is unique; prove that the 1 element in any Boolean algebra is unique.

6. Prove that no Boolean algebra can have an odd number of elements. (Note that in the definition of a Boolean algebra, 0 and 1 are distinct elements of B, so B has at least two elements. Arrange the remaining elements of B so that each element is paired with its complement.)

7. Show that $M_2(\mathbb{Z})$ under matrix multiplication is associative.

8. Show that the set $M_2(\mathbb{Z})$ under matrix addition is a commutative monoid.

★ 9. (a) Write the equation that subtraction on $\mathbb{Z}$ must satisfy to be associative. Is subtraction on $\mathbb{Z}$ associative?
   (b) Write the equation that subtraction on $\mathbb{Z}$ must satisfy to be commutative. Is subtraction on $\mathbb{Z}$ commutative?

★ 10. Each case below defines a binary operation, denoted by $\cdot$, on a given set. Which are associative? Which are commutative?

(a) on $\mathbb{Z}$, $x \cdot y = \begin{cases} x & \text{if } x \text{ is even} \\ x + 1 & \text{if } x \text{ is odd} \end{cases}$

(b) on $\mathbb{N}$, $x \cdot y = (x + y)^2$

(c) on $\mathbb{R}^+$, $x \cdot y = x^4$

(d) on $\mathbb{Q}$, $x \cdot y = \dfrac{xy}{2}$

(e) on $\mathbb{R}^+$, $x \cdot y = \dfrac{1}{(x + y)}$

Section 3.2   *MORPHISMS—SIMULATION II*

In the preceding section we explored several cases where structures were defined to simulate common properties found in various situations (Simulation I). Now we want to investigate what it might mean for one instance of a particular structure to simulate another instance of that structure (Simulation II). Such a "second order simulation" occurs frequently in computer solutions of problems. A real world situation may be modeled by a continuous function (Simulation I), but the computer deals with a second function that is a discrete approximation of the first (Simulation II).

EXAMPLES OF ISOMORPHISMS

We will start with a very simple case.

3.19 Example   In one of the exercises in Chapter 2, you were asked to draw the graphs of each of two partially ordered sets:

(a) $S = \{1, 2, 3, 5, 6, 10, 15, 30\}$; $x \rho y \Leftrightarrow x$ divides y
(b) $S = \mathscr{P}(\{1, 2, 3\})$; $A \rho B \Leftrightarrow A \subseteq B$

The two graphs appear in Figure 3.1.

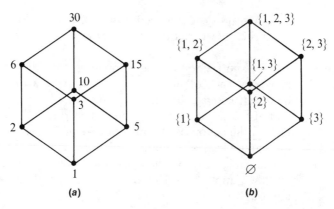

Figure 3.1

The striking thing about these two graphs is that if the node labels were removed, the graphs would be identical even though they arose as representatives of two different, partially ordered sets. Let's define a one-to-one function f from the set of nodes of Graph a onto the set of nodes of Graph b:

$$f(1) = \varnothing \qquad f(6) = \{1, 2\}$$
$$f(2) = \{1\} \qquad f(10) = \{1, 3\}$$
$$f(3) = \{2\} \qquad f(15) = \{2, 3\}$$
$$f(5) = \{3\} \qquad f(30) = \{1, 2, 3\}$$

This function equates the two sets of nodes so that the relationships of Graph a are preserved in Graph b. For example, if x and y are nodes of Graph a with x an immediate predecessor of y, then in Graph b, $f(x)$ is an immediate predecessor of $f(y)$. If in Graph a nodes x and y are unrelated, then their images under f, $f(x)$ and $f(y)$, are unrelated in Graph b. The inverse function preserves the relationships of Graph b in Graph a.

Each graph simulates the other in the following sense: if we want to find, for example, a successor of node x in Graph a, we can consult node $f(x)$ in Graph b where we see that some node y' is a successor of $f(x)$. But y' is $f(y)$ for some node y in Graph a; so we apply the inverse function to find y. Then y will be a successor of x in Graph a. In short, Graph b is a mirror image of Graph a. We obtain information about Graph a by "looking in the mirror," which amounts to applying f, working in Graph b, and then applying f^{-1}. In the same way, Graph a is a mirror image of Graph b, and we obtain information about Graph b by applying f^{-1}, working in Graph a, and then applying f.

The two partially ordered sets these graphs represent are said to be **isomorphic**. □

A graph is not a very dynamic structure. We will look next at instances of a structure where there is a binary operation available.

3.20   Example $[\mathbb{R}, +]$ and $[\mathbb{R}^+, \cdot]$ are both monoids. Can these two monoids be mirror images of each other? If so, there must be a one-to-one correspondence between the sets $\mathbb{R}$ and $\mathbb{R}^+$; so we want a bijection $f: \mathbb{R} \to \mathbb{R}^+$. More than that, if $[\mathbb{R}^+, \cdot]$ is to act as a mirror for $[\mathbb{R}, +]$, then f has to preserve in $[\mathbb{R}^+, \cdot]$ the action of the operation $+$ in $\mathbb{R}$. Thus, if we want to find $x + y$ for x and y in $\mathbb{R}$, we should "look in the mirror"—take $f(x)$ and $f(y)$, which puts us in $[\mathbb{R}^+, \cdot]$, do the operation available there, that is, find $f(x) \cdot f(y)$, and then apply f^{-1}. This process should yield $x + y$. Thus, we want, for any x and y in $\mathbb{R}$,

$$x + y = f^{-1}(f(x) \cdot f(y))$$

or

$$f(x + y) = f(x) \cdot f(y)$$

This equation says the following. We want to be able to operate in $\mathbb{R}$ with $+$ and then map to $\mathbb{R}^+$, or to map to $\mathbb{R}^+$ and then operate there with $\cdot$; the result should be the same in either case. We can illustrate this with a commutative diagram (Figure 3.2). In Figure 3.2, $+$ is a binary operation

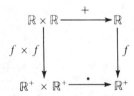

Figure 3.2

taking members of $\mathbb{R} \times \mathbb{R}$ (ordered pairs of real numbers) to $\mathbb{R}$, $\cdot$ is a binary operation taking members of $\mathbb{R}^+ \times \mathbb{R}^+$ to $\mathbb{R}^+$, and $f \times f$ applies f to each component of an element of $\mathbb{R} \times \mathbb{R}$. A function $f: \mathbb{R} \to \mathbb{R}^+$ that does the job is given by $f(x) = e^x$. Then f is a one-to-one, onto function (a bijection). Also, for x and y in $\mathbb{R}$,

$$f(x + y) = e^{x+y} = e^x \cdot e^y = f(x) \cdot f(y)$$ □

3.21   Practice   Use the function f defined in Example 3.20. Choose -4 and 7 as elements of $\mathbb{R}$.

(a) What is $f(-4 + 7)$? What paths in the commutative diagram of Figure 3.2 does this computation trace?

(b) What is $f(-4) \cdot f(7)$? What paths in the commutative diagram of Figure 3.2 does this computation trace?   (*Answer, page A-10*)

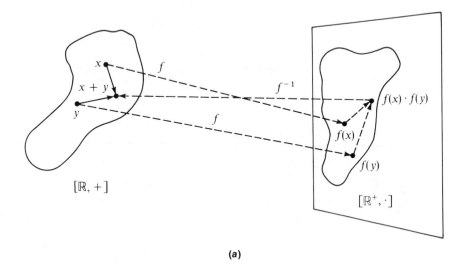

(a)

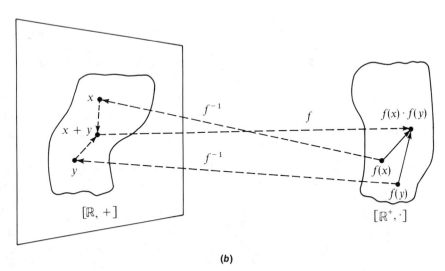

(b)

Figure 3.3

Figure 3.3a illustrates how $[\mathbb{R}^+, \cdot]$ simulates $[\mathbb{R}, +]$ in that $x + y$ in $[\mathbb{R}, +]$ can be done by mapping to $[\mathbb{R}^+, \cdot]$, operating there, and then taking f^{-1}. Figure 3.3b shows how $[\mathbb{R}, +]$ simulates $[\mathbb{R}^+, \cdot]$ in that $f(x) \cdot f(y)$ in $\mathbb{R}^+$ (remember that f is onto) can be done by using f^{-1} to get to $[\mathbb{R}, +]$, operating there, and then applying f. Each monoid simulates the other, and $[\mathbb{R}, +]$ and $[\mathbb{R}^+, \cdot]$ are said to be **isomorphic** monoids.

Two instances of a structure will be **isomorphic** whenever there is a bijection (called an **isomorphism**) between the elements of sets involved that preserves the effects of operations or functions on the sets. More specific definitions will be given for particular structures, but the general idea is that of the mirror pictures of Figure 3.3. Isomorphic structures are essentially the same since the elements are simply relabeled from one to the other. Therefore, we can use the idea of isomorphism to classify instances of a structure, lumping together those that are isomorphic.

ISOMORPHIC BOOLEAN ALGEBRAS

Now let's try to understand isomorphism for instances of a structure with which we are relatively familiar, a Boolean algebra. Suppose we have two Boolean algebras, $[B, +, \cdot, ', 0, 1]$ and $[b, \&, *, '', \emptyset, I]$. This notation means that, for example, if x is in B, x' is the result of performing on x the unary operation defined in B, and if z is an element of b, z'' is the result of performing on z the unary operation defined in b. How would we define isomorphism between these two Boolean algebras? First, we would need a bijection f from B onto b. Then f must preserve in b the effects of the various operations in B. There are three operations, so we will need three equations. To preserve the operation $+$, we want to be able to operate in B and then map to b, or to map to b and operate there. (Think of the commutative diagram of Example 3.20.) Thus, for x and y in B, we require

$$f(x + y) = f(x) \mathbin{\&} f(y)$$

3.22 Practice
(a) Write the equation requiring f to preserve the effect of the binary operation $\cdot$.
(b) Write the equation requiring f to preserve the effect of the unary operation $'$. *(Answer, page A-10)*

Here is the definition of isomorphic Boolean algebras.

3.23 Definition
Let $[B, +, \cdot, ', 0, 1]$ and $[b, \&, *, '', \emptyset, I]$ be Boolean algebras. A function $f: B \to b$ is an **isomorphism** from $[B, +, \cdot, ', 0, 1]$ to $[b, \&, *, '', \emptyset, I]$ if

(a) f is a bijection
(b) $f(x + y) = f(x) \mathbin{\&} f(y)$
(c) $f(x \cdot y) = f(x) * f(y)$
(d) $f(x') = [f(x)]''$ ☐

3.24 Practice
Illustrate equations (b), (c), and (d) in Definition 3.23 by a commutative diagram. *(Answer, page A-10)*

We already know (see the beginning of Section 3.1) that for any set S, $\mathscr{P}(S)$ under the operations of union, intersection, and complementation is a Boolean algebra. If we pick $S = \{1, 2\}$, then the elements of $\mathscr{P}(S)$ are $\varnothing$, $\{1\}$, $\{2\}$, and $\{1, 2\}$. The operations are given by the tables

$\cup$	$\varnothing$	$\{1, 2\}$	$\{1\}$	$\{2\}$
$\varnothing$	$\varnothing$	$\{1, 2\}$	$\{1\}$	$\{2\}$
$\{1, 2\}$	$\{1, 2\}$	$\{1, 2\}$	$\{1, 2\}$	$\{1, 2\}$
$\{1\}$	$\{1\}$	$\{1, 2\}$	$\{1\}$	$\{1, 2\}$
$\{2\}$	$\{2\}$	$\{1, 2\}$	$\{1, 2\}$	$\{2\}$

$\cap$	$\varnothing$	$\{1, 2\}$	$\{1\}$	$\{2\}$
$\varnothing$	$\varnothing$	$\varnothing$	$\varnothing$	$\varnothing$
$\{1, 2\}$	$\varnothing$	$\{1, 2\}$	$\{1\}$	$\{2\}$
$\{1\}$	$\varnothing$	$\{1\}$	$\{1\}$	$\varnothing$
$\{2\}$	$\varnothing$	$\{2\}$	$\varnothing$	$\{2\}$

$'$	
$\varnothing$	$\{1, 2\}$
$\{1, 2\}$	$\varnothing$
$\{1\}$	$\{2\}$
$\{2\}$	$\{1\}$

In Exercise 1 of the last section, a Boolean algebra was defined on the set $B = \{0, 1, a, a'\}$ where the tables defining the operations of $+$, $\cdot$, and $'$ were

$+$	0	1	a	a'
0	0	1	a	a'
1	1	1	1	1
a	a	1	a	1
a'	a'	1	1	a'

$\cdot$	0	1	a	a'
0	0	0	0	0
1	0	1	a	a'
a	0	a	a	0
a'	0	a'	0	a'

$'$	
0	1
1	0
a	a'
a'	a

We claim that the mapping $f : B \to \mathscr{P}(S)$ given by

$$f(0) = \varnothing$$

$$f(1) = \{1, 2\}$$

$$f(a) = \{1\}$$

$$f(a') = \{2\}$$

is an isomorphism. Certainly it is a bijection. For $x, y \in B$, we can verify each of the equations

$$f(x + y) = f(x) \cup f(y)$$

$$f(x \cdot y) = f(x) \cap f(y)$$

$$f(x') = [f(x)]'$$

by examining all possible cases. Thus, for example,

$$f(a \cdot 1) = f(a) = \{1\} = \{1\} \cap \{1, 2\} = f(a) \cap f(1)$$

3.25   Practice   Verify the following equations:

(a) $f(0 + a) = f(0) \cup f(a)$
(b) $f(a + a') = f(a) \cup f(a')$
(c) $f(a \cdot a') = f(a) \cap f(a')$
(d) $f(1') = [f(1)]'$

(*Answer, page A-10*)

The remaining cases also hold. Even without testing all cases, it is pretty clear here that f is going to work because it merely relabels the entries in the tables for B so that they resemble the tables for $\mathcal{P}(S)$. In general, it may not be so easy to decide whether a given f is an isomorphism between two instances of a structure. Even harder to answer is the question of whether or not two given instances of a structure are isomorphic; in this case, we must either think up a function that works or show that no such function exists. One case where no such function exists is if the sets involved are not the same size; we cannot have a 4-element Boolean algebra isomorphic to an 8-element Boolean algebra.

We just showed that a particular 4-element Boolean algebra is isomorphic to $\mathcal{P}\{1, 2\}$. It turns out that any finite Boolean algebra is isomorphic to the Boolean algebra of a power set. We will state this as a theorem, but we will not prove it.

3.26   Theorem   Let B be any Boolean algebra with n elements. Then $n = 2^m$ for some m, and B is isomorphic to $\mathcal{P}\{1, 2, \cdots, m\}$.   □

Theorem 3.26 gives us two pieces of information. We know by Exercise 6 of the last section that a finite Boolean algebra must have an even number of elements; now we find that the even number must be a power of 2. Also we learn that finite Boolean algebras that are power sets, under the operations of union, intersection, and complementation, are—in our lumping together of isomorphic things—really the only kinds of finite Boolean algebras. In a sense we have come full circle. We defined a Boolean algebra to represent many kinds of situations, now we find that (for the finite case) the situations, except for the labels of objects, are the same anyway!

A HOMOMORPHISM EXAMPLE

When A is isomorphic to B, A and B can simulate each other. The success of the simulation depends on two properties of the isomorphism $f : A \rightarrow B$. First, f is a bijection (we use both the onto property and the existence of f^{-1}). Second, f preserves in B the effect of the operations in A (commutative diagrams hold). We get simulations of a less perfect nature by relaxing the one-to-one requirement on the function f. Before we look at an example, we need one more monoid to work with.

3.27 Example   The set $\mathbb{Z}$ of integers under multiplication is a monoid denoted by $[\mathbb{Z}, \cdot]$. The product of two integers exists and is a unique integer, so multiplication is a binary operation on $\mathbb{Z}$. Multiplication is associative because $(a \cdot b) \cdot c = a \cdot (b \cdot c)$ for all $a, b, c \in \mathbb{Z}$. The number $1 \in \mathbb{Z}$ is an identity element because $1 \cdot a = a \cdot 1 = a$ for all $a \in \mathbb{Z}$.   ☐

Now let's consider the monoids $[M_2^D(\mathbb{Z}), \cdot]$ and $[\mathbb{Z}, \cdot]$. Recall that $M_2^D(\mathbb{Z})$ is the set of all 2×2 diagonal matrices with integer entries. For $\begin{bmatrix} a_{11} & 0 \\ 0 & a_{22} \end{bmatrix}$ in $M_2^D(\mathbb{Z})$, we define f by

$$f\left(\begin{bmatrix} a_{11} & 0 \\ 0 & a_{22} \end{bmatrix}\right) = a_{11}$$

Then $f: M_2^D(\mathbb{Z}) \to \mathbb{Z}$ and f is onto. Clearly, f is not one-to-one. Now we claim that f preserves in $[\mathbb{Z}, \cdot]$ the effect of the operation in $M_2^D(\mathbb{Z})$:

$$f\left(\begin{bmatrix} a_{11} & 0 \\ 0 & a_{22} \end{bmatrix} \cdot \begin{bmatrix} b_{11} & 0 \\ 0 & b_{22} \end{bmatrix}\right) = f\left(\begin{bmatrix} a_{11} & 0 \\ 0 & a_{22} \end{bmatrix}\right) \cdot f\left(\begin{bmatrix} b_{11} & 0 \\ 0 & b_{22} \end{bmatrix}\right)$$

Notice that there are two different multiplications in this equation. The left side of the equation is matrix multiplication, the right, integer multiplication.

3.28 Practice   Verify the above equation.   (*Answer, page A-10*)

The function f is not an isomorphism because it is not one-to-one; f is a **homomorphism** (it preserves the operation).

Because f is not one-to-one, f^{-1} does not exist. That means the mirror pictures don't work. A direct simulation of $[M_2^D(\mathbb{Z}), \cdot]$ by $[\mathbb{Z}, \cdot]$ is not possible using f. However, there is still a type of simulation possible. We will define a relation ρ on $M_2^D(\mathbb{Z})$ by saying that two members of $M_2^D(\mathbb{Z})$ are related if they map under f to the same member of $\mathbb{Z}$. Thus,

$$\begin{bmatrix} a_{11} & 0 \\ 0 & a_{22} \end{bmatrix} \rho \begin{bmatrix} b_{11} & 0 \\ 0 & b_{22} \end{bmatrix} \Leftrightarrow f\left(\begin{bmatrix} a_{11} & 0 \\ 0 & a_{22} \end{bmatrix}\right) = f\left(\begin{bmatrix} b_{11} & 0 \\ 0 & b_{22} \end{bmatrix}\right)$$

$$\Leftrightarrow a_{11} = b_{11}$$

ρ is an equivalence relation—reflexive, symmetric, and transitive—on $M_2^D(\mathbb{Z})$. Therefore, $M_2^D(\mathbb{Z})$ is partitioned into equivalence classes, and each class can be associated with an integer, namely the integer that is the 1–1 element of every member of the class.

We can use $[M_2^D(\mathbb{Z}), \cdot]$ to simulate $[\mathbb{Z}, \cdot]$ as follows. Let v and w belong to $\mathbb{Z}$; we want to compute $v \cdot w$ by simulation. The integer v is associated with a whole class of matrices in $M_2^D(\mathbb{Z})$, all those with 1–1 element v. Pick any member of this class, and name it V. Similarly, w is associated with a class of matrices; pick any member W of this class. Compute $V \cdot W$ in

$[M_2^D(\mathbb{Z}), \cdot]$. It will belong to a class of matrices; the integer in $\mathbb{Z}$ corresponding to this class is $v \cdot w$.

3.29 *Example*   Suppose we want to compute $3 \cdot 4$. One matrix from the class for 3 is

$$\begin{bmatrix} 3 & 0 \\ 0 & 5 \end{bmatrix}$$

one matrix from the class for 4 is

$$\begin{bmatrix} 4 & 0 \\ 0 & -18 \end{bmatrix}$$

Then

$$\begin{bmatrix} 3 & 0 \\ 0 & 5 \end{bmatrix}\begin{bmatrix} 4 & 0 \\ 0 & -18 \end{bmatrix} = \begin{bmatrix} 12 & 0 \\ 0 & -90 \end{bmatrix}$$

The corresponding integer is 12. □

So far we have f a homomorphism from $M_2^D(\mathbb{Z})$ onto $\mathbb{Z}$, and we have been able to simulate $[\mathbb{Z}, \cdot]$ by $[M_2^D(\mathbb{Z}), \cdot]$. The only difference between this case and Figure 3.3b is that we don't have f^{-1}, and we must choose preimages of v and w under f. Still, given that we can choose such preimages, all goes well. It is the other direction that is less than perfect. We want to simulate $[M_2^D(\mathbb{Z}), \cdot]$ by $[\mathbb{Z}, \cdot]$. We pick two matrices, X and Y, in $M_2^D(\mathbb{Z})$. We find $f(X)$ and $f(Y)$ and compute $f(X) \cdot f(Y)$. So far so good. But here is the problem. When we try to go from $f(X) \cdot f(Y)$, which is some integer z, back to $M_2^D(\mathbb{Z})$, we only know that z is associated with a class of matrices. And although we can choose a member of the class, we may not choose the one that is really $X \cdot Y$.

3.30 *Example*   Suppose we want to compute

$$\begin{bmatrix} 2 & 0 \\ 0 & -4 \end{bmatrix} \cdot \begin{bmatrix} -6 & 0 \\ 0 & 7 \end{bmatrix}$$

We find

$$f\left(\begin{bmatrix} 2 & 0 \\ 0 & -4 \end{bmatrix}\right) = 2, \quad f\left(\begin{bmatrix} -6 & 0 \\ 0 & 7 \end{bmatrix}\right) = -6$$

and compute $(2)(-6) = -12$. The class associated with -12 is the set of all matrices with -12 in the 1–1 position. But only one member of this class, namely

$$\begin{bmatrix} -12 & 0 \\ 0 & -28 \end{bmatrix}$$

is the answer we want. □

Thus, the simulation of $[M_2^D(\mathbb{Z}), \cdot]$ by $[\mathbb{Z}, \cdot]$ is imperfect; we can only obtain the answer to within an equivalence class. Actually, this result is not so surprising; $[M_2^D(\mathbb{Z}), \cdot]$ is somehow more complex than $[\mathbb{Z}, \cdot]$. It is possible to show that we can treat the classes of $M_2^D(\mathbb{Z})$ as objects, with an operation of multiplication defined on them. This new structure will be isomorphic to $[\mathbb{Z}, \cdot]$. Thus, in a way, a picture of $[\mathbb{Z}, \cdot]$ is somehow embedded within $[M_2^D(\mathbb{Z}), \cdot]$, and it makes sense that $[M_2^D(\mathbb{Z}), \cdot]$ simulates $[\mathbb{Z}, \cdot]$, but that $[\mathbb{Z}, \cdot]$ only imperfectly simulates $[M_2^D(\mathbb{Z}), \cdot]$.

What has been done with these two particular monoids will be done in a more general setting in Chapter 5. Meanwhile, let's summarize the ideas. Suppose A and B are two examples of a structure. Simulation must involve a function that preserves operations, a homomorphism. If there is in fact an isomorphism (a bijective homomorphism) from A onto B, then A simulates B and B simulates A; thus, we think of A and B as being essentially the same. If there is a homomorphism but not an isomorphism from A onto B, then A simulates B, but B only imperfectly simulates A; that is, B is isomorphic to a structure composed of classes of A.

We will also see in Chapter 5 that the nature of computer arithmetic is governed by these ideas of simulation and homomorphism.

✔ CHECKLIST

Definition

Isomorphism from one Boolean algebra to another (*p. 118*)

Techniques

Write the equation meaning that a function f preserves an operation from one instance of a structure to another; verify or disprove such an equation.

Main Ideas

If there is an isomorphism (bijection preserving operations) from A to B, where A and B are instances of a structure, then A and B simulate each other (Simulation II), and, except for labels, A and B are the same.

If there is a homomorphism (function preserving operations) from A onto B, where A and B are instances of a structure, then A simulates B and B imperfectly simulates A.

All finite Boolean algebras are isomorphic to Boolean algebras that are power sets.

EXERCISES, SECTION 3.2

★ 1. Let $(S, \leq)$ and $(S', \leq')$ be two partially ordered sets. $(S, \leq)$ is *isomorphic* to $(S', \leq')$ if there is a bijection $f : S \to S'$ such that for x, y in S, $x \prec y \Rightarrow f(x) \prec' f(y)$ and $f(x) \prec' f(y) \Rightarrow x \prec y$.

    (a) Show that there are exactly two nonisomorphic, partially ordered sets with two elements (use graphs).

    (b) Show that there are exactly five nonisomorphic, partially ordered sets with three elements.

    (c) How many nonisomorphic, partially ordered sets with four elements exist?

★ 2. Find an example of two partially ordered sets $(S, \leq)$ and $(S', \leq')$ and a bijection $f : S \to S'$ where, for x, y in S, $x \prec y \Rightarrow f(x) \prec' f(y)$, but $f(x) \prec' f(y) \not\Rightarrow x \prec y$.

3. Let $S = \{0, 1\}$ and $\cdot$ be defined on S by

$\cdot$	0	1
0	1	0
1	0	1

Let $T = \{5, 7\}$ and $+$ be defined on T by

$+$	5	7
5	7	5
7	5	7

Then $[S, \cdot]$ and $[T, +]$ are both monoids.

    (a) If a function f is an isomorphism from $[S, \cdot]$ to $[T, +]$, what two properties must f satisfy?

    (b) Define a function f and prove it is an isomorphism from $[S, \cdot]$ to $[T, +]$.

4. On the set B of all functions mapping $\{0, 1\}^2$ to $\{0, 1\}$, we can define operations of $+, \cdot,$ and $'$ by

$$(f_1 + f_2)(x, y) = \max(f_1(x, y), f_2(x, y))$$
$$(f_1 \cdot f_2)(x, y) = \min(f_1(x, y), f_2(x, y))$$
$$f'_1(x, y) = \begin{cases} 1 & \text{if } f_1(x, y) = 0 \\ 0 & \text{if } f_1(x, y) = 1 \end{cases}$$

Then, according to Exercise 3 of Section 3.1, $[B, +, \cdot, ', 0, 1]$ is a Boolean algebra of 16 elements. The following table assigns names to these 16 functions.

(x, y)	0	1	f_1	f_2	f_3	f_4	f_5	f_6	f_7	f_8	f_9	f_{10}	f_{11}	f_{12}	f_{13}	f_{14}
$(0, 0)$	0	1	1	1	1	1	1	1	0	0	0	1	0	0	0	0
$(0, 1)$	0	1	0	1	1	0	1	0	1	1	1	0	0	1	0	0
$(1, 0)$	0	1	1	0	1	0	0	1	1	1	0	0	1	0	1	0
$(1, 1)$	0	1	0	0	0	0	1	1	1	0	1	1	1	0	0	1

By Theorem 3.26, this Boolean algebra is isomorphic to $[\mathscr{P}\{1, 2, 3, 4\},$ $\cup, \cap, {}', \varnothing, \{1, 2, 3, 4\}]$. Complete the following definition of an isomorphism from B to $\mathscr{P}\{1, 2, 3, 4\}$:

$$0 \to \varnothing$$

$$1 \to \{1, 2, 3, 4\}$$

$$f_4 \to \{1\}$$

$$f_{12} \to \{2\}$$

$$f_{13} \to \{3\}$$

$$f_{14} \to \{4\}$$

★ 5. Suppose that $[B, +, \cdot, {}', 0, 1]$ and $[b, \&, *, {}'', \emptyset, I]$ are isomorphic Boolean algebras and that f is an isomorphism from B to b.
   (a) Prove that $f(0) = \emptyset$
   (b) Prove that $f(1) = I$

★ 6. Show that the function $f : M_2(\mathbb{Z}) \to M_2^D(\mathbb{Z})$ given by

$$f\left(\begin{bmatrix} a_{11} & a_{12} \\ a_{21} & a_{22} \end{bmatrix}\right) = \begin{bmatrix} a_{11} & 0 \\ 0 & a_{22} \end{bmatrix}$$

is *not* a homomorphism from the monoid $[M_2(\mathbb{Z}), \cdot]$ to the monoid $[M_2^D(\mathbb{Z}), \cdot]$.

7. (a) Let $\mathbb{R}^*$ denote the set of nonzero real numbers. Show that $[\mathbb{R}^*, \cdot]$ is a monoid.
   (b) Define a function $f : \mathbb{R}^* \to \mathbb{R}^+$ by $f(a) = |a|$. Show that f is a homomorphism from $[\mathbb{R}^*, \cdot]$ to $[\mathbb{R}^+, \cdot]$.
   (c) Show that f is an onto function but is not one-to-one.
   (d) Define a relation ρ on $\mathbb{R}^*$ by $x \rho y \Leftrightarrow f(x) = f(y)$. Prove that ρ is an equivalence relation on $\mathbb{R}^*$ and describe the equivalence classes.
   (e) Use $[\mathbb{R}^*, \cdot]$ (and the classes of part (d)) to simulate the computation $3 \cdot 5$ in $[\mathbb{R}^+, \cdot]$.
   (f) Use $[\mathbb{R}^+, \cdot]$ to simulate the computation $(-2) \cdot 6$ in $[\mathbb{R}^*, \cdot]$ to within an equivalence class.

For Further Reading

A theoretical development of Boolean algebras is given in [1]. However, [2] is more down-to-earth and has an electronics orientation akin to what we will do in Chapter 4. Algebraic structures such as monoids and groups can be studied further in any abstract algebra text such as [3] or [4], but such study is best deferred until after Chapter 5.

1. James C. Abbott, *Sets, Lattices, and Boolean Algebras.* Boston: Allyn and Bacon, 1969.

2. Gerald E. Williams, *Boolean Algebra with Computer Applications.* New York: McGraw-Hill, 1970.

3. John B. Fraleigh, *A First Course in Abstract Algebra*, 2nd ed. Reading, Mass.: Addison-Wesley, 1976.

4. Joseph E. Kuczkowski and Judith L. Gersting, *Abstract Algebra, A First Look.* New York: Marcel Dekker, 1977.

Chapter 4

BOOLEAN ALGEBRA AND COMPUTER LOGIC

This chapter establishes a relationship between the Boolean algebra structure and the wiring diagrams for the electronic circuits in computers, calculators, industrial control devices, telephone systems, and so forth.

Indeed, we will see that truth functions, expressions made up of variables and the operations of Boolean algebra, and these wiring diagrams are all related in such a way that we can effectively pass from one formulation to another and still preserve characteristic behavior. We will find that we can simplify the wiring diagrams by using properties of Boolean algebras (as discussed in Chapter 3). We will also look at two other simplification procedures.

Section 4.1   LOGIC NETWORKS

COMBINATIONAL NETWORKS

Basic Logic Elements

The concept of a Boolean algebra was first formulated by George Boole, an English mathematician, around 1850 to model statement logic. In 1938 the American mathematician Claude Shannon perceived the parallel between statement logic and computer logic, and realized that Boolean algebra could play a part in systematizing this new realm of electronics.

Recall that a statement is either true or false. Two statement letters, A and B, can be combined by one of two binary operations. If the operation is disjunction, for example, then $A \vee B$ is true or false as a function of the truth or falsity of A and B, according to the truth table in Figure 4.1. We

A	B	$A \vee B$
T	T	T
T	F	T
F	T	T
F	F	F

Figure 4.1

can view A and B as inputs to a device that performs the disjunction operation. Thus, we have Figure 4.2.

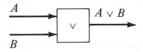

Figure 4.2

Any Boolean algebra operation can be associated with an electronic device; the inputs and outputs of the device represent statements, which may be complex rather than just single statement letters. Figure 4.3 represents a device paralleling the behavior of the Boolean operation $+$. In Figure 4.3 an input or output wire is assigned the value 0 or 1, depending

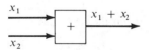

x_1	x_2	$x_1 + x_2$
1	1	1
1	0	1
0	1	1
0	0	0

Figure 4.3

on its voltage. There are two technical details to consider here. First, the voltages carried along wires x_1, x_2 and $x_1 + x_2$ may fluctuate somewhat, so actually, we associate one range of voltage values with 0 and another with 1. Second, we assume that the output of the device at a given instant is a function of its inputs at that instant; that is, we assume that the time required for the device to perform its actions is negligible. This assumption helps us avoid timing difficulties in connecting devices together.

Let's consider for a moment what might be inside the box in Figure 4.3.

If we try to produce the required behavior for $x_1 + x_2$ using mechanical switches, the earliest technology used, we would wire a switch controlled by x_1 and one controlled by x_2 in parallel, so that the circuit would be broken only if both switches were open. Later technologies included vacuum tubes and then transistors. However, the actual physical implementation does not concern us here; we shall simply label our device to identify its behavior. We will have one device for each of the three Boolean operations. Each has a standard symbol. The **OR gate**, Figure 4.4a, behaves like the Boolean operation $+$. The **AND gate**, Figure 4.4b, represents the Boolean operation $\cdot$. Figure 4.4c shows an **inverter** corresponding to the unary Boolean operation $'$. Because of the associativity property for $+$ and $\cdot$, the OR and AND gates can have more than two inputs.

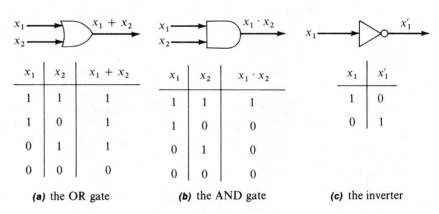

x_1	x_2	$x_1 + x_2$
1	1	1
1	0	1
0	1	1
0	0	0

x_1	x_2	$x_1 \cdot x_2$
1	1	1
1	0	0
0	1	0
0	0	0

x_1	x_1'
1	0
0	1

(a) the OR gate   (b) the AND gate   (c) the inverter

Figure 4.4

Boolean Expressions

Each device in Figure 4.4 represents a single Boolean operation acting upon some variables; the corresponding truth table is also shown.

4.1 Definition A **Boolean expression** in n variables, $x_1, \ldots, x_n$, is any finite string of symbols formed by application of the following rules:

1. $x_1, x_2, \ldots, x_n$ are Boolean expressions.
2. If P and Q are Boolean expressions, so are $(P + Q), (P \cdot Q)$ and $(P)'$.

□

(Definition 4.1 is an example of a recursive, or inductive, definition. Rule 1 gives us a base step, tells us some definite objects that are Boolean expressions; then rule 2 tells us how to build up more complex expressions from already existing expressions. Many constructs in computer science, including elements of many programming languages, are defined recursively.) When there is no chance of confusion, we omit the parentheses introduced by rule 2. In addition, $\cdot$ takes precedence over $+$, so that $x_1 + x_2 \cdot x_3$ stands for $x_1 + (x_2 \cdot x_3)$; this convention also allows removal of parentheses. Finally, we will generally omit the symbol $\cdot$ and use juxtaposition, so that $x_1 \cdot x_2$ is written $x_1 x_2$.

4.2 Example $x_3, (x_1 + x_2)'x_3, (x_1 x_3 + x_4')x_2,$ and $(x_1' x_2)'x_1$ are all Boolean expressions. $\square$

Truth Functions

4.3 Definition   A **truth function** is a function f such that $f : \{0, 1\}^n \to \{0, 1\}$ for some integer $n \geq 1$. $\square$

$\{0, 1\}^n$ is the set of all n-tuples of 0's and 1's. A truth function thus associates a value of 0 or 1 with each such n-tuple.

4.4 Example   The truth table for the Boolean operation $+$ describes a truth function f with $n = 2$. The domain of f is $\{(1, 1), (1, 0), (0, 1), (0, 0)\}$, and $f((1, 1)) = 1$, $f((1, 0)) = 1$, $f((0, 1)) = 1$, $f((0, 0)) = 0$. Similarly, the Boolean operation $\cdot$ describes a different truth function with $n = 2$, and the Boolean operation $'$ describes a truth function for $n = 1$. $\square$

4.5 Practice   (a) If we are writing a truth function $f : \{0, 1\}^n \to \{0, 1\}$ in tabular form (like a truth table), how many rows will the table have?
(b) How many different truth functions are there that take $\{0, 1\}^2 \to \{0, 1\}$?
(c) How many different truth functions are there that take $\{0, 1\}^n \to \{0, 1\}$?   (*Answer, page A-10*)

Any Boolean expression defines a unique truth function, just as do the simple Boolean expressions $x_1 + x_2, x_1 x_2,$ and x_1'.

4.6 Example   The Boolean expression $x_1 x_2' + x_3$ defines the truth function given by the table in Figure 4.5.

x_1	x_2	x_3	$x_1 x_2' + x_3$
1	1	1	1
1	1	0	0
1	0	1	1
1	0	0	1
0	1	1	1
0	1	0	0
0	0	1	1
0	0	0	0

Figure 4.5 $\square$

Networks and Expressions

By combining AND gates, OR gates, and inverters, we can construct a logic network representing a given Boolean expression and producing the same truth function.

4.7 Example

The logic network for the Boolean expression $x_1 x_2' + x_3$ is shown in Figure 4.6. □

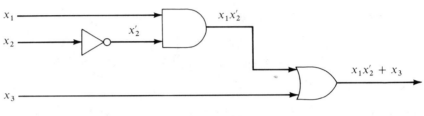

Figure 4.6

4.8 Practice

Design the logic network for the Boolean expressions

(a) $x_1 + x_2'$
(b) $x_1(x_2 + x_3)'$

(*Answer, page A-10*)

Conversely, if we have a logic network, we can write a Boolean expression with the same truth function.

4.9 Example

A Boolean expression for the logic network of Figure 4.7 is $(x_1 x_2 + x_3)' x_3$. □

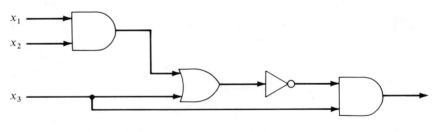

Figure 4.7

4.10 Practice

(a) Write a Boolean expression for the logic network of Figure 4.8.
(b) Write the truth function (in table form) for the network (and expression) of part (a).

(*Answer, page A-11*)

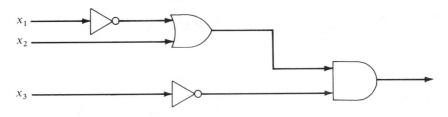

Figure 4.8

The logic networks we have been looking at are also called **combinational networks**. They have several features we should note. First, input or output lines are not tied together except by passing through gates. The lines can be split, however, to serve as input to more than one device. There are no loops where the output of an element can become part of the input to that element. And, finally, network output is an instantaneous function of input; there are no delay elements that capture and remember input signals. Notice also that the picture of any network is, in effect, a directed graph.

Canonical Form

Here is the situation so far. Arrows indicate a procedure we can carry out.

truth function ← Boolean expression ↔ logic network

We can write a unique truth function from either a network or an expression. Given an expression, we can find a network with the same truth function, and conversely. The last part of the puzzle concerns how to get from an arbitrary truth function to an expression (and hence a network) with that truth function. We want a procedure applicable to any truth function, and having precise instructions that can be mechanically carried out, always terminating with a correct expression. In short, we want an algorithm. Algorithms are the very heart of computer science, and we will have much to say about them in Chapter 9. No doubt you are already aware that one of the major facets of computer programming concerns formulation of an algorithm to solve the problem at hand. At any rate, our algorithm for this particular problem is an easy one and is explained in the next example.

4.11 Example   Suppose we want to find a Boolean expression for the truth function f of Figure 4.9. There are four rows in the table, 1, 3, 4, and 7, for which f is 1. The basic form of our expression will be a sum of four terms

$$(\ ) + (\ ) + (\ ) + (\ )$$

such that the first term has the value 1 for the input values of row 1, the second term has the value 1 for the input values of row 3, and so on. Thus,

x_1	x_2	x_3	$f(x_1, x_2, x_3)$
1	1	1	1
1	1	0	0
1	0	1	1
1	0	0	1
0	1	1	0
0	1	0	0
0	0	1	1
0	0	0	0

Figure 4.9

the entire expression has the value 1 for these inputs and no others—precisely what we want. (Other inputs cause each term in the sum, hence the sum itself, to be 0.)

Each individual term in the sum will be a product of the form $\alpha\beta\gamma$ where α is either x_1 or x_1', β is either x_2 or x_2', and γ is either x_3 or x_3'. If the input value of x_i, $i = 1, 2, 3$, in the row we are working on is 1, then x_i itself is used; if the input value of x_i in the row we are working on is 0, then x_1' is used. These values will force $\alpha\beta\gamma$ to be 1 for that row and 0 for all other rows. Thus, we have

row 1: $x_1x_2x_3$

row 3: $x_1x_2'x_3$

row 4: $x_1x_2'x_3'$

row 7: $x_1'x_2'x_3$

The final expression is

$$(x_1x_2x_3) + (x_1x_2'x_3) + (x_1x_2'x_3') + (x_1'x_2'x_3) \qquad \square$$

The procedure described in Example 4.11 always leads to an expression that is a sum of products, called the **canonical sum-of-products form**, or the **disjunctive normal form**, for the given truth function. The only case not covered by this procedure is when the function has a 0 value everywhere. Then we use an expression such as

$$x_1x_1'$$

that is also a sum (one term) of products. Therefore, we can find a sum-of-products expression to represent any truth function. The algorithm described is illustrated by the flowchart in Figure 4.10.

Because any expression has a corresponding network, we see that any truth function has a logic network representation. Furthermore, the AND gate, OR gate, and inverter are the only devices needed to construct the network. Thus, we can build a network for any truth function with only

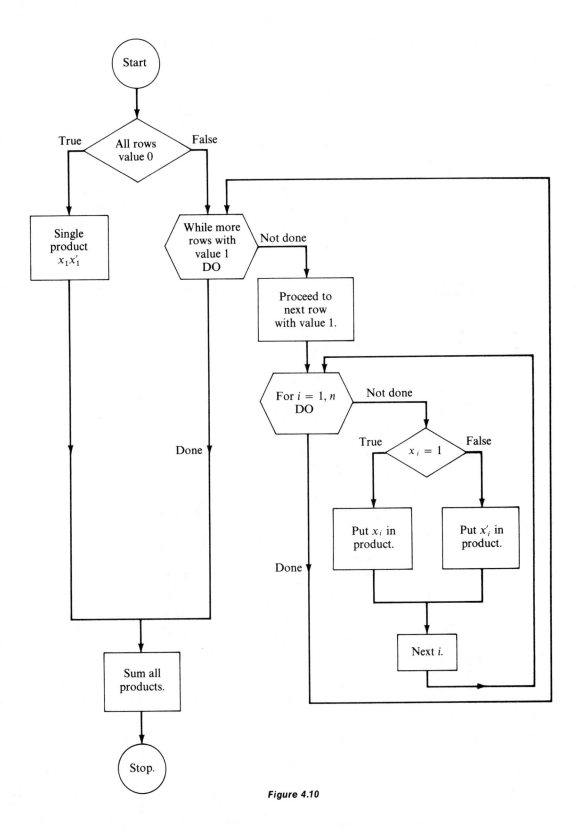

Figure 4.10

three kinds of parts—and lots of wire! Later we will see that it is only necessary to stock one kind of part.

Given a truth function, the canonical sum-of-products form just described is one expression having this truth function, but it is not the only possible one. A method for obtaining a different expression for any truth function is given in Exercise 7 at the end of this section.

4.12 *Example* The network for the canonical sum-of-products form of Example 4.11 is shown in Figure 4.11. We have drawn the inputs to each AND gate separately because it looks neater, but actually a single $x_1, x_2,$ or x_3 input can be split as needed. □

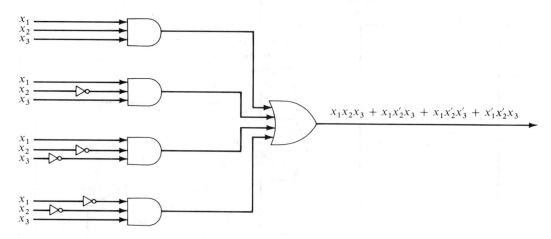

Figure 4.11

4.13 *Practice* (a) Find the canonical sum-of-products form for the truth function of Figure 4.12a.

(b) Draw the network for the expression of part (a). *(Answer, page A-11)*

x_1	x_2	x_3	$f(x_1, x_2, x_3)$
1	1	1	1
1	1	0	0
1	0	1	1
1	0	0	1
0	1	1	0
0	1	0	0
0	0	1	1
0	0	0	1

Figure 4.12(a)

Minimization

A given truth function may, as already noted, be represented by more than one Boolean expression, hence, more than one logic network composed of AND gates, OR gates, and inverters.

4.14 Example   The Boolean expression

$$x_1 x_3 + x_2'$$

has the truth function of Figure 4.12a. The logic network corresponding to this expression is given by Figure 4.12b. Compare this with your network in Practice 4.13b!   □

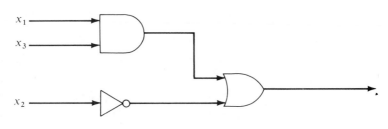

Figure 4.12(b)

4.15 Definition   Two Boolean expressions are **equivalent** if they have the same truth function.   □

We know that

$$x_1 x_2 x_3 + x_1 x_2' x_3 + x_1 x_2' x_3' + x_1' x_2' x_3 + x_1' x_2' x_3'$$

and

$$x_1 x_3 + x_2'$$

for example, are equivalent Boolean expressions.

Clearly, equivalence of Boolean expressions is an equivalence relation on the set of all Boolean expressions in *n* variables. Each equivalence class is associated with a distinct truth function. Given a truth function, the procedure of Example 4.11 produces one particular member of the class associated with that function, namely, the canonical sum-of-products form. However, if we are trying to design the logic network for that function, we are interested in finding a member of the class that is as simple as possible. We would rather build the network of Figure 4.12b than the one for Practice 4.13b. (Using the algorithm for the canonical sum-of-products form is rather like solving a quadratic equation by the quadratic formula; it always works, but it may not always be the easiest way.)

How can we reduce a Boolean expression to an equivalent, simpler expression? We can use the properties of a Boolean algebra because they express equivalence of Boolean expressions. If P is a Boolean expression containing the subexpression $x_1 + (x_2 x_3)$, for example, and Q is the expression obtained from P by replacing $x_1 + (x_2 x_3)$ with the equivalent expression $(x_1 + x_2)(x_1 + x_3)$, then P and Q are equivalent.

4.16   Example   Using the properties of Boolean algebra, we can reduce

$$x_1 x_2 x_3 + x_1 x_2' x_3 + x_1 x_2' x_3' + x_1' x_2' x_3 + x_1' x_2' x_3'$$

to

$$x_1 x_3 + x_2'$$

as follows:

$$
\begin{aligned}
x_1 x_2 x_3 &+ x_1 x_2' x_3 + x_1 x_2' x_3' + x_1' x_2' x_3 + x_1' x_2' x_3' \\
&= x_1 x_2 x_3 + x_1 x_2' x_3 + x_1 x_2' x_3' + x_1 x_2' x_3' + x_1' x_2' x_3 + x_1' x_2' x_3' \\
&= x_1 x_3 x_2 + x_1 x_3 x_2' + x_1 x_2' x_3 + x_1 x_2' x_3' + x_1' x_2' x_3 + x_1' x_2' x_3' \\
&= x_1 x_3 (x_2 + x_2') + x_1 x_2' (x_3 + x_3') + x_1' x_2' (x_3 + x_3') \\
&= x_1 x_3 \cdot 1 + x_1 x_2' \cdot 1 + x_1' x_2' \cdot 1 \\
&= x_1 x_3 + x_1 x_2' + x_1' x_2' \\
&= x_1 x_3 + x_2' x_1 + x_2' x_1' \\
&= x_1 x_3 + x_2' (x_1 + x_1') \\
&= x_1 x_3 + x_2' \cdot 1 \\
&= x_1 x_3 + x_2' \qquad\qquad \square
\end{aligned}
$$

Unfortunately, one must be fairly clever to apply Boolean algebra properties to simplify an expression. In the next section we will discuss more systematic approaches to this minimization problem requiring less ingenuity. For now, we should say a bit more about why we would want to minimize. When logic networks were built from separate gates and inverters, the cost of these elements was a considerable factor in the design of the network, and it was desirable to have as few elements as possible. But now, many networks are built using integrated circuit technology, a development that began in the early 1960s. An integrated circuit is itself a logic network representing a certain truth function, just as if some gates and inverters had been combined in the appropriate arrangement inside this package; the package, however, is extremely small and inexpensive. These integrated circuits are then combined as needed to produce the desired result. The integrated circuits themselves are in fact so inexpensive that we might think it pointless to bother minimizing a network. However, minimization is still important because reliability of the final network is a function of the number of connections occurring between integrated circuit packages.

A USEFUL NETWORK

We can design a network to carry out addition of binary numbers, a basic operation a computer must be able to perform.

First, we recall the rules for adding two binary digits (Figure 4.13a).

x_1	x_2	sum
1	1	10
1	0	1
0	1	1
0	0	0

x_1	x_2	s
1	1	0
1	0	1
0	1	1
0	0	0

x_1	x_2	c
1	1	1
1	0	0
0	1	0
0	0	0

Figure 4.13(a)   *Figure 4.13(b)*

Next, we express the sum as a single sum digit s (the right-hand digit of the actual sum) together with a single carry digit c giving us the two truth functions of Figure 4.13b. The canonical sum-of-products form for each truth function is

$$s = x_1'x_2 + x_1x_2'$$

$$c = x_1x_2$$

An equivalent Boolean expression for s is

$$s = (x_1 + x_2)(x_1x_2)'$$

Figure 4.14a shows a network with inputs x_1 and x_2 and outputs of s and c. This device, for reasons which will be clear shortly, is called a **half-adder**.

To actually add two n-digit binary numbers, we begin at the right-hand or low-order end; the ith column has as input its two binary digits x_1 and x_2 together with the carry digit from the addition of column $i - 1$ to its right. We need a device incorporating the previous carry digit as

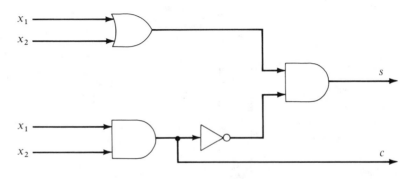

Figure 4.14(a)

input, which can be accomplished by adding x_1 and x_2, using a half-adder, and then adding the previous carry digit c_{i-1} (using another half-adder) to the result. Again, a sum digit and final carry digit c_i are output where c_i should be 1 if either half-adder produces a 1 as its carry digit. The **full-adder** is shown in Figure 4.14b. The full-adder is thus composed of two half-adders and an additional OR gate. To add two n-digit binary numbers, the carry signal must be propagated through n full-adders, or $2n$ half-adders. Although we have assumed that gates output instantaneously, there is in fact a small time delay that can be appreciable for large n. Circuitry that speeds up the addition process is available in today's computers although a greater time savings can be realized by a clever representation of the numbers to be added.

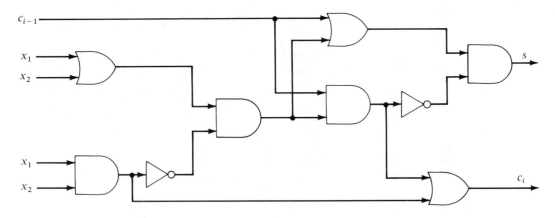

Figure 4.14(b)

OTHER LOGIC ELEMENTS

The basic elements used in integrated circuits are not AND and OR gates and inverters, but NAND and NOR gates. Figure 4.15 shows the standard

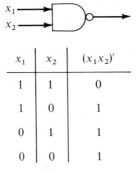

x_1	x_2	$(x_1x_2)'$
1	1	0
1	0	1
0	1	1
0	0	1

Figure 4.15

symbol for the NAND gate (the NOT AND gate) and the truth function it implements. The NAND gate alone is sufficient to realize any truth function because we can construct networks using only NAND gates that do the job of inverters, OR gates, and AND gates. Figure 4.16 shows these networks.

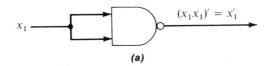

(a)

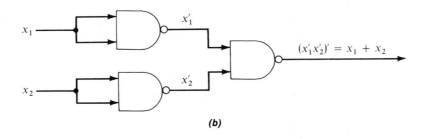

(b)

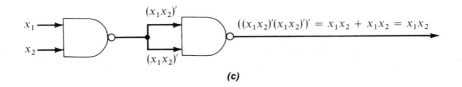

(c)

Figure 4.16

The NOR (NOT OR) gate and its truth function appear in Figure 4.17.

x_1	x_2	$(x_1 + x_2)'$
1	1	0
1	0	0
0	1	0
0	0	1

Figure 4.17

An exercise at the end of this section has you construct networks using only NOR gates for inverters, OR gates, and AND gates.

Although we can construct a network for a truth function using NAND elements by replacing AND gates, OR gates, and inverters in the canonical form or a minimized form with the appropriate NAND networks, we can often obtain a simpler network by using the properties of NAND elements directly.

4.17   Practice

(a) Rewrite the network of Figure 4.12b with NAND elements by directly replacing the AND gate, OR gate, and inverter, as in Figure 4.16.
(b) Rewrite the Boolean expression $x_1x_3 + x_2'$ for Figure 4.12b using DeMorgan's Laws, and then construct a network using only two NAND elements.   (*Answer, page A-11*)

CONSTRUCTING TRUTH FUNCTIONS

We know how to write a Boolean expression and construct a network from a given truth function. Often the truth function itself must first be deduced from the description of the actual problem.

4.18   Example

An automatic control device supervises packaging of orders received by a mail-order cosmetics firm. Lipstick, perfume, makeup, and nail polish can be ordered. As a bonus item, shampoo is included with any order that includes perfume, or with any lipstick-makeup-nail polish order. How can we design the logic network that controls whether shampoo is packaged with an order?

The inputs to the network will represent the four items that can be ordered. We label these

$x_1 = $ lipstick

$x_2 = $ perfume

$x_3 = $ makeup

$x_4 = $ nail polish

The value of x_i will be 1 when that item is included in the order, 0 otherwise. The output from the network should be 1 if shampoo is to be packaged with the order, 0 otherwise. The truth table for the circuit appears in Figure 4.18a. The canonical sum-of-products form for this truth function is lengthy, but the expression $x_1x_3x_4 + x_2$ also represents the function. Figure 4.18b shows the logic network for this expression.   □

x_1	x_2	x_3	x_4	$f(x_1, x_2, x_3, x_4)$
1	1	1	1	1
1	1	1	0	1
1	1	0	1	1
1	1	0	0	1
1	0	1	1	1
1	0	1	0	0
1	0	0	1	0
1	0	0	0	0
0	1	1	1	1
0	1	1	0	1
0	1	0	1	1
0	1	0	0	1
0	0	1	1	0
0	0	1	0	0
0	0	0	1	0
0	0	0	0	0

(a)

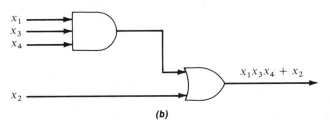

(b)

Figure 4.18

4.19 Practice A hall light is controlled by two light switches, one at each end. Find (a) a truth function, (b) a Boolean expression, and (c) a logic network that allows the light to be switched on or off by either switch. (*Answer, page A-12*)

The problem might be such that the corresponding truth function has certain undefined values because certain combinations of input cannot occur (see Exercise 11 at the end of this section). Under these "don't care" conditions, any value may be assigned to the output.

When using a programming language where the Boolean operators AND, OR, and NOT are available, designing the logic of a computer program often depends on deciding upon appropriate truth functions and finding corresponding Boolean expressions (see Exercise 6 of Section 1.1).

✓ CHECKLIST

Definitions

OR gate (*p. 130*)
AND gate (*p. 130*)
inverter (*p. 130*)
Boolean expression (*p. 130*)
truth function (*p. 131*)
combinational network (*p. 133*)
canonical sum-of-products form (disjunctive normal form) (*p. 134*)
equivalent Boolean expressions (*p. 137*)
NAND gate (*p. 140*)
NOR gate (*p. 141*)

Techniques

Find the truth function corresponding to a given Boolean expression or logic network.

Construct a logic network with the same truth function as a given Boolean expression.

Write a Boolean expression with the same truth function as a given logic network.

Write the Boolean expression in canonical sum-of-products form for a given truth function.

Using only NAND gates, find a network having the same truth function as a given network with AND gates, OR gates, and inverters.

Find a truth function satisfying the description of a particular problem.

Main Ideas

We can effectively convert information from any of the three forms below to any other form:

$$\text{truth function} \leftrightarrow \text{Boolean expression} \leftrightarrow \text{logic network}$$

A Boolean expression can sometimes be converted to a simpler, equivalent expression using the properties of Boolean algebra, thus producing a simpler network for a given truth function.

EXERCISES, SECTION 4.1

1. Construct logic networks for the following Boolean expressions, using AND gates, OR gates, and inverters.
   (a) $(x_1' + x_2)x_3$
   (b) $(x_1 + x_2)' + x_1'x_3$

2. Write a Boolean expression and a truth function for each of the logic networks in Figures 4.19 and 4.20.

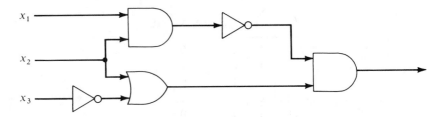

Figure 4.19

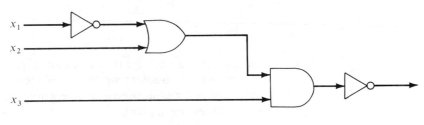

Figure 4.20

★ 3. Find the canonical sum-of-products form for the truth functions in Figures 4.21 and 4.22.

(a)

x_1	x_2	$f(x_1, x_2)$
1	1	1
1	0	0
0	1	1
0	0	0

Figure 4.21

(b)

x_1	x_2	x_3	$f(x_1, x_2, x_3)$
1	1	1	0
1	1	0	1
1	0	1	1
1	0	0	0
0	1	1	1
0	1	0	0
0	0	1	0
0	0	0	1

Figure 4.22

★ 4. (a) Find the canonical sum-of-products form for the truth function in Figure 4.23.

x_1	x_2	x_3	$f(x_1, x_2, x_3)$
1	1	1	0
1	1	0	1
1	0	1	0
1	0	0	1
0	1	1	0
0	1	0	0
0	0	1	0
0	0	0	0

Figure 4.23

(b) Draw the logic network for the expression of part (a).
(c) Use properties of a Boolean algebra to reduce the expression of part (a) to an equivalent expression whose network requires only two logic elements. Draw the network.

5. (a) Find the canonical sum-of-products form for the truth function in Figure 4.24.

x_1	x_2	x_3	$f(x_1, x_2, x_3)$
1	1	1	1
1	1	0	0
1	0	1	0
1	0	0	0
0	1	1	1
0	1	0	1
0	0	1	0
0	0	0	0

Figure 4.24

(b) Draw the logic network for the expression of part (a).
(c) Use properties of a Boolean algebra to reduce the expression of part (a) to an equivalent expression whose network requires only three logic elements. Draw the network.

6. (a) Show that the two Boolean expressions

$$(x_1 + x_2)(x_1' + x_3)(x_2 + x_3)$$

and

$$(x_1 x_3) + (x_1' x_2)$$

are equivalent by writing the truth table for each.

(b) Write the canonical sum-of-products form equivalent to the two expressions of part (a).

(c) Use properties of a Boolean algebra to reduce one of the expressions of part (a) to the other.

★ 7. There is also a **canonical product-of-sums form (conjunctive normal form)** for any truth function. This expression has the form

$$(\ )(\ ) \cdots (\ )$$

with each factor a sum of the form

$$\alpha + \beta + \cdots + \gamma$$

where $\alpha = x_1$ or x_1', $\beta = x_2$ or x_2', and so on. Each factor is constructed to have a 0 value for the input values of one of the rows of the truth function having value 0. Thus, the entire expression has value 0 for these inputs and no others. Find the canonical product-of-sums form for the truth functions of Exercise 3 above.

8. (a) Construct a network for the following expression using only NAND elements. Replace the AND and OR gates and inverters with the appropriate NAND networks.

$$x_3'x_1 + x_2'x_1 + x_3'$$

(b) Use the properties of a Boolean algebra to reduce the expression of part (a) to one whose network would require only three NAND gates. Draw the network.

★ 9. Replace the network of Figure 4.25 with an equivalent network using one AND gate, one OR gate, and one inverter.

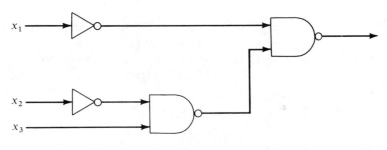

Figure 4.25

10. Using only NOR elements, construct networks that can replace (a) an inverter; (b) an OR gate; and (c) an AND gate.

★ 11. You have just been hired at Mercenary Motors. Your job is to design a logic network that will allow a car to be started only when it is in neutral or park, and the driver's seat belt is fastened. Find a truth function, a

Boolean expression, and a logic network. (There is a "don't care" condition to the truth function since the car cannot be in both neutral and park.)

12. Mercenary Motors has expanded into the calculator business. You need to design the circuitry for the display readout on a new calculator. This design involves a two-step process.

   (a) Any digit 0, 1, . . . , 9 put into the calculator is first converted to binary form. Figure 4.26 describes this conversion, which involves the design of four separate networks. Each network has 10 inputs, but only 1 input can be "on" at any given moment. Write a Boolean expression and then draw a network for x_2.

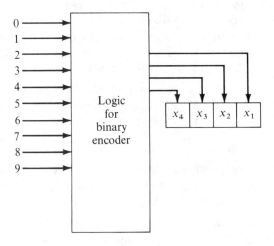

Figure 4.26

   (b) The binary form of the digit is then converted into visual display form by activating a pattern of 7 outputs arranged as shown in Figure 4.27a. To display the digit 3, for example, requires that y_1, y_2, y_3, y_5, and y_7 be "on," as in Figure 4.27b. Thus, the second step of the process

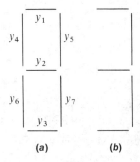

(a)   (b)

Figure 4.27

can be represented by Figure 4.28, which involves the design of 7 separate networks, each with 4 inputs. Write a truth function, a Boolean expression, and a network for y_5 and for y_6.

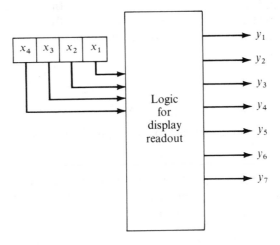

Figure 4.28

Section 4.2   MINIMIZATION

THE MINIMIZATION PROCESS

Remember from the last section that a given truth function is associated with an equivalence class of Boolean expressions. If we want to design a logic network for the function, we would *ideally* like to have a procedure that would choose a "simplest" Boolean expression from the class. What we consider simple will depend upon the technology employed in building the network, what kind of logic elements are available, and so on. At any rate, we probably want to minimize the total number of connections that must be made and the total number of logic elements used. (As we discuss minimization procedures, keep in mind that other factors may influence the economics of the situation. If a network is to be built only once, the time spent in minimization effort is more costly than building the network. But if the network is to be mass produced, then the cost of minimization time may be worthwhile.)

We have had some experience with simplifying Boolean expressions by applying the properties of Boolean algebra. However, we had no procedure to use; we simply had to guess—to attack each problem individually. What we want now is a procedure (algorithm) that we can use mechanically without having to be clever or insightful. As a matter of fact, we won't develop such a procedure to solve our ideal problem. But we already know how to select the canonical sum-of-products form from the equivalence class of expressions for a given truth function. And in this section, we will discuss two procedures to reduce a canonical sum-of-products form to a minimal sum-of-products form. Therefore, we can minimize within this framework and reduce, if not actually achieve the absolute minimum, the number of elements and connections required.

4.20 Example   The Boolean expression

$$x_1 x_2 x_3 + x_1' x_2 x_3 + x_1' x_2 x_3'$$

is in sum-of-products form. An equivalent minimal sum-of-products form is

$$x_2 x_3 + x_2 x_1'$$

Implementing a network for this form would require two AND gates, one OR gate, and an inverter. Using one of the distributive laws of Boolean algebra, this expression reduces to

$$x_2(x_3 + x_1')$$

which requires only one AND gate, one OR gate, and an inverter, but is no longer in sum-of-products form. Thus, a minimal sum-of-products form may not be minimal in an absolute sense.   □

There are two extremely useful equivalencies in minimizing a sum-of-products form. They are

$$x_1 x_2 + x_1' x_2 = x_2$$

and

$$x_1 + x_1' x_2 = x_1 + x_2$$

4.21 Practice   Use properties of Boolean algebra

(a) to reduce

$$x_1 x_2 + x_1' x_2 \text{ to } x_2$$

and

(b) to reduce

$$x_1 + x_1' x_2 \text{ to } x_1 + x_2 \qquad \text{(Answer, page A-12)}$$

The equivalency $x_1 x_2 + x_1' x_2 = x_2$ means that the expression $x_1' x_2 x_3' x_4 + x_1' x_2' x_3' x_4$ reduces to $x_1' x_3' x_4$. Thus, when we have a sum of two products that differ in only one factor, we can eliminate that factor. But the canonical sum-of-products form for a truth function of four variables, for example, might be quite long and require some searching to locate two product terms differing by only one factor. The **Karnaugh map** provides us with a visual way to represent the truth function so that terms in the canonical sum-of-products form differing by only one factor can be quickly matched.

THE KARNAUGH MAP

Figure 4.29

In the canonical sum-of-products form for a truth function, we are interested in values of the input variables producing outputs of 1. The Karnaugh map records the 1's of the function in an array that forces products of inputs differing by only one factor to be adjacent. The array form for a two-variable function is given in Figure 4.29. Notice that the square corresponding to x_1x_2, the upper left-hand square, is adjacent to squares $x_1'x_2$ and x_1x_2', which differ in one factor from x_1x_2; but it is not adjacent to the $x_1'x_2'$ square, which differs in two factors from x_1x_2.

4.22 Example   The truth function of Figure 4.30 is represented by the Karnaugh map of Figure 4.31. At once we can observe 1's in two adjacent squares, so there are two terms in the canonical sum-of-products form differing by one variable; again from the graph, we see that the variable that changes is x_1. It can be eliminated. We conclude that the function can be represented by x_2. Indeed, the canonical sum-of-products form for the function is $x_1x_2 + x_1'x_2$, which, by our basic reduction rule, reduces to x_2. But we did not have to write the canonical form, only look at the map.   □

x_1	x_2	$f(x_1, x_2)$
1	1	1
1	0	0
0	1	1
0	0	0

Figure 4.30

Figure 4.31

4.23 Practice   Draw the Karnaugh map and use it to find a reduced expression for the function of Figure 4.32.   (*Answer, page A-12*)

x_1	x_2	$f(x_1, x_2)$
1	1	0
1	0	0
0	1	1
0	0	1

Figure 4.32

Maps for Three and Four Variables

The array forms for functions of three variables and four variables are shown in Figure 4.33a and b. In these arrays, adjacent squares also differ by only one variable. However, in Figure 4.33a, the leftmost and rightmost squares in a row also differ by one variable, so we consider them adjacent. (They would in fact be adjacent in the usual sense if we wrapped the map around a cylinder and glued the left and right edges together.) In Figure 4.33b, the leftmost and rightmost squares in a row are adjacent (differ by exactly one variable) and so are the top and bottom squares of a column.

In three-variable maps, two adjacent squares marked with 1 allow elimination of one variable, and four adjacent squares marked with 1 (either in a single row or arranged in a square) allow elimination of two variables.

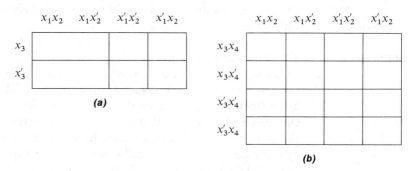

(a)

(b)

Figure 4.33

4.24 Example   In the map of Figure 4.34, the squares that combine for a reduction are looped. These four adjacent squares reduce to x_3 (eliminate the changing

	x_1x_2	x_1x_2'	$x_1'x_2'$	$x_1'x_2$
x_3	1	1	1	1
x_3'				

Figure 4.34

variables x_1 and x_2). The reduction uses our basic reduction rule more than once, namely,

$$x_1x_2x_3 + x_1x_2'x_3 + x_1'x_2'x_3 + x_1'x_2x_3$$
$$= x_1x_3(x_2 + x_2') + x_1'x_3(x_2' + x_2)$$
$$= x_1x_3 + x_1'x_3$$
$$= x_3(x_1 + x_1')$$
$$= x_3 \qquad \qquad \square$$

In four-variable maps, two adjacent squares marked with 1 allow elimination of one variable, four adjacent squares marked with 1 allow elimination of two variables, and eight adjacent squares marked with 1 allow elimination of three variables.

Figure 4.35 illustrates some ways in which two adjacent marked squares could occur. Figure 4.36 illustrates some ways in which four adjacent marked

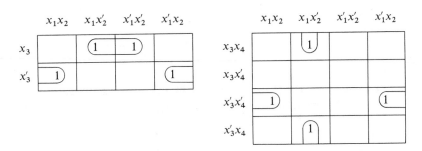

Figure 4.35

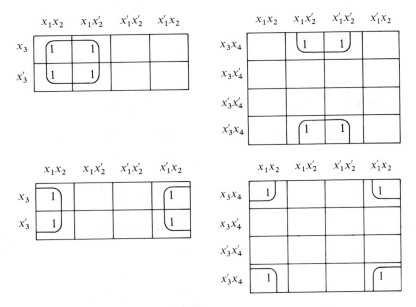

Figure 4.36

squares could occur, and Figure 4.37 shows possibilities for eight adjacent marked squares.

	x_1x_2	x_1x_2'	$x_1'x_2'$	$x_1'x_2$
x_3x_4	1	1	1	1
x_3x_4'				
$x_3'x_4'$				
$x_3'x_4$	1	1	1	1

	x_1x_2	x_1x_2'	$x_1'x_2'$	$x_1'x_2$
x_3x_4			1	1
x_3x_4'			1	1
$x_3'x_4'$			1	1
$x_3'x_4$			1	1

Figure 4.37

4.25 Example   In the map of Figure 4.38, the four outside corners reduce to x_2x_4 and the inside square reduces to $x_2'x_4'$.   □

	x_1x_2	x_1x_2'	$x_1'x_2'$	$x_1'x_2$
x_3x_4	1			1
x_3x_4'		1	1	
$x_3'x_4'$		1	1	
$x_3'x_4$	1			1

Figure 4.38

4.26 Practice   Find the two terms represented by the map in Figure 4.39.   (*Answer, page A-12*)

	x_1x_2	x_1x_2'	$x_1'x_2'$	$x_1'x_2$
x_3x_4	1	1		
x_3x_4'	1	1		
$x_3'x_4'$	.			1
$x_3'x_4$				1

Figure 4.39

Using the Karnaugh Map

How do we find a minimal sum-of-products form from a Karnaugh map (or from a truth function or a canonical sum-of-products form)? We must include every marked square of the map. And we want to include every marked square as part of the largest combination of marked squares possible since doing so provides maximum reduction of the expression. However, we cannot simply begin by looking for the largest blocks of marked squares on the map.

4.27 Example   In the Karnaugh map of Figure 4.40, if we simply look for the largest block of

	x_1x_2	x_1x_2'	$x_1'x_2'$	$x_1'x_2$
x_3x_4		1	1	
x_3x_4'			1	1
$x_3'x_4'$		1	1	
$x_3'x_4$			1	1

Figure 4.40

marked squares, we would use the column of 1's and reduce it to $x_1'x_2'$, still leaving us four marked squares unaccounted for. Each of these marked squares can be combined into a two-square block in only one way (see Figure 4.41), and each of these blocks has to be included. But when this is done, each

	x_1x_2	x_1x_2'	$x_1'x_2'$	$x_1'x_2$
x_3x_4		1	1	
x_3x_4'			1	1
$x_3'x_4'$		1	1	
$x_3'x_4$			1	1

Figure 4.41

square in the column of 1's is used, and the term $x_1'x_2'$ is superfluous. The minimal sum-of-products form for this map becomes

$$x_2'x_3x_4 + x_1'x_3x_4' + x_2'x_3'x_4' + x_1'x_3'x_4$$

☐

To avoid the redundancy illustrated by Example 4.27, we analyze the map as follows. First, we form terms for those marked squares that cannot be combined with anything. Then we use the remaining marked squares to find those that can only be combined into two-square blocks and in only one way. Then among the unused marked squares, that is, those not already assigned to a block, we find those that can only be combined into four-square blocks and in only one way; then we look for any unused squares that go uniquely into eight-square blocks. At each step, if an unused marked square has a choice of membership in blocks, we do nothing with it. Finally, we take any unused marked squares that are left (for which there was a choice of blocks) and arbitrarily select the largest blocks that include them.

4.28 Example   In Figure 4.42a, we have shown the only square that cannot be combined into a larger block. In Figure 4.42b, we have formed the unique two-square block

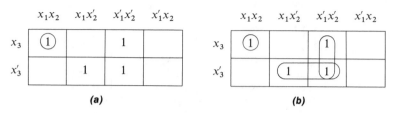

Figure 4.42

for the $x_1x_2'x_3'$ square, and the unique two-square block for the $x_1'x_2'x_3$ square. All marked squares are covered. The minimal sum-of-products expression is

$$x_1x_2x_3 + x_2'x_3' + x_1'x_2'$$

We got this sum by expanding $x_1'x_2'x_3'$ into $x_1'x_2'x_3' + x_1x_2'x_3'$, and then combining it with each of its neighbors.   □

4.29 Example   Figure 4.43a shows the unique two-square blocks for the $x_1'x_2x_3x_4$ square and the $x_1x_2'x_3'x_4'$ square. In Figure 4.43b the two unused squares have been

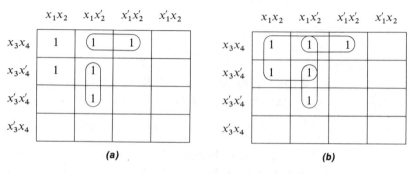

Figure 4.43

combined into a unique four-square block. The minimal sum-of-products expression is

$$x_1x_3 + x_2'x_3x_4 + x_1x_2'x_4'$$ □

4.30 Example   In Figure 4.44a the unique two-square blocks are shown. We can assign the remaining unused marked square to two different two-square blocks; these

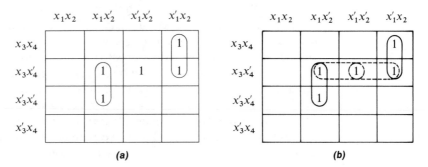

Figure 4.44

blocks are shown in Figure 4.44b. There are two minimal sum-of-products forms,

$$x_1x_2'x_4' + x_1'x_2x_3 + x_2'x_3x_4'$$

and

$$x_1x_2'x_4' + x_1'x_2x_3 + x_1'x_3x_4'$$ □

4.31 Practice   Write the minimal sum-of-products expression for the map shown in Figure 4.45.   *(Answer, page A-12)*

	x_1x_2	x_1x_2'	$x_1'x_2'$	$x_1'x_2$
x_3x_4		1		
x_3x_4'				
$x_3'x_4'$		1	1	
$x_3'x_4$	1	1	1	

Figure 4.45

We have used Karnaugh maps for functions of two, three, and four variables. It is possible, by using three-dimensional drawings, to construct Karnaugh maps for functions of five, six, or even more variables, but the visualization gets too complicated to be worthwhile.

If the Karnaugh map corresponds to a function with don't care conditions, then the don't care squares on the map can be left blank or assigned the value 1, whichever aids the minimization process.

THE QUINE–MCCLUSKEY PROCEDURE

Remember that the key to reducing the canonical sum-of-products form for a truth function lies in recognizing terms of the sum that differ in only one factor. In the Karnaugh map, we see where such terms occur. The second method of reduction we will discuss, the Quine–McCluskey procedure, organizes information from the canonical sum-of-products form into a table so that the search for terms differing by only one factor is easy to carry out. The procedure is a two-step process paralleling the use of the Karnaugh map. First, we find the groupings of terms (as we looped together marked squares in the Karnaugh map); then we eliminate redundant groupings and make choices for terms that can belong to several groups.

4.32 Example   Let's illustrate the Quine–McCluskey procedure by using the truth function for Example 4.27. We did not write the actual truth function there, but the information is contained in the Karnaugh map. The truth function is shown in Figure 4.46. In Figure 4.47 the eight 4-tuples of 0's and 1's producing a function value of 1 are listed in a table, separated into groupings according

x_1	x_2	x_3	x_4	$f(x_1, x_2, x_3, x_4)$
1	1	1	1	0
1	1	1	0	0
1	1	0	1	0
1	1	0	0	0
1	0	1	1	1
1	0	1	0	0
1	0	0	1	0
1	0	0	0	1
0	1	1	1	0
0	1	1	0	1
0	1	0	1	1
0	1	0	0	0
0	0	1	1	1
0	0	1	0	1
0	0	0	1	1
0	0	0	0	1

Figure 4.46

	x_1	x_2	x_3	x_4
(three 1's)	1	0	1	1
	0	1	1	0
(two 1's)	0	1	0	1
	0	0	1	1
	1	0	0	0
(one 1)	0	0	1	0
	0	0	0	1
(no 1's)	0	0	0	0

Figure 4.47

to the number of 1's. Note that terms of the canonical sum-of-products form differing by only one factor must be in adjacent groupings, which simplifies the search for such terms. Thus, we compare the first term, 1011, with the three terms of the second group, 0110, 0101, and 0011, to locate terms differing by only one factor. Such a term is 0011. The combination 1011 and 0011 reduces to –011 where the changing variable x_1 has been eliminated, indicated by a dash in the x_1 position. We form a second column in the table consisting of the reduced terms, again grouped according to number of 1's (see Figure 4.48). The two terms 1011 and 0011 are marked with the number 1 as a pointer to the reduced term of column two that they form (numbering terms corresponds to putting loops in the Karnaugh map). We continue this process with all the terms. A numbered term may still be used in other combinations, just as a marked square in a Karnaugh map can be in more than one loop. Thus, we arrive at Figure 4.48.

x_1	x_2	x_3	x_4			x_1	x_2	x_3	x_4
1	0	1	1	1		–	0	1	1
0	1	1	0	2		0	–	1	0
0	1	0	1	3		0	–	0	1
0	0	1	1	1,4,5		0	0	1	–
						0	0	–	1
1	0	0	0	6					
0	0	1	0	2,4,7		–	0	0	0
0	0	0	1	3,5,8		0	0	–	0
						0	0	0	–
0	0	0	0	6,7,8					

Figure 4.48

We continue this reduction process on the second column of Figure 4.48. Here, not only the groupings but also the dashes help organize the search process since terms differing by only one variable must have dashes in the same location. Numbers on terms that combine serve as pointers to the reduced terms in column three. When the process cannot be continued, we reach the reduction table of Figure 4.49. The unnumbered terms are irreducible, so they represent the possible maximum-sized loops on a Karnaugh map.

x_1	x_2	x_3	x_4		x_1	x_2	x_3	x_4		x_1	x_2	x_3	x_4
1	0	1	1	1	–	0	1	1		0	0	–	–
0	1	1	0	2	0	–	1	0					
0	1	0	1	3	0	–	0	1					
0	0	1	1	1,4,5	0	0	1	–	1				
					0	0	–	1	1				
1	0	0	0	6									
0	0	1	0	2,4,7	–	0	0	0					
0	0	0	1	3,5,8	0	0	–	0	1				
					0	0	0	–	1				
0	0	0	0	6,7,8									

Figure 4.49

For the second step of the process, we compare the original terms with the irreducible terms. A check in the comparison table in Figure 4.50 indicates that the original term in that column eventually led to the irreducible term in that row, which can be determined by following the pointers.

	1011	0110	0101	0011	1000	0010	0001	0000
–011	✓			✓				
0–10		✓				✓		
0–01			✓				✓	
–000					✓			✓
00––				✓		✓	✓	✓

Figure 4.50

If a column in the comparison table has a check in only one row, the irreducible term for that row is the only one covering the original term, so it is an essential term and must appear in the final sum-of-products form. Thus, we see from Figure 4.50 that the terms -011, $0-10$, $0-01$, and -000 are essential and must be in the final expression. We also note that all columns with a check in row five also have checks in another row and so are covered by an essential, reduced term already in the expression. Thus, $00--$ is redundant. As in Example 4.27, the minimal sum-of-products form is

$$x'_2x_3x_4 + x'_1x_3x'_4 + x'_1x'_3x_4 + x'_2x_3x'_4 \qquad \square$$

In situations where there is more than one minimal sum-of-products form, the comparison table will have nonessential, nonredundant, reduced terms. A selection must be made from these reduced terms to cover all columns not covered by essential terms.

4.33 Example

We will use the Quine–McCluskey procedure on the problem presented in Example 4.30. The reduction table is given in Figure 4.51. The comparison table appears in Figure 4.52. We see from the comparison table that $011-$ and $10-0$ are essential, reduced terms and that there are no redundant terms. Column four is the only original term not covered by essential terms, and the choice of the reduced term for row two or for row four will cover it.

x_1	x_2	x_3	x_4			x_1	x_2	x_3	x_4
0	1	1	1	1		0	1	1	–
1	0	1	0	2,3		–	0	1	0
0	1	1	0	1,4		1	0	–	0
0	0	1	0	2,4		0	–	1	0
1	0	0	0	3					

Figure 4.51

	0111	1010	0110	0010	1000
$011-$	✓		✓		
-010		✓		✓	
$10-0$		✓			✓
$0-10$			✓	✓	

Figure 4.52

Thus, the minimal sum-of-products form is

$$x_1' x_2 x_3 + x_1 x_2' x_4' + x_2' x_3 x_4'$$

or

$$x_1' x_2 x_3 + x_1 x_2' x_4' + x_1' x_3 x_4' \qquad \square$$

4.34 Practice   Use the Quine–McCluskey procedure to find a minimal sum-of-products form for the truth function of Figure 4.53.   (*Answer, page A-13*)

x_1	x_2	x_3	$f(x_1, x_2, x_3)$
1	1	1	1
1	1	0	1
1	0	1	0
1	0	0	1
0	1	1	0
0	1	0	0
0	0	1	1
0	0	0	1

Figure 4.53

The Quine–McCluskey procedure applies to truth functions with any number of input variables, but for a large number of variables, the procedure would be extremely tedious to do by hand. However, it is exactly the kind of systematic, mechanical process that lends itself to a computerized solution.

If the truth function f has few 0 values and a large number of 1 values, it may be simpler to implement the Quine–McCluskey procedure for the complement of the function, f', which will have 1 values where f has 0 values, and vice versa. Once a minimal sum-of-products expression is obtained for f', it can be complemented to obtain an expression for f although the new expression will not be in sum-of-products form. (In fact, by DeMorgan's Laws, it will be equivalent to a product-of-sums form.) We can obtain the network for f from the sum-of-products network for f' by tacking an inverter on the end.

The whole object of minimizing a network is to simplify the internal configuration while preserving the external behavior. In Chapter 8 we will attempt the same sort of minimization on finite-state machine structures.

⟜ CHECKLIST

Techniques

Minimize the canonical sum-of-products form for a truth function by using a Karnaugh map.

Minimize the canonical sum-of-products form for a truth function by using the Quine–McCluskey procedure.

Main Ideas

Algorithms exist to reduce a canonical sum-of-products form to a minimized sum-of-products form.

EXERCISES, SECTION 4.2

★ 1. Write the minimal sum-of-products form for the Karnaugh maps of Figure 4.54.

	$x_1 x_2$	$x_1 x_2'$	$x_1' x_2'$	$x_1' x_2$
x_3	1	1	1	1
x_3'	1			1

(a)

	$x_1 x_2$	$x_1 x_2'$	$x_1' x_2'$	$x_1' x_2$
$x_3 x_4$		1		
$x_3 x_4'$		1	1	1
$x_3' x_4'$	1	1	1	
$x_3' x_4$		1		

(b)

	$x_1 x_2$	$x_1 x_2'$	$x_1' x_2'$	$x_1' x_2$
$x_3 x_4$				1
$x_3 x_4'$	1	1		
$x_3' x_4'$		1	1	
$x_3' x_4$			1	

(c)

Figure 4.54

2. Use a Karnaugh map to find the minimal sum-of-products form for the truth functions of Figure 4.55.

x_1	x_2	x_3	$f(x_1, x_2, x_3)$
1	1	1	1
1	1	0	1
1	0	1	0
1	0	0	0
0	1	1	1
0	1	0	0
0	0	1	0
0	0	0	0

(a)

x_1	x_2	x_3	x_4	$f(x_1, x_2, x_3, x_4)$
1	1	1	1	1
1	1	1	0	1
1	1	0	1	1
1	1	0	0	1
1	0	1	1	0
1	0	1	0	1
1	0	0	1	0
1	0	0	0	1
0	1	1	1	1
0	1	1	0	1
0	1	0	1	1
0	1	0	0	1
0	0	1	1	0
0	0	1	0	0
0	0	0	1	0
0	0	0	0	0

★ (b)

Figure 4.55

3. Use a Karnaugh map to find the minimal sum-of-products form for the following Boolean expressions.
   (a) $x_1'x_2'x_3x_4 + x_1x_2x_3'x_4 + x_1'x_2'x_3'x_4 + x_1x_2'x_3x_4' + x_1'x_2x_3x_4$
   $+ x_1'x_2x_3'x_4 + x_1'x_2'x_3x_4'$
   ★ (b) $x_1'x_2'x_3'x_4' + x_1x_2x_3'x_4 + x_1'x_2'x_3'x_4 + x_1x_2x_3'x_4' + x_1'x_2x_3x_4$
   $+ x_1x_2'x_3x_4'$

4. Use a Karnaugh map to find a minimal sum-of-products expression for the network of three variables shown in Figure 4.56. Sketch the new network.

★ 5. Use a Karnaugh map to find a minimal sum-of-products form for the truth function in Figure 4.57. Don't care conditions are shown by dashes.

★ 6. Use the Quine–McCluskey procedure to find a minimal sum-of-products form for the truth function illustrated by the map in Figure 4.54a.

7. Use the Quine–McCluskey procedure to find a minimal sum-of-products form for the network in Figure 4.58 (p. 166). Sketch the new network.

8. Use the Quine–McCluskey procedure to find the minimal sum-of-products form for the truth functions in Figure 4.59 (p. 167).

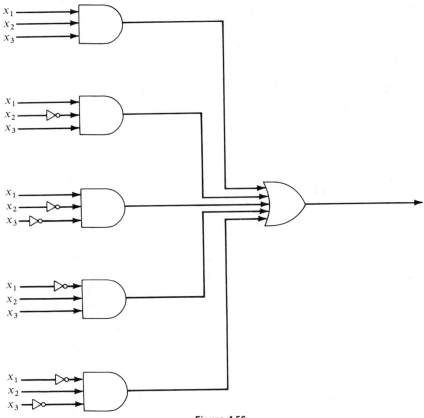

Figure 4.56

x_1	x_2	x_3	x_4	$f(x_1, x_2, x_3, x_4)$
1	1	1	1	0
1	1	1	0	1
1	1	0	1	0
1	1	0	0	–
1	0	1	1	0
1	0	1	0	–
1	0	0	1	0
1	0	0	0	0
0	1	1	1	0
0	1	1	0	1
0	1	0	1	0
0	1	0	0	1
0	0	1	1	1
0	0	1	0	0
0	0	0	1	–
0	0	0	0	0

Figure 4.57

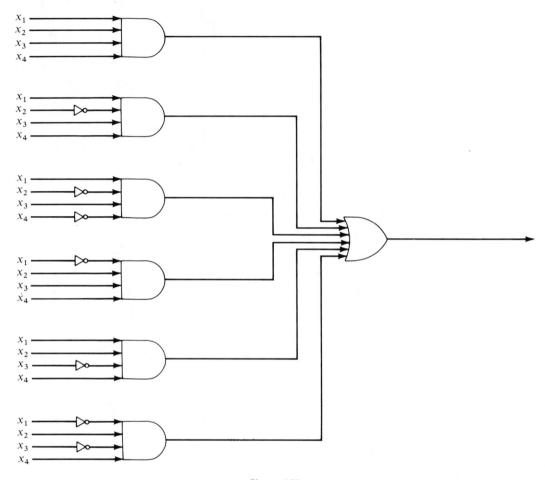

Figure 4.58

9. Use the Quine–McCluskey procedure to find the minimal sum-of-products form for the following Boolean expressions.

★ (a) $x_1 x_2' x_3 x_4' + x_1' x_2' x_3 x_4 + x_1' x_2 x_3 x_4 + x_1' x_2' x_3' x_4' + x_1' x_2 x_3' x_4'$
$+ x_1' x_2' x_3' x_4$

(b) $x_1 x_2 x_3 x_4 + x_1 x_2' x_3 x_4 + x_1 x_2 x_3 x_4' + x_1 x_2' x_3 x_4' + x_1' x_2 x_3 x_4'$
$+ x_1 x_2 x_3' x_4' + x_1' x_2 x_3' x_4' + x_1 x_2 x_3' x_4 + x_1' x_2 x_3' x_4$

(c) $x_1 x_2 x_3 x_4 + x_1 x_2 x_3 x_4' + x_1' x_2 x_3 x_4' + x_1 x_2 x_3' x_4' + x_1' x_2' x_3' x_4'$
$+ x_1' x_2 x_3' x_4' + x_1 x_2' x_3' x_4 + x_1' x_2' x_3' x_4 + x_1 x_2 x_3' x_4$

(d) $x_1' x_2 x_3' x_4 x_5' + x_1' x_2 x_3 x_4' x_5 + x_1 x_2 x_3 x_4 x_5 + x_1' x_2' x_3 x_4' x_5$
$+ x_1 x_2' x_3 x_4 x_5 + x_1' x_2' x_3' x_4' x_5 + x_1 x_2' x_3 x_4' x_5 + x_1 x_2 x_3 x_4' x_5'$
$+ x_1 x_2 x_3' x_4 x_5 + x_1' x_2' x_3' x_4 x_5'$

x_1	x_2	x_3	x_4	$f(x_1, x_2, x_3, x_4)$		x_1	x_2	x_3	x_4	$f(x_1, x_2, x_3, x_4)$
1	1	1	1	0		1	1	1	1	1
1	1	1	0	1		1	1	1	0	0
1	1	0	1	0		1	1	0	1	1
1	1	0	0	0		1	1	0	0	0
1	0	1	1	0		1	0	1	1	1
1	0	1	0	1		1	0	1	0	0
1	0	0	1	1		1	0	0	1	1
1	0	0	0	1		1	0	0	0	1
0	1	1	1	0		0	1	1	1	1
0	1	1	0	0		0	1	1	0	0
0	1	0	1	0		0	1	0	1	1
0	1	0	0	1		0	1	0	0	1
0	0	1	1	1		0	0	1	1	1
0	0	1	0	1		0	0	1	0	0
0	0	0	1	0		0	0	0	1	1
0	0	0	0	1		0	0	0	0	1

(a)   (b)

Figure 4.59

FOR FURTHER READING

A brief but thorough treatment of the material of this chapter is given in [1]. Another good general reference is [2]; [3] takes a more modern look at logic networks in the light of newer technology. More specific applications of logic design to computers appear in [4] and [5].

1. Elliott Mendelson, *Boolean Algebra and Switching Circuits*. New York: McGraw-Hill, 1970.

2. Michael A. Harrison, *Introduction to Switching Theory and Automata Theory*. New York: McGraw-Hill, 1965.

3. Arthur D. Friedman and P. R. Menon, *Theory and Design of Switching Circuits*. Woodland Hills, Calif.: Computer Science Press, 1975.

4. Taylor L. Booth, *Digital Networks and Computer Systems*, 2nd ed. New York: Wiley, 1978.

5. Morris Mano, *Computer Logic Design*. Englewood Cliffs, N.J.: Prentice-Hall, 1972.

Chapter 5

ALGEBRAIC STRUCTURES

From Chapter 3 we have a general notion of what a mathematical structure is. Here we will consider some algebraic structures, those which simulate (among other things) various types of arithmetic. Besides looking at numerous examples, we will also develop some of the basic theorems about these well-studied structures. The Simulation II ideas will be used to classify some instances of these structures. Finally, we will discuss further the conditions under which one instance of an algebraic structure can simulate another.

To balance our emphasis on algebraic structures, we should inject a realistic note. When we notice that a certain concrete situation S is an instance of an algebraic structure, say a group, we may hope to reap all sorts of benefits from this observation. In particular, it would be nice if some theorem from group theory were to reveal a property of S difficult to discover had we not noticed that we have a group. Often the results are not that spectacular, and the observation serves merely to classify S or clarify some of its properties. Sometimes, however, we gain a great deal from the knowledge that S has a certain algebraic structure—such will be the case with the group codes of Chapter 6, and in the general loop-free decomposition of finite-state machines in Section 4 of Chapter 8.

Section 5.1   SEMIGROUPS, MONOIDS, GROUPS—SIMULATION I

DEFINITIONS OF ALGEBRAIC STRUCTURES

If we wanted to parallel the historical development of a structure, we could look at many different situations and try to abstract common properties, the process we called Simulation I in Chapter 3 and used there to define a Boolean algebra and a monoid. But we really don't have time to pursue this approach. Instead, we will jump ahead and define those structures proven to be the most interesting, and then give some examples.

Algebraic structures are often arithmetic in nature, so we are concerned with binary operations on sets and with properties of these binary operations. We have already mentioned that the associative property of a binary operation means that grouping of elements does not matter, and the commutative property of a binary operation means the order of the elements does not matter. Identity elements also appeared in Chapter 3. Let's give the formal definitions of these terms as applied to algebraic structures.

5.1 Definition   Let S be a set and $\cdot$ denote a binary operation on S. Then $\cdot$ is **associative** if for all x, y, z in S,

$$(x \cdot y) \cdot z = x \cdot (y \cdot z) \qquad \qquad \square$$

If $\cdot$ is associative on S, then we can write an expression such as $x \cdot y \cdot z$ without parentheses.

5.2 Definition   Let S be a set and $\cdot$ denote a binary operation on S. Then $\cdot$ is **commutative** if for all x, y, in S,

$$x \cdot y = y \cdot x \qquad \qquad \square$$

5.3 Definition   Let S be a set and $\cdot$ denote a binary operation on S. Then an element $i \in S$ is an **identity element** if for all $x \in S$,

$$x \cdot i = i \cdot x = x \qquad \qquad \square$$

A monoid is an algebraic structure; we recall the definition.

5.4 Definition $[S, \cdot]$ is a **monoid** if S is a nonempty set, $\cdot$ is a binary operation on S, and

  (a) $\cdot$ is associative
  (b) There is an identity element $i \in S$; that is,

$$x \cdot i = i \cdot x = x$$

  for all $x \in S$. $\qquad \qquad \square$

5.5 Example   From Chapter 3 we already have the following examples of monoids: $[M_2(\mathbb{Z}), \cdot]$, $[M_2^D(\mathbb{Z}), \cdot]$, $[\mathbb{R}, +]$, $[\mathbb{R}^+, \cdot]$, $[\mathbb{Z}, \cdot]$. And if we review the properties of a Boolean algebra, we see that for any Boolean algebra $[B, +, \cdot, ', 0, 1]$, $[B, +]$ is a monoid and $[B, \cdot]$ is a monoid. Hence, for any sets, $[\mathscr{P}(S), \cup]$ and $[\mathscr{P}(S), \cap]$ are monoids. (Can you name an identity element for each of these monoids?) $\qquad \qquad \square$

As far as complexity is concerned, the monoid is in the middle of the three structures considered in this section. To have a *semigroup* we drop one of the requirements for a monoid, and to get a *group* we add another requirement to the monoid definition.

5.6   Definition $[S, \cdot]$ is a **semigroup** if S is a nonempty set, $\cdot$ is a binary operation on S, and $\cdot$ is associative.   □

A semigroup $[S, \cdot]$ in which $\cdot$ is commutative is called (what else?) a **commutative semigroup**.

Naturally every structure that is a monoid is also a semigroup, so we already have some examples of semigroups. Here are additional easy ones.

5.7   Practice   Verify that $[\mathbb{N}, +]$, $[\mathbb{N}, \cdot]$, $[\mathbb{Z}, +]$, $[\mathbb{R}, \cdot]$, $[\mathbb{R}^+, +]$ and $[M_2(\mathbb{Z}), +]$ are semigroups.   *(Answer, page A-13)*

5.8   Practice   (a) Which of the semigroups of Practice 5.7 are commutative?
   (b) Which of the semigroups of Practice 5.7 are monoids? Name the identities.

(Answer, page A-13)

Now for the last structure.

5.9   Definition $[S, \cdot]$ is a **group** if S is a nonempty set, $\cdot$ is a binary operation on S, and

   (a) $\cdot$ is associative
   (b) There is an identity element $i \in S$; that is,

$$x \cdot i = i \cdot x = x$$

   for all $x \in S$.
   (c) Each $x \in S$ has an **inverse element** x^{-1} in S; that is,

$$x \cdot x^{-1} = x^{-1} \cdot x = i$$ □

5.10   Example   The monoid $[\mathbb{N}, +]$ fails to be a group because 2, for instance, has no inverse element. There is no $n \in \mathbb{N}$ such that $2 + n = n + 2 = 0$. The only element having an inverse is 0.   □

A group $[S, \cdot]$ in which $\cdot$ is commutative is called a **commutative** or **abelian group** (after the Norwegian mathematician Niels Abel). Note that although we use the multiplication symbol to denote the binary operation on an abstract group, in any particular group, the binary operation could be addition of real numbers, addition of matrices, and so on. Also, we may refer to a structure as "the group S," rather than the group $[S, \cdot]$, when it is clear what the binary operation is.

Because the requirements that must be satisfied in going from semi-group to monoid to group keep getting stiffer, we expect some examples to drop out, but those remaining should have richer and more interesting personalities.

5.11 Practice   Which of the following monoids are groups? $[M_2(\mathbb{Z}), \cdot], [M_2^D(\mathbb{Z}), \cdot], [\mathbb{R}, +],$ $[\mathbb{R}^+, \cdot], [\mathbb{Z}, \cdot], [\mathscr{P}(S), \cup], [\mathscr{P}(S), \cap], [\mathbb{N}, \cdot], [\mathbb{Z}, +], [\mathbb{R}, \cdot], [M_2(\mathbb{Z}), +]$

(Answer, page A-13)

EXAMPLES OF ALGEBRAIC STRUCTURES

Next we will look at some other examples of semigroups, monoids, and groups.

5.12 Example   An expression of the form

$$a_n x^n + a_{n-1}x^{n-1} + \cdots + a_0$$

where $a_i \in \mathbb{R}$, $i = 0, \ldots, n$, and $n \in \mathbb{N}$ is a **polynomial in x with real number coefficients** (or a **polynomial in x over** $\mathbb{R}$.) For each i, a_i is the **coefficient** of x^i. If i is the largest integer greater than 0 for which $a_i \neq 0$, the polynomial is of **degree** i; if no such i exists, the polynomial is of **zero degree**. Terms with zero coefficients are generally not written. Thus, $\pi x^4 - 2/3x^2 + 5$ is a polynomial of degree 4, and the constant polynomial 6 is of zero degree. The set of all polynomials in x over $\mathbb{R}$ is denoted by $\mathbb{R}[x]$.

We define binary operations of $+$ and $\cdot$ in $\mathbb{R}[x]$ to be the familiar operations of polynomial addition and multiplication. For polynomials $f(x)$ and $g(x)$ members of $\mathbb{R}[x]$, the products $f(x) \cdot g(x)$ and $g(x) \cdot f(x)$ are equal because the coefficients are real numbers, and we can use all the properties of real numbers under multiplication and addition (properties such as commutativity and associativity). Similarly, for $f(x)$, $g(x)$, $h(x)$ members of $\mathbb{R}[x], (f(x) \cdot g(x)) \cdot h(x) = f(x) \cdot (g(x) \cdot h(x))$. The constant polynomial 1 is an identity because $1 \cdot f(x) = f(x) \cdot 1 = f(x)$ for every $f(x) \in \mathbb{R}[x]$. Thus, $[\mathbb{R}[x], \cdot]$ is a commutative monoid. It fails to be a group because only the nonzero constant polynomials have inverses. For example, there is no polynomial $g(x)$ such that $g(x) \cdot x = x \cdot g(x) = 1$, so the polynomial x has no inverse. However, $[\mathbb{R}[x], +]$ is a commutative group.   □

5.13 Practice   (a) For $f(x)$, $g(x)$, $h(x) \in \mathbb{R}[x]$, write the equations saying that $\mathbb{R}[x]$ under $+$ is commutative and associative.
(b) What is the identity element in $[\mathbb{R}[x], +]$?
(c) What is the inverse of $7x^4 - 2x^3 + 4$ in $[\mathbb{R}[x], +]$?

(Answer, page A-14)

Polynomials play a special part in the history of group theory (the study of groups) because much group theory research was prompted by the very practical problem of solving polynomial equations of the form $f(x) = 0$, $f(x) \in \mathbb{R}[x]$. The quadratic formula provides an algorithm for finding solutions for every $f(x)$ of degree 2, and the algorithm uses only the algebraic operations of addition, subtraction, multiplication, division, and taking roots. Other such algorithms exist for polynomials of degree 3 and 4. One of the highlights of abstract algebra is the proof that no algorithm exists using only these operations that works for every $f(x)$ of degree 5. (Notice that this statement is much stronger than simply saying that no algorithm has yet been found; it says to stop looking for one.)

The next example introduces **modular arithmetic**, an important idea in computer science because of the finite nature of a computer. (More on this notion later in this chapter.)

5.14 Example   Let $\mathbb{Z}_5 = \{0, 1, 2, 3, 4\}$ and define **addition modulo 5**, denoted by $+_5$, on $\mathbb{Z}_5$ by $x +_5 y = r$, where r is the remainder when $x + y$ is divided by 5. For example, $1 +_5 2 = 3$ and $3 +_5 4 = 2$. **Multiplication modulo 5** is defined by $x \cdot_5 y = r$, where r is the remainder when $x \cdot y$ is divided by 5. Thus, $2 \cdot_5 3 = 1$ and $3 \cdot_5 4 = 2$. Then $[\mathbb{Z}_5, +_5]$ is a commutative group, and $[\mathbb{Z}_5, \cdot_5]$ is a commutative monoid.   □

5.15 Practice   (a) Complete the tables defining $+_5$ and $\cdot_5$ on $\mathbb{Z}_5$.

$+_5$	0	1	2	3	4
0					
1			3		
2					
3				2	
4					

$\cdot_5$	0	1	2	3	4
0					
1					
2				1	
3					2
4					

(b) What is an identity in $[\mathbb{Z}_5, +_5]$? in $[\mathbb{Z}_5, \cdot_5]$?
(c) What is an inverse of 2 in $[\mathbb{Z}_5, +_5]$?
(d) Which elements in $[\mathbb{Z}_5, \cdot_5]$ have inverses?   (*Answer, page A-14*)

Notice that when we use a table to define an operation on a finite set, it is easy to check for commutativity by looking for symmetry around the main diagonal. It is also easy to find an identity element because its row looks like the top of the table and its column looks like the side. And it is easy to locate an inverse of an element. Look along the row until you find a column where the identity appears; then check to see that changing the order of the elements still gives the identity. But associativity (or the lack of it) is not immediately apparent from the table.

Similar to $\mathbb{Z}_5$, we can define operations of **addition modulo n** and **multiplication modulo n** on the set $\mathbb{Z}_n = \{0, 1, \ldots, n - 1\}$ where n is any positive integer. Again $[\mathbb{Z}_n, +_n]$ is a commutative group and $[\mathbb{Z}_n, \cdot_n]$ is a commutative monoid.

5.16 Practice

(a) Give the table for $\cdot_6$ on $\mathbb{Z}_6$.

(b) Which elements in $[\mathbb{Z}_6, \cdot_6]$ have inverses?   *(Answer, page A-14)*

The next two examples give us algebraic structures where the elements are functions.

5.17 Example

Let A be a set and consider the set S of all functions f such that $f : A \to A$. The binary operation is function composition (note that S is closed under $\circ$). Then $[S, \circ]$ is a semigroup, called the **semigroup of transformations on A**. Actually $[S, \circ]$ is a monoid because the identity function i_A that takes each member of A to itself has the property that for any $f \in S$,

$$f \circ i_A = i_A \circ f = f$$ □

5.18 Practice

Prove that function composition on the set S defined above is associative.

          (Answer, page A-14)

5.19 Example

Again let A be a set and consider the set S_A of all bijections f such that $f : A \to A$ (permutations of A). Bijectiveness is preserved under function composition, the identity function i_A is a permutation, and for any $f \in S_A$, the inverse function f^{-1} exists and is a permutation. Furthermore,

$$f \circ f^{-1} = f^{-1} \circ f = i_A$$

Thus, $[S_A, \circ]$ is a group, called the **group of permutations on A**.   □

If $A = \{1, 2, \ldots, n\}$ for some positive integer n, then S_A is called the **symmetric group of degree n** and denoted by S_n. Thus, S_3, for example, is the set of all permutations on $\{1, 2, 3\}$. There are six such permutations, which we will name as follows (using the cycle notation of Section 2.2):

$$\alpha_1 = i$$

$$\alpha_2 = (1, 2)$$

$$\alpha_3 = (1, 3)$$

$$\alpha_4 = (2, 3)$$

$$\alpha_5 = (1, 2, 3)$$

$$\alpha_6 = (1, 3, 2)$$

Then $\alpha_2 \circ \alpha_3 = (1, 2) \circ (1, 3) = (1, 3, 2) = \alpha_6$.

5.20 *Practice*   (a) Complete the group table for $[S_3, \circ]$.

$\circ$	α_1	α_2	α_3	α_4	α_5	α_6
α_1						
α_2			α_6			
α_3						
α_4						
α_5						
α_6						

(b) Is $[S_3, \circ]$ a commutative group?   *(Answer, page A-14)*

$[S_3, \circ]$ is our first example of a noncommutative group ($[M_2(\mathbb{Z}), \cdot]$ was a noncommutative monoid).

The next example is very simple but particularly appropriate because it appears in several areas of computer science, including formal language theory and automata theory.

5.21 *Example*   Let A be a set; its elements are called **symbols** and A itself an **alphabet**. A **string**, or **word**, over A is a finite sequence of symbols from A. Thus, if $A = \{a, b\}$, then *abbaa*, *bbbbba*, and *a* are all strings over A. The number of symbols in a string is called its **length**. We also allow a 0-length string, a string with no symbols, denoted by λ. We let A^* denote the set of all strings over A. (The empty string λ should not be confused with the empty set $\emptyset$; even if A itself is $\emptyset$, then $A^* = \{\lambda\}$.) The binary operation $\cdot$ is **concatenation** (or catenation), meaning juxtaposition. Therefore, *abbaa* $\cdot$ *a* gives the string *abbaaa*. The operation of concatenation on A^* is associative, and the empty string λ is an identity because for any string $x \in A^*$,

$$x \cdot \lambda = \lambda \cdot x = x$$

Therefore, $[A^*, \cdot]$ is a monoid, called the **free monoid generated by** A. Note that even if A contains only one element, say a, A^* is an infinite set because λ, a, aa, aaa, and so on, all belong to A^*.   □

5.22 *Practice*   For $A = \{a, b\}$

(a) Is $[A^*, \cdot]$ a commutative monoid?
(b) Is $[A^*, \cdot]$ a group?   *(Answer, page A-14)*

Our final example simply gives us a procedure to construct new groups from already existing groups, which will be helpful in producing groups with certain properties.

5.23 *Example*   Let $[G, \circ]$ and $[H, *]$ be two groups. We can form a new group whose elements come from the set $G \times H$. For the binary operation $\cdot$ of the new group, we must define $(g_1, h_1) \cdot (g_2, h_2)$. We define $\cdot$ to be a componentwise operation, acting on the first components of the ordered pairs—those from G—with the binary operation available in G and acting on the second components of the ordered pairs with the binary operation available in H. Thus,

$$(g_1, h_1) \cdot (g_2, h_2) = (g_1 \circ g_2, h_1 * h_2)$$

Then $[G \times H, \cdot]$ is a group, called the **direct product** of groups G and H.   □

5.24 Practice

(a) What is $(2, 3) \cdot (5, 6)$ in the direct product of $[\mathbb{R}^+, \cdot]$ and $[\mathbb{Z}, +]$?
(b) What is $(2, 3) \cdot (5, 6)$ in the direct product of $[\mathbb{Z}_6, +_6]$ and $[\mathbb{Z}_8, +_8]$?
(c) What is the inverse of $(1, 4)$ in the direct product of $[\mathbb{Z}_4, +_4]$ and $[\mathbb{Z}_6, +_6]$?   (*Answer, page A-14*)

5.25 Practice

(a) Let 1_G be an identity in the group $[G, \circ]$ and 1_H be an identity in the group $[H, *]$. Show that $(1_G, 1_H)$ is an identity in the direct product $[G \times H, \cdot]$.
(b) For (g, h) an element of the direct product $[G \times H, \cdot]$, what is $(g, h)^{-1}$?
(c) If $[G, \circ]$ and $[H, *]$ are both commutative groups, is $[G \times H, \cdot]$ commutative?   (*Answer, page A-14*)

BASIC RESULTS ABOUT GROUPS

We will now prove some basic theorems about groups. There are hundreds of theorems about groups and many books devoted exclusively to group theory, so here we are barely scratching the surface. The results we will prove follow almost immediately from the definitions involved.

By definition, a group $[G, \cdot]$ (or a monoid) has an identity element, and we have tried to be careful to refer to *an* identity element rather than *the* identity element. But, in fact, it is legal to say *the* identity because there is only one. To prove that the identity element is unique, suppose that i_1 and i_2 are both identity elements. Then

$$i_1 = i_1 \cdot i_2 = i_2$$

5.26 Practice   Justify the above equality signs.   (*Answer, page A-14*)

Because $i_1 = i_2$, the identity element is unique. Thus, we have proved Theorem 5.27.

5.27 Theorem   In any group (or monoid) $[G, \cdot]$, the identity element i is unique.   □

Each element x in a group $[G, \cdot]$ has an inverse element, x^{-1}. Therefore, G contains many different inverse elements, but for each x, its inverse is unique.

5.28 Practice   Prove Theorem 5.29. (Hint: assume two inverses for x, namely y and z, and let i be the identity. Then $y = y \cdot i = y(xz) =$ etc.)   *(Answer, page A-14)*

5.29 Theorem   For each x in a group $[G, \cdot]$, x^{-1} is unique.   □

If x and y belong to a group $[G, \cdot]$, then $x \cdot y$ belongs to G and must have an inverse element in G. Naturally, we expect that inverse to have some connection with x^{-1} and y^{-1}, which we know exist in G. We can show that $(x \cdot y)^{-1} = y^{-1} \cdot x^{-1}$, so that the inverse of a product is the product of the inverses in reverse order.

5.30 Theorem   For x and y members of a group $[G, \cdot]$, $(x \cdot y)^{-1} = y^{-1} \cdot x^{-1}$.

Proof:
We must show that $y^{-1} \cdot x^{-1}$ has the two properties required of $(x \cdot y)^{-1}$.

$$(x \cdot y) \cdot (y^{-1} \cdot x^{-1}) = x \cdot (y \cdot y^{-1}) \cdot x^{-1}$$
$$= x \cdot i \cdot x^{-1}$$
$$= x \cdot x^{-1}$$
$$= i$$

Similarly, $(y^{-1} \cdot x^{-1}) \cdot (x \cdot y) = i$. Notice how associativity and the meaning of i and inverses all come into play in this proof.   □

5.31 Practice   Write 10 as $7 +_{12} 3$ and use Theorem 4.33 to find $(10)^{-1}$ in the group $[\mathbb{Z}_{12}, +_{12}]$.   *(Answer, page A-14)*

Many familiar number systems such as $[\mathbb{Z}, +]$ and $[\mathbb{R}, +]$ are groups, and we make use of group properties when we do arithmetic or algebra in these systems. In $[\mathbb{Z}, +]$, for example, if we see the equation $x + 5 = y + 5$, we conclude that $x = y$. We are making use of the right cancellation law that, we will soon see, holds in any group.

5.32 Definition   A set S with a binary operation $\cdot$ satisfies the **right cancellation law** if for $x, y, z \in S$, $x \cdot z = y \cdot z$ implies $x = y$. It satisfies the **left cancellation law** if $z \cdot x = z \cdot y$ implies $x = y$.   □

Now suppose that x, y, and z are members of a group $[G, \cdot]$ and that $x \cdot z = y \cdot z$. To conclude that $x = y$, we take advantage of z^{-1}. Thus,

$$x \cdot z = y \cdot z$$

implies

$$(x \cdot z) \cdot z^{-1} = (y \cdot z) \cdot z^{-1}$$
$$x \cdot (z \cdot z^{-1}) = y \cdot (z \cdot z^{-1})$$
$$x \cdot i = y \cdot i$$
$$x = y$$

Hence, G satisfies the right cancellation law.

5.33 Practice   Show that any group $[G, \cdot]$ satisfies the left cancellation law.

(Answer, page A-14)

We have proved Theorem 5.34.

5.34 Theorem   Any group $[G, \cdot]$ satisfies the left and right cancellation laws.   □

5.35 Example   We know that $[\mathbb{Z}_6, \cdot_6]$ is not a group. Here the equation

$$4 \cdot_6 2 = 1 \cdot_6 2$$

holds, but of course $4 \neq 1$.   □

Again, working in $[\mathbb{Z}, +]$, we would solve the equation $x + 6 = 13$ by adding -6 to both sides, producing a unique answer of $x = 13 + (-6) = 7$. The property of being able to solve linear equations for unique solutions holds in all groups. Consider the equation $a \cdot x = b$ in the group $[G, \cdot]$ where a and b belong to G and x is to be found. Then $x = a^{-1} \cdot b$ is an element of G satisfying the equation. Should x_1 and x_2 both be solutions to the equation $ax = b$, then $a \cdot x_1 = a \cdot x_2$ and, by left cancellation, $x_1 = x_2$. Similarly, the unique solution to $x \cdot a = b$ is $x = b \cdot a^{-1}$.

5.36 Theorem   Let a and b be any members of a group $[G, \cdot]$. Then the linear equations $a \cdot x = b$ and $x \cdot a = b$ have unique solutions in G.   □

5.37 Practice   Solve the equation $x +_8 3 = 1$ in $[\mathbb{Z}_8, +_8]$.   *(Answer, page A-15)*

Theorem 5.36 tells us something about tables for finite groups. As we look along row a of the table, does element b appear twice? If so, then the table says there are two distinct elements x_1 and x_2 of the group such that $a \cdot x_1 = b$ and $a \cdot x_2 = b$. But by Theorem 5.36, this double occurrence can't happen. Thus, a given element of a finite group appears once at most in a given row of the group table. But to complete the row, each element must appear at least once. A similar result holds for columns. Therefore, in a group table, each element appears exactly once in each row and each column. This property alone, however, is not sufficient to insure that a table represents a group; the operation must also be known to be associative (see Exercise 14 at the end of this section).

5.38 Practice   Assume that ∘ is an associative binary operation on $\{1, a, b, c, d\}$. Complete the table in Figure 5.1 to define a group with identity 1.   *(Answer, page A-15)*

∘	1	a	b	c	d
1	1				
a			c	d	1
b		c	d		
c		d		a	
d				b	c

Figure 5.1

If $[S, \cdot]$ is a semigroup where S is finite with n elements, then n is said to be the **order of the semigroup**, denoted by $|S|$. If S is an infinite set, the semigroup is of infinite order.

5.39 Practice   (a) Name a commutative group of order 18.
(b) Name a noncommutative group of order 6.

(Answer, page A-15)

More properties of groups appear in the Exercises at the end of this section.

✔ CHECKLIST

Definitions

associative binary operation (*p. 169*)
commutative binary operation (*p. 169*)
identity element (*p. 169*)
inverse element (*p. 170*)
monoid (*p. 169*)
semigroup (*p. 170*)
group (*p. 170*)
abelian group (*p. 170*)
polynomial in x over $\mathbb{R}$ (*p. 171*)
degree of a polynomial (*p. 171*)

addition modulo n (*p. 173*)
multiplication modulo n (*p. 173*)
semigroup of transformations on a set A (*p. 173*)
group of permutations on a set A (*p. 173*)
symmetric group of degree n (*p. 173*)
concatenation of strings (*p. 174*)
direct product of groups (*p. 175*)
cancellation laws (*p. 176*)
order of a semigroup (*p. 178*)

Techniques

Test whether a given set and operation have the properties necessary to form a semigroup, monoid, or group structure.

Main Ideas

There are many instances of semigroups, monoids, and groups.

In any group structure, the identity and inverse elements are unique, cancellation laws hold, and linear equations are solvable. These and other properties of groups follow from the definitions involved.

EXERCISES, SECTION 5.1

★ 1. (a) A binary operation · is defined on the set $\{a, b, c, d\}$ by the table in Figure 5.2. Is · commutative? Is · associative?

·	a	b	c	d
a	a	c	d	a
b	b	c	a	d
c	c	a	b	d
d	d	b	a	c

Figure 5.2

(b) Let $S = \{p, q, r, s\}$. An associative operation · is partly defined on S by the table in Figure 5.3. Complete the table to preserve associativity. Is · commutative?

	p	q	r	s
p	p	q	r	s
q	q	r	s	p
r			p	
s	s		q	r

Figure 5.3

2. Define binary operations on the set $\mathbb{N}$ that are:
   (a) Commutative but not associative.
   (b) Associative but not commutative.
   (c) Neither associative nor commutative.
   (d) Both associative and commutative.

3. Rate the following structures $[S, \cdot]$ as semigroups, monoids, groups, or none of these. Name the identity element in any monoid or group structure.
   ★ (a) $S = \mathbb{N}; x \cdot y = \min(x, y)$
   ★ (b) $S = \mathbb{R}; x \cdot y = (x + y)^2$
   ★ (c) $S = \{a\sqrt{2} \mid a \in \mathbb{N}\}; \cdot = $ multiplication

⋆(d) $S = \{a + b\sqrt{2} \mid a, b \in \mathbb{Z}\}; \cdot = $ multiplication

⋆(e) $S = \{a + b\sqrt{2} \mid a, b \in \mathbb{Q}, a$ and b not both $0\}; \cdot = $ multiplication

⋆(f) $S = \{1, -1, i, -i\}; \cdot = $ multiplication (where $i^2 = -1$)

⋆(g) $S = \{1, 2, 4\}; \cdot = \cdot_6$

(h) $S = \mathbb{N} \times \mathbb{N}; (x_1, y_1) \cdot (x_2, y_2) = (x_1, y_2)$

(i) $S = \mathbb{N} \times \mathbb{N}; (x_1, y_1) \cdot (x_2, y_2) = (x_1 + x_2, y_1 y_2)$

(j) $S = $ set of even integers; $\cdot = $ addition

(k) $S = $ set of odd integers; $\cdot = $ addition

(l) $S = $ set of all polynomials in $\mathbb{R}[x]$ of degree $\leq 3; \cdot = $ polynomial addition

(m) $S = $ set of all polynomials in $\mathbb{R}[x]$ of degree $\leq 3; \cdot = $ polynomial multiplication

(n) $S = \left\{ \begin{bmatrix} 1 & z \\ 0 & 1 \end{bmatrix} \middle| z \in \mathbb{Z} \right\}; \cdot = $ matrix multiplication

(o) $S = \{1, 2, 3, 4\}; \cdot = \cdot_5$

(p) $S = \{f \mid f : \mathbb{N} \to \mathbb{N}\}; \cdot = $ function addition, that is, $(f + g)(x) = f(x) + g(x)$

4. Write a computer program that uses the definition of a group to determine whether a given 10×10 composition table of integers from 1 to 10 represents a group.

5. (a) Show that any group of order 2 is commutative by constructing a group table on the set $\{1, a\}$ with 1 as the identity.

   (b) Show that any group of order 3 is commutative by constructing a group table on the set $\{1, a, b\}$ with 1 as the identity. (You may assume associativity.)

   (c) Show that any group of order 4 is commutative by constructing a group table on the set $\{1, a, b, c\}$ with 1 as the identity. (You may assume associativity.) There will be four such tables, but three of them are isomorphic because the elements have simply been relabeled from one to the other. Find these three groups and indicate the relabeling. Thus, there are two essentially different groups of order 4, and both of these are commutative.

⋆ 6. Given an equilateral triangle, there are six permutations that can be performed on the triangle that will leave its image in the plane unchanged. Three of these permutations are clockwise rotations in the plane of $120°$, $240°$, and $360°$ about the center of the triangle; these permutations are denoted R_1, R_2, R_3, respectively. The triangle can also be flipped about any of the axes 1, 2, and 3 (see Figure 5.4); these permutations are denoted F_1, F_2, F_3, respectively. Composition of permutations is a binary operation on the set D_3 of all six permutations. For example, $F_3 \circ R_2 = F_2$. The set D_3 under composition is a group, called the **group of symmetries of an equilateral triangle**. Complete the group table (Figure 5.5)

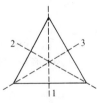

Figure 5.4

for $[D_3, \circ]$. What is an identity element in $[D_3, \circ]$? What is an inverse element for F_1? for R_2?

$\circ$	R_1	R_2	R_3	F_1	F_2	F_3
R_1						
R_2						
R_3						
F_1						
F_2						
F_3	F_2					

Figure 5.5

7. S_3, the symmetric group of degree 3, is isomorphic to D_3, the group of symmetries of an equilateral triangle (see Exercise 6). Find a bijection from the elements of S_3 to the elements of D_3 that preserves the operation. (Hint: R_1 of D_3 may be considered a permutation in S_3 sending 1 to 2, 2 to 3, and 3 to 1.)

★ 8. Let $[S, \cdot]$ be a semigroup. An element $i_L \in S$ is a **left identity element** if for all $x \in S$, $i_L \cdot x = x$. An element $i_R \in S$ is a **right identity element** if for all $x \in S$, $x \cdot i_R = x$.
   (a) Show that if a semigroup $[S, \cdot]$ has both a left identity element and a right identity element, then $[S, \cdot]$ is a monoid.
   (b) Give an example of a finite semigroup with two left identities and no right identity.
   (c) Give an example of a finite semigroup with two right identities and no left identity.
   (d) Give an example of a semigroup with neither a right nor a left identity.

9. Let $[S, \cdot]$ be a monoid with identity i, and let $x \in S$. An element x_L^{-1} in S is a **left inverse** of x if $x_L^{-1} \cdot x = i$. An element x_R^{-1} in S is a **right inverse** of x if $x \cdot x_R^{-1} = i$.
   (a) Show that if every element in a monoid $[S, \cdot]$ has both a left inverse and a right inverse, then $[S, \cdot]$ is a group.
   (b) Let S be the set of all functions f such that $f : \mathbb{N} \to \mathbb{N}$. Then S under function composition is a monoid. Define a function $f \in S$ by $f(x) = 2x$, $x \in \mathbb{N}$. Then define a function $g \in S$ by

$$g(x) = \begin{cases} x/2 & \text{if } x \in \mathbb{N}, x \text{ even} \\ 1 & \text{if } x \in \mathbb{N}, x \text{ odd} \end{cases}$$

   Show that g is a left inverse for f. Also show that f has no right inverse.
   (c) Given an example of a monoid where at least one element has neither a right nor a left inverse.

★ 10. Let $[S, \cdot]$ be a semigroup. An element $0_R \in S$ is a **right zero** in the semigroup if for all $x \in S$, $x \cdot 0_R = 0_R$. An element $0_L \in S$ is a **left zero** if for all $x \in S$, $0_L \cdot x = 0_L$.

(a) Show that if a semigroup $[S, \cdot]$ has both a left zero 0_L and a right zero 0_R, then $0_L = 0_R$; this element is called a **zero**.

(b) Name any zero elements that exist in the semigroups of Exercise 3.

11. Let $[S, \cdot]$ be a monoid and let $x \in S$ be a right zero. If $|S| > 1$, show that x has no right inverse (see Exercises 9 and 10).

12. An element x of a semigroup $[S, \cdot]$ is **idempotent** if $x \cdot x = x$. Prove that a group has only one **idempotent** element.

13. Let $[S, \cdot]$ be a semigroup having a left identity i_L (see Exercise 8) and the property that for every $x \in S$, x has a left inverse y such that $y \cdot x = i_L$. Prove that $[S, \cdot]$ is a group. (Hint: y also has a left inverse in S.)

14. Show that if $[S, \cdot]$ is a semigroup in which the linear equations $a \cdot x = b$ and $x \cdot a = b$ are solvable for any $a, b \in S$, then $[S, \cdot]$ is a group. (Hint: use Exercise 13.)

15. Prove that a finite semigroup satisfying the left and right cancellation laws is a group. (Hint: use Exercise 13.)

16. Show that a group $[G, \cdot]$ in which $x \cdot x = i$ for each $x \in G$ is commutative.

Section 5.2   ***SUBSTRUCTURES***

We know what structures are and we know what subsets are, so it should not be hard to guess what a substructure would be. But we will look at an example before we give any definitions. Addition is an associative binary operation on $\mathbb{R}^+$, the set of positive real numbers; thus, $[\mathbb{R}^+, +]$ is a semigroup. Now let A be any nonempty subset of $\mathbb{R}^+$. For any x and y in A, x and y are also in $\mathbb{R}^+$, so that $x + y$ exists and is unique. The set A "inherits" a well-defined operation, $+$, from $[\mathbb{R}^+, +]$. Also, the associative property of $+$ is inherited because for $x, y, z \in A$, $x, y, z \in \mathbb{R}^+$, and the equation $(x + y) + z = x + (y + z)$ is true. Perhaps A under the inherited operation has all of the structure of $[\mathbb{R}^+, +]$ and is itself a semigroup. The only remaining property necessary for $[A, +]$ to be a semigroup is closure of A under $+$. This property is not inherited and depends upon the set A. For $E = \{2, 4, 6, 8, \ldots\}$, closure holds and $[E, +]$ is a semigroup; for $O = \{1, 3, 5, 7, \ldots\}$, closure does not hold and $[O, +]$ is not a semigroup.

SUBSEMIGROUPS

5.40   Definition   Let $[S, \cdot]$ be a semigroup and $A \subseteq S$, $A \neq \varnothing$. Then $[A, \cdot]$ is a **subsemigroup** of $[S, \cdot]$ if $[A, \cdot]$ is itself a semigroup.   □

Clearly, the only test necessary to determine whether we have a subsemigroup is closure of the nonempty subset under the inherited operation.

5.41 Theorem   For $[S, \cdot]$ a semigroup and $A \subseteq S$, $A \neq \varnothing$, $[A, \cdot]$ is a subsemigroup of $[S, \cdot]$ if and only if A is closed under $\cdot$. $\square$

5.42 Example   Let A be the nonempty subset of $M_2(\mathbb{Z})$ consisting of all matrices of the form $\begin{bmatrix} a & b \\ 0 & c \end{bmatrix}$ where $a, b, c \in \mathbb{Z}$. Then $[A, +]$ is a subsemigroup of the semigroup $[M_2(\mathbb{Z}), +]$ because matrix addition of any two elements of A gives an element of A. $\square$

5.43 Practice   Which of the following are subsemigroups of the direct product of $[\mathbb{N}, \cdot]$ and $[\mathbb{N}, \cdot]$?

   (a) $[A, \cdot]$ where $A = \{(x, y) | (x, y) \in \mathbb{N} \times \mathbb{N}$ and $y = 1\}$
   (b) $[B, \cdot]$ where $B = \{(x, y) | (x, y) \in \mathbb{N} \times \mathbb{N}$ and $x \leq 2\}$
   (c) $[C, \cdot]$ where $C = \{(x, y) | (x, y) \in \mathbb{N} \times \mathbb{N}$ and $x + y \leq 1\}$
   (Answer, page A-15)

5.44 Practice   If $[A, \cdot]$ is a subsemigroup of a commutative semigroup $[S, \cdot]$, will $[A, \cdot]$ be commutative?   *(Answer, page A-15)*

5.45 Example   Let $I = \{f | f: \mathbb{N} \to \mathbb{N}$ and $f(x) \geq x\}$. Then $[I, \circ]$ is a subsemigroup of the semigroup of transformations on $\mathbb{N}$. To prove this, we must show closure of I under function composition. Let $f, g \in I$; then $(g \circ f)(x) = g(f(x)) \geq f(x) \geq x$, and $g \circ f \in I$. $\square$

We should mention some rather confusing terminology. The set of *all* functions on a set A into itself under function composition is called, as we said in the last section, *the semigroup of transformations on A.* Any subsemigroup of this (such as $[I, \circ]$ in the above example) is called a **transformation semigroup.** The distinction is that *a* transformation semigroup may not include all the functions on A into A but *the* semigroup of transformations on A does. Transformation semigroups will arise in connection with finite-state machines in Chapter 8.

SUBMONOIDS

We would expect any subset of a monoid that is itself a monoid under the inherited operation to be called a submonoid. It is possible to live with this definition, but it turns out to make life slightly less complicated if we require that the identity element of the submonoid agree with the identity of the original monoid. Thus, we make the following definition.

5.46 Definition   Let $[S, \cdot]$ be a monoid with identity i, and $A \subseteq S$. Then $[A, \cdot]$ is a **submonoid** of $[S, \cdot]$ if $[A, \cdot]$ is itself a monoid with identity i. (Note this assures that $i \in A$, so we have dropped the remark that A be nonempty.)   □

To test whether $[A, \cdot]$ is a submonoid, we test for closure (to get a subsemigroup) and whether $i \in A$ (to get a submonoid).

5.47 Example   (a) $[M_2^D(\mathbb{Z}), \cdot]$ is a commutative submonoid of the noncommutative monoid $[M_2(\mathbb{Z}), \cdot]$ (closure holds and $\begin{bmatrix} 1 & 0 \\ 0 & 1 \end{bmatrix}$ belongs to $M_2^D(\mathbb{Z})$.

(b) The set of all polynomials in x over $\mathbb{R}$ with constant term equal to 1 forms a submonoid of the monoid $[\mathbb{R}[x], \cdot]$ (closure holds and the constant polynomial 1 belongs to the set).   □

5.48 Practice   Which of the sets on the left form submonoids of the monoids on the right?

(a) $\{0, 2, 4\}$; $[\mathbb{Z}_6, +_6]$
(b) $\{1, 2, 4\}$; $[\mathbb{Z}_6, \cdot_6]$
(c) $\{(0, x) \mid x \in \mathbb{N}\}$; direct product of $[\mathbb{N}, \cdot]$ and $[\mathbb{N}, \cdot]$ *(Answer, page A-15)*

SUBGROUPS

Now for the last substructure. The definition of a subgroup, for reasons we shall see immediately, does not have to include a requirement about the identity.

5.49 Definition   Let $[G, \cdot]$ be a group, and $A \subseteq G$. Then $[A, \cdot]$ is a **subgroup** of $[G, \cdot]$ if $[A, \cdot]$ is itself a group.   □

Suppose $[A, \cdot]$ is a subgroup of $[G, \cdot]$ and that the identity of $[A, \cdot]$ is i_A, and the identity of $[G, \cdot]$ is i_G. We want to show that $i_A = i_G$. Notice that this equation does *not* follow from the uniqueness of a group identity because the element i_A, as far as we know, may not be an identity for all of G, and we cannot yet say that i_G is an element of A. However, $i_A = i_A \cdot i_A$ because i_A is the identity for $[A, \cdot]$, and $i_A = i_A \cdot i_G$ because i_G is the identity for $[G, \cdot]$. Because of the left cancellation law holding in the group $[G, \cdot]$, it follows that $i_A = i_G$. Hence, we did not need to specify in the definition that the subgroup have the same identity as the group.

To test whether $[A, \cdot]$ is a subgroup of $[G, \cdot]$, we make sure $[A, \cdot]$ is a submonoid and then test for the one additional property necessary to make $[A, \cdot]$ a subgroup.

5.50 Practice   Complete the following theorem.   *(Answer, page A-15)*

5.51   Theorem   For $[G, \cdot]$, a group with identity i and $A \subseteq G$, $[A, \cdot]$ is a subgroup of $[G, \cdot]$ if

(a) ———
(b) ———
(c) ———   ☐

5.52   Example   (a) $[\mathbb{Z}, +]$ is a subgroup of the group $[\mathbb{R}, +]$.
(b) $[\{1, 4\}, \cdot_5]$ is a subgroup of the group $[\{1, 2, 3, 4\}, \cdot_5]$ (closure holds; $1 \in \{1, 4\}$, $1^{-1} = 1$, $4^{-1} = 4$).   ☐

5.53   Practice   (a) Show that $[\{0, 2, 4, 6\}, +_8]$ is a subgroup of the group $[\mathbb{Z}_8, +_8]$.
(b) Show that $[\{1, 2, 4\}, \cdot_7]$ is a subgroup of the group $[\{1, 2, 3, 4, 5, 6\}, \cdot_7]$.
(*Answer, page A-15*)

If $[G, \cdot]$ is a group with identity i, then it is true that $[\{i\}, \cdot]$ and $[G, \cdot]$ are subgroups of $[G, \cdot]$. These somewhat trivial subgroups of $[G, \cdot]$ are called **improper subgroups**. Any other subgroups of $[G, \cdot]$ are **proper subgroups**.

5.54   Practice   Find all the proper subgroups of S_3, the symmetric group of degree 3. (You can find them by looking at the group table—see Practice 5.20.)
(*Answer, page A-15*)

Another point of confusing terminology: the set of *all* bijections on a set A into itself under function composition is called *the group of permutations on A*, and any subgroup of this set (such as those in the above Practice) is called **a permutation group**. Again, the distinction is that the group of permutations on a set A includes all bijections on A into itself, but a permutation group may not. Permutation groups are of particular importance, not only because they were the first groups to be studied, but also because they are, if we consider isomorphic structures to be the same, the only groups. We will see this result of Arthur Cayley's in the next section.

There is an interesting subgroup we can always find in the symmetric group S_n for $n > 1$. We know that every member of S_n can be written as a composition of cycles. But it is also true that each cycle can be written as the composition of cycles of length two, called **transpositions**. In S_7, for example, $(5, 1, 7, 2, 3, 6) = (5, 6) \circ (5, 3) \circ (5, 2) \circ (5, 7) \circ (5, 1)$, or $(1, 5) \circ (1, 6) \circ (1, 3) \circ (1, 2) \circ (2, 4) \circ (1, 7) \circ (4, 2)$. For any $n > 1$, the identity permutation i in S_n can be written as $i = (a, b) \circ (a, b)$ for any two elements a and b in the set $\{1, 2, \ldots, n\}$. This equation also shows that the inverse of the transposition (a, b) in S_n is (a, b). Now we borrow (without proof) one more fact: even though there are various ways to write a cycle as the composition of transpositions, for a given cycle, the number of transpositions will always be *even*, or it will always be *odd*. Consequently, we classify any permutation in

S_n, $n > 1$, as **even** or **odd** according to the number of transpositions in any representation of that permutation. For example, in S_7, $(5, 1, 7, 2, 3, 6)$ is odd. If we denote by A_n the set of all even permutations in S_n, then A_n determines a subgroup of $[S_n, \circ]$. Composition of even permutations produces an even permutation, and $i \in A_n$. If $\alpha \in A_n$ and α as a product of transpositions is $\alpha = \alpha_1 \circ \alpha_2 \circ \cdots \circ \alpha_k$, then $\alpha^{-1} = \alpha_k^{-1} \circ \cdots \circ \alpha_1^{-1}$. Each inverse of a transposition is a transposition, so α^{-1} is also even.

The order of the group $[S_n, \circ]$ (the number of elements) is $n!$ What is the order of the subgroup $[A_n, \circ]$? We might expect half the permutations in S_n to be even and half to be odd. Indeed, this is the case. If we let O_n denote the set of odd transpositions in S_n (which is not even closed under function composition), then the mapping $f: A_n \to O_n$ defined by $f(\alpha) = \alpha \circ (1, 2)$ is a bijection.

5.55 Practice   Prove that $f: A_n \to O_n$ given by $f(\alpha) = \alpha \circ (1, 2)$ is one-to-one and onto.

(*Answer, page A*-15)

Because there is a bijection from A_n onto O_n, each set has the same number of elements. But $A_n \cap O_n = \varnothing$ and $A_n \cup O_n = S_n$, so $|A_n| = |S_n|/2 = n!/2$.

5.56 Theorem   For $n \in \mathbb{N}$, $n > 1$, the set A_n of even permutations determines a subgroup, called the **alternating group**, of $[S_n, \circ]$ of order $n!/2$.   □

Another interesting situation occurs when we consider subgroups of the group $[\mathbb{Z}, +]$. For n any fixed element of $\mathbb{N}$, the set $n\mathbb{Z}$ is defined as the set of all integral multiples of n; $n\mathbb{Z} = \{nz \mid z \in \mathbb{Z}\}$. Thus, for example, $3\mathbb{Z} = \{0, \pm3, \pm6, \pm9, \ldots\}$.

5.57 Practice   Show that for any $n \in \mathbb{N}$, $[n\mathbb{Z}, +]$ is a subgroup of $[\mathbb{Z}, +]$. (*Answer, page A-15*)

Not only is $[n\mathbb{Z}, +]$ a subgroup of $[\mathbb{Z}, +]$ for any fixed n, but sets of the form $n\mathbb{Z}$ are the only subgroups of $[\mathbb{Z}, +]$. To illustrate, let $[S, +]$ be any subgroup of $[\mathbb{Z}, +]$. If $S = \{0\}$, then $S = 0\mathbb{Z}$. If $S \neq \{0\}$, let m be a member of S, $m \neq 0$. Either m is positive or, if m is negative, then $-m \in S$ and $-m$ is positive. S, therefore, contains at least one positive integer. Let n be the smallest positive integer in S. We will now see that $S = n\mathbb{Z}$.

First, since 0 and $-n$ are members of S and S is closed under $+$, $n\mathbb{Z} \subseteq S$. To obtain inclusion in the other direction, let $s \in S$. Now we divide the integer s by the integer n to get an integer quotient q and an integer remainder r with $0 \leq r < n$. Thus, $s = nq + r$. Solving for r, $r = s + (-nq)$. But $nq \in S$, therefore $-nq \in S$, and $s \in S$, so by closure of S under $+$, $r \in S$. If r is positive, we have a contradiction of the definition of n as the smallest positive number in S. Therefore, $r = 0$ and $s = nq + r = nq$. We now have $S \subseteq n\mathbb{Z}$, and thus $S = n\mathbb{Z}$, which completes the proof of Theorem 5.58.

5.58   Theorem   Subgroups of the form $[n\mathbb{Z}, +]$ for $n \in \mathbb{N}$ are the only subgroups of $[\mathbb{Z}, +]$.

                                                                            □

        Our last subgroup example concerns the direct product of two groups $|G, \cdot|$ and $[H, \cdot]$ and shows why the direct product is useful for producing groups with certain properties. Let i_G denote the identity of G and i_H the identity of H. Then $[G \times \{i_H\}, \cdot]$ is a subgroup of the group $[G \times H, \cdot]$ because if (g_1, i_H) and (g_2, i_H) are two members of $G \times \{i_H\}$, then $(g_1, i_H) \cdot (g_2, i_H) = (g_1 \cdot g_2, i_H \cdot i_H) = (g_1 \cdot g_2, i_H) \in G \times \{i_H\}$, and closure holds. The identity of $G \times H$, (i_G, i_H), belongs to $G \times \{i_H\}$, and for $(g, i_H) \in G \times \{i_H\}$, its inverse is (g^{-1}, i_H), which also belongs to $G \times \{i_H\}$. Similarly, $[\{i_G\} \times H, \cdot]$ is a subgroup of $[G \times H, \cdot]$.

        There is an obvious bijection from G onto $G \times \{i_H\}$ given by $f(g) = (g, i_H)$. In fact, this bijection preserves the operation. Thus, for g_1 and g_2 members of G,

$$f(g_1 \cdot g_2) = (g_1 \cdot g_2, i_H) = (g_1, i_H) \cdot (g_2, i_H) = f(g_1) \cdot f(g_2)$$

Figure 5.6 illustrates this equation by a commutative diagram. It says that we can operate and then map, or map and then operate, and the result will be the same. Thus, G and $G \times \{i_H\}$ are isomorphic and essentially the same, except for relabeling. Any properties of $[G, \cdot]$ relating to how elements act under $\cdot$ will carry over to the mirror image $G \times \{i_H\}$. Therefore, the direct product $[G \times H, \cdot]$ has a subgroup that acts like G. Let's make use of this fact to force construction of an infinite noncommutative group, something we have not yet seen.

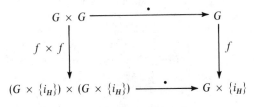

Figure 5.6

5.59   Example   Take the noncommutative group $[S_3, \circ](\alpha_2 \circ \alpha_3 \neq \alpha_3 \circ \alpha_2)$ and form its direct product with the group $[\mathbb{Z}, +]$. Because $\mathbb{Z}$ is infinite, $S_3 \times \mathbb{Z}$ is infinite. The subgroup $[S_3 \times \{0\}, \cdot]$ of $[S_3 \times \mathbb{Z}, \cdot]$ is structurally identical to $[S_3, \circ]$, so it will be noncommutative. In particular, $(\alpha_2, 0)$ and $(\alpha_3, 0)$ are members of $[S_3 \times \{0\}, \cdot]$, and thus of $[S_3 \times \mathbb{Z}, \cdot]$, that do not commute. $[S_3 \times \mathbb{Z}, \cdot]$ is thus an infinite noncommutative group.   □

✔ CHECKLIST

Definitions

subsemigroup (*p. 182*)

transformation semigroup (*p. 183*)

submonoid (*p. 184*)

subgroup (*p. 184*)

proper subgroup (*p. 185*)

permutation group (*p. 185*)

transposition (*p. 185*)

even and odd permutations
 (*p. 186*)

alternating group (*p. 186*)

Techniques

Test whether a given subset of a semigroup, monoid, or group is a subsemigroup, submonoid, or subgroup.

Main Ideas

Subsets of a given algebraic structure may themselves be structures of the same type under the inherited operation.

The only subgroups of the group $[\mathbb{Z}, +]$ are of the form $[n\mathbb{Z}, +]$ where $n\mathbb{Z}$ is the set of all integral multiples of a fixed $n \in \mathbb{N}$.

EXERCISES, SECTION 5.2

1. Let $A = \{p, q, r\}$. Show that the set of all strings over A with an even number of q's is a submonoid of the free monoid $[A^*, \cdot]$.

★ 2. In each case, decide whether the structure on the left is a subgroup of the group on the right. If not, why not? (Note: here S^* denotes $S - \{0\}$.)
   (a) $[\mathbb{Z}_5^*, \cdot_5]; [\mathbb{Z}_5, +_5]$
   (b) $[P, +]; [\mathbb{R}[x], +]$ where P is the set of all polynomials in x over $\mathbb{R}$ of degree ≥ 3.
   (c) $[\mathbb{Z}^*, \cdot]; [\mathbb{Q}^*, \cdot]$
   (d) $[A, \circ]; [S, \circ]$ where S is the set of all bijections on $\mathbb{N}$ and A is the set of all bijections on $\mathbb{N}$ mapping 3 to 3.
   (e) $[\mathbb{Z}, +]; [M_2(\mathbb{Z}), +]$
   (f) $[K, +]; [\mathbb{R}[x], +]$ where K is the set of all polynomials in x over $\mathbb{R}$ of degree $\leq k$ for some fixed k.
   (g) $[\{0, 3, 6\}, +_8]; [\mathbb{Z}_8, +_8]$

★ 3. Find all the distinct subgroups of $[\mathbb{Z}_{12}, +_{12}]$.

4. (a) Show that the subset

$$\alpha_1 = i$$
$$\alpha_2 = (1, 2) \circ (3, 4)$$
$$\alpha_3 = (1, 4) \circ (2, 3)$$
$$\alpha_4 = (1, 3) \circ (2, 4)$$

forms a subgroup of the symmetric group S_4.

(b) Show that the subset

$$\alpha_1 = i$$
$$\alpha_2 = (1, 2, 3, 4)$$
$$\alpha_3 = (1, 3) \circ (2, 4)$$
$$\alpha_4 = (1, 4, 3, 2)$$
$$\alpha_5 = (1, 2) \circ (3, 4)$$
$$\alpha_6 = (1, 4) \circ (2, 3)$$
$$\alpha_7 = (2, 4)$$
$$\alpha_8 = (1, 3)$$

forms a subgroup of the symmetric group S_4.

5. Find the elements of the alternating group A_4.

6. (a) Let $(G, \cdot]$ be a group and let $[S, \cdot]$ and $[T, \cdot]$ be subgroups of $[G, \cdot]$. Show that $[S \cap T, \cdot]$ is a subgroup of $[G, \cdot]$.
   (b) Will $[S \cup T, \cdot]$ be a subgroup of $[G, \cdot]$? Prove or give a counter-example.

★ 7. In any semigroup $[S, \cdot]$, an element $a \in S$ with $a^2 = a$ is called **idempotent**. Prove that the set of idempotent elements in a commutative monoid forms a submonoid.

8. Let $[G, \cdot]$ be a commutative group with subgroups $[S, \cdot]$ and $[T, \cdot]$. Let $ST = \{s \cdot t \mid s \in S, t \in T\}$. Show that $[ST, \cdot]$ is a subgroup of $[G, \cdot]$.

★ 9. Let $[G, \cdot]$ be a commutative group with identity i. For a fixed positive integer k, let $B_k = \{x \mid x \in G, x^k = i\}$. Show that $[B_k, \cdot]$ is a subgroup of $[G, \cdot]$.

10. For any group $[G, \cdot]$, the **center** of the group is $A = \{x \in G \mid x \cdot g = g \cdot x$ for all $g \in G\}$.
    (a) Prove that $[A, \cdot]$ is a subgroup of $[G, \cdot]$.
    (b) Find the center of the group of symmetries of an equilateral triangle, $[D_3, \circ]$ (see Exercise 6 of Section 5.1).
    (c) Show that G is commutative if and only if $G = A$.
    (d) Let x and y be members of G with $x \cdot y^{-1} \in A$. Show that $x \cdot y = y \cdot x$.

★ 11. Let $[G, \cdot]$ be a group and $A = \{(x, x) \mid x \in G\}$. Show that $[A, \cdot]$ is a subgroup of $[G \times G, \cdot]$.

12. (a) Let S_A denote the group of permutations on a set A, and let a be a fixed element of A. Show that the set H_a of all permutations in S_A leaving a fixed forms a subgroup of S_A.
    (b) If A has n elements, what is $|H_a|$?

13. (a) Let $[G, \cdot]$ be a group and $A \subseteq G$, $A \neq \emptyset$. Show that $[A, \cdot]$ is a subgroup of $[G, \cdot]$ if for each $x, y \in A$, $x \cdot y^{-1} \in A$. This subgroup

test is sometimes more convenient to use than Theorem 5.51.

(b) Use the test of part (a) to work Exercise 9 above.

14. (a) Let $[G, \cdot]$ be any group, with identity i. For a fixed $a \in G$, a^0 denotes i and a^{-n} means $(a^n)^{-1}$. Let $A = \{a^z \mid z \in \mathbb{Z}\}$. Show that $[A, \cdot]$ is a subgroup of G.

(b) The group $[G, \cdot]$ is a **cyclic group** if for some $a \in G$, $A = \{a^z \mid z \in \mathbb{Z}\}$ is the entire group G. In this case, a is a **generator** of $[G, \cdot]$. For example, 1 is a generator of the group $[\mathbb{Z}, +]$; remember that the operation is addition. Thus, $1^0 = 0$, $1^1 = 1$, $1^2 = 1 + 1 = 2$, $1^3 = 1 + 1 + 1 = 3, \ldots$, $1^{-1} = (1)^{-1} = -1$, $1^{-2} = (1^2)^{-1} = -2$, $1^{-3} = (1^3)^{-1} = -3, \ldots$. Every integer can be written as an integral "power" of 1, and $[\mathbb{Z}, +]$ is cyclic with generator 1. Show that the group $[\mathbb{Z}_7, +_7]$ is cyclic with generator 2.

(c) Show that 5 is also a generator of the cyclic group $[\mathbb{Z}_7, +_7]$.

(d) Let $\mathbb{Z}_{11}^* = \mathbb{Z}_{11} - \{0\}$. Then $[\mathbb{Z}_{11}^*, \cdot_{11}]$ is a cyclic group; show that 2 is a generator.

15. Let $[G, \cdot]$ be a cyclic group with generator a (see Exercise 14). Show that G is commutative.

Section 5.3   **MORPHISMS—SIMULATION II**

HOMOMORPHISMS

The ideas of homomorphism and isomorphism were introduced in Chapter 3. Now we want to apply these ideas to the algebraic structures of semigroups, monoids, and groups. Recall that if A and B are two instances of a structure, a homomorphism from A to B is a function $f : A \to B$ that *preserves the operations* of A in B. When we have particular structures to work with, we express the property of preserving an operation by means of an equation. In effect, the equation always says, "operate and then map, or map and then operate," and it corresponds to a commutative diagram. Let's write the equation defining a homomorphism from one semigroup to another.

5.60 Practice   Complete the following definition.   (*Answer, page A-15*)

5.61 Definition   Let $[S, \cdot]$ and $[T, +]$ be semigroups. A function $f : S \to T$ is a **homomorphism** from $[S, \cdot]$ to $[T, +]$ if for all $x, y \in S$, $f(x \cdot y) = $ _____.   □

The commutative diagram corresponding to the equation $f(x \cdot y) = f(x) + f(y)$ appears in Figure 5.7. The left half of the equation, $f(x \cdot y)$, says "operate and then map" and traces the top and right paths of the diagram.

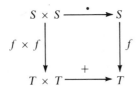

Figure 5.7

The right half of the equation, $f(x) + f(y)$, says "map and then operate" and traces the left and bottom paths of the diagram. Notice that the operation on x and y (in S) is $\cdot$, but the operation on $f(x)$ and $f(y)$ (in T) is $+$. We used two different symbols to emphasize this distinction. When working with homomorphisms, you must pay attention to the structure you are in, and what the appropriate operation is for that structure.

5.62 Example   Let $\mathbb{N}^+$ denote the positive integers; then $[\mathbb{N}^+, +]$ is a semigroup. Let E be the set of positive even integers; then $[E, \cdot]$ is a semigroup. The function $f : \mathbb{N}^+ \to E$ defined by $f(x) = 2^x$ is a homomorphism. To prove this, note that for x and y in $\mathbb{N}^+$,

$$f(x + y) = 2^{x+y} = 2^x \cdot 2^y = f(x) \cdot f(y)$$

(Again, we must keep track of the appropriate operations. In comparing this example with the general definition, $[\mathbb{N}^+, +]$ corresponds to $[S, \cdot]$, so the generic $\cdot$ of the definition becomes $+$ for this case. $[E, \cdot]$ corresponds to $[T, +]$, so the generic $+$ of the definition becomes $\cdot$.)   ☐

Because monoids and groups are also semigroups, Definition 5.61 applies equally well to monoids or groups.

5.63 Example   Consider the semigroup (actually a group) $[\mathbb{R}, +]$. The function $f : \mathbb{R} \to \mathbb{R}$ given by $f(x) = x^2$ is not a homomorphism. The required equation $f(x + y) = f(x) + f(y)$ becomes $(x + y)^2 = x^2 + y^2$, which is true only if $x = 0$ or $y = 0$.   ☐

5.64 Practice   In each case, decide whether the given function is a homomorphism from the semigroup on the left to the one on the right.

(a) $[\mathbb{N}, +], [2\mathbb{N}, +]$ (where $2\mathbb{N} = \{2n \mid n \in \mathbb{N}\}$); $f(x) = 2x$

(b) $[\mathbb{N}, +] [2\mathbb{N}, \cdot]$; $f(x) = 2x$

(c) $[M_2(\mathbb{Z}), +], [\mathbb{Z}, +]; f\left(\begin{bmatrix} a & b \\ c & d \end{bmatrix}\right) = a$

(d) $[\mathbb{N}^2[x], +]$ (where $\mathbb{N}^2[x] = \{a_2 x^2 + a_1 x + a_0 \mid a_2, a_1, a_0 \in \mathbb{N}\}$), $[\mathbb{N}, +]; f(a_2 x^2 + a_1 x + a_0) = a_2$ *(Answer, page A-15)*

Preserving Properties

Suppose that f is a homomorphism from a semigroup (monoid, group) $[S, \cdot]$ to a semigroup (monoid, group) $[T, +]$. The range of f, $f(S)$, is a subset of T. A homomorphism need not be an onto mapping, so we may have $f(S) \subset T$. The definition of a homomorphism requires that f preserve in its range the effect of the operation $\cdot$ in $[S, \cdot]$. Consequently, many of the properties of $[S, \cdot]$ are also preserved in the range, for example, the structure of $[S, \cdot]$. Thus, if $[S, \cdot]$ is a semigroup, then $f(S)$ is a semigroup under $+$. To prove this, we must verify closure (associativity is inherited from $[T, +]$). Let $f(x)$ and $f(y)$ belong to $f(S)$. We must prove $f(x) + f(y)$ belongs to $f(S)$, that is, that $f(x) + f(y)$ is $f(s)$ for some $s \in S$. But $f(x) + f(y) = f(x \cdot y)$ and, because x and y belong to S and S is closed under $\cdot$, $x \cdot y \in S$. This procedure verifies closure, and $[f(S), +]$ is a semigroup, a subsemigroup of $[T, +]$.

5.65 Practice Let $[S, \cdot]$ and $[T, +]$ be monoids with identities i_S and i_T, respectively. Let f be a homomorphism from S to T. Show that $f(i_S)$ is an identity for $f(S)$ under $+$. *(Answer, page A-16)*

Now we know that if f is a homomorphism from a monoid $[S, \cdot]$ to a monoid $[T, +]$, the image $f(S)$ is a monoid under $+$. Note, however, that $[f(S), +]$ may not be a submonoid of $[T, +]$ because $f(i_S)$ may differ from i_T.

5.66 Practice Let $[S, \cdot]$ and $[T, +]$ be groups with identities i_S and i_T, respectively. Let f be a homomorphism from S to T. For $f(x) \in f(S)$, show that

$$f(x^{-1}) + f(x) = f(x) + f(x^{-1}) = f(i_S)$$

(Note that $x \in S$, a group, so x^{-1} exists.) *(Answer, page A-16)*

When we have a homomorphism from a group $[S, \cdot]$ to a group $[T, +]$, we know that $[f(S), +]$ is a semigroup; from Practice 5.65, $f(i_S)$ ia an identity for $f(S)$, and from Practice 5.66 every element in $f(S)$ has an inverse element (with respect to the identity $f(i_S)$) in $f(S)$. Thus, the range $f(S)$ is a subgroup of $[T, +]$, and it also follows that $f(i_S) = i_T$. To summarize:

5.67 Theorem (a) Let f be a homomorphism from a semigroup $[S, \cdot]$ to a semigroup $[T, +]$. Then the range $f(S)$ is a semigroup under $+$.
(b) Let f be a homomorphism from a monoid $[S, \cdot]$ to a monoid $[T, +]$. Then the range $f(S)$ is a monoid under $+$.
(c) Let f be a homomorphism from a group $[S, \cdot]$ with identity i_S to a group $[T, +]$ with identity i_T. Then the range $f(S)$ is a group under $+$. Further, $f(i_S) = i_T$ and $f(x^{-1}) = -f(x)$ for all $x \in S$. $\square$

Theorem 5.67 says that a homomorphism preserves the structure of the domain in the range. Other properties of the domain structure relating entirely to the behavior of elements under the binary operation, called **algebraic properties**, are also preserved.

5.68 Practice   Let f be a homomorphism from a semigroup $[S, \cdot]$ to a semigroup $[T, +]$. Show that if $[S, \cdot]$ is commutative, then so is the semigroup $[f(S), +]$.

(Answer, page A-16)

Nonalgebraic properties need not be preserved by a homomorphism. Thus, $f: \mathbb{Z} \to \{0\}$ given by $f(z) = 0$ for all $z \in \mathbb{Z}$ is a homomorphism from the infinite group $[\mathbb{Z}, +]$ to the one-element group $[\{0\}, +]$. Order of a group is an example of a nonalgebraic property.

Homomorphic Images and Simulation

Because a homomorphism preserves algebraic properties, we begin to think in terms of one instance of a structure simulating another instance of that structure. A homomorphism from $[S, \cdot]$ to $[T, +]$ need not be onto or one-to-one. As we add these conditions one at a time to the homomorphism, we see what sorts of simulations are possible. In the following discussion, $[S, \cdot]$ and $[T, +]$ are two semigroups or two monoids or two groups.

Suppose f is a homomorphism from S to T and that f is *onto*, that is, $f(S) = T$. T is called the **homomorphic image** of S under f. We know that algebraic properties of S must also appear in T. From our discussion of Simulation II at the end of Chapter 3, we expect that S will simulate T and T will imperfectly simulate S. We will consider this situation more thoroughly in the next section; for now, we will simply look at another example.

5.69 Example   Let $f: \mathbb{Z} \to \mathbb{Z}_4$ be defined by $f(x) = x \cdot_4 1$. (Here $x \cdot_4 1$ denotes the *nonnegative* remainder when x is divided by 4.) Then we claim that f is a homomorphism from the group $[\mathbb{Z}, +]$ onto the group $[\mathbb{Z}_4, +_4]$. Clearly f is onto. For $x, y \in \mathbb{Z}, f(x + y) = (x + y) \cdot_4 1 = x \cdot_4 1 +_4 y \cdot_4 1 = f(x) +_4 f(y)$; so f is a homomorphism.

The simulation of $[\mathbb{Z}_4, +_4]$ by $[\mathbb{Z}, +]$ is accomplished by the use of equivalence classes of $\mathbb{Z}$. We define a relation ρ on $\mathbb{Z}$ by $x \, \rho \, y \Leftrightarrow f(x) = f(y) \Leftrightarrow x \cdot_4 1 = y \cdot_4 1$. Then ρ is an equivalence relation on $\mathbb{Z}$ and partitions $\mathbb{Z}$ into the equivalence classes

$$[0] = \{0, 4, 8, 12, 16, \ldots, -4, -8, -12, -16, \ldots\}$$

$$[1] = \{1, 5, 9, 13, \ldots, -3, -7, -11, -15, \ldots\}$$

$$[2] = \{2, 6, 10, 14, \ldots, -2, -6, -10, -14, \ldots\}$$

$$[3] = \{3, 7, 11, 15, \ldots, -1, -5, -9, -13, \ldots\}$$

(ρ is the relation of **congruence modulo 4**—see Example 2.31). There is an obvious one-to-one correspondence between the equivalence classes of $\mathbb{Z}$ and the elements of $\mathbb{Z}_4$.

Suppose now that n and m belong to $\mathbb{Z}_4$; we wish to compute $n +_4 m$ by simulation in $[\mathbb{Z}, +]$. We choose any integer z_n in the class of $\mathbb{Z}$ associated with n; we choose any integer z_m in the class of $\mathbb{Z}$ associated with m. (We choose preimages of n and m under f; remember f^{-1} does not exist.) We add $z_n + z_m$ in $[\mathbb{Z}, +]$; the sum belongs to a class of $\mathbb{Z}$, and we choose the member of $\mathbb{Z}_4$ corresponding to that class. This element of $\mathbb{Z}_4$ will be $n +_4 m$. For example, we compute $3 +_4 2$ in $[\mathbb{Z}_4, +_4]$ by choosing an element of $[3]$ and an element of $[2]$, say 11 and -2, respectively. We add these in $[\mathbb{Z}, +]$ $(11 + (-2) = 9)$, and then note that $9 \in [1]$, so $3 +_4 2 = 1$.

The simulation of $[\mathbb{Z}, +]$ by $[\mathbb{Z}_4, +_4]$ is much less satisfactory and can give us answers only to within an equivalence class. Thus, if we try to add $z_n + z_m$ in $[\mathbb{Z}, +]$ by using $[\mathbb{Z}_4, +_4]$, we compute $f(z_n) +_4 f(z_m)$; this is some member p of $\mathbb{Z}_4$. But to get back to $\mathbb{Z}$, the best we can find is the class containing p. Thus, to add $13 + 22$, for example, we use $f(13) +_4 f(22)$, or $1 +_4 2 = 3$. But in going back to $\mathbb{Z}$, $[3]$ contains an infinite number of values, only one of which is the correct answer of 35.   □

This example has implications for computer arithmetic. When we add integers on a computer, we would really like to do the arithmetic of $[\mathbb{Z}, +]$, but we cannot because $\mathbb{Z}$ is an infinite set and the capacity of the computer is finite (although large). We need a finite arithmetic to simulate $[\mathbb{Z}, +]$. If we use $[\mathbb{Z}_n, +_n]$ for some (large) n, then, just as in the case of $\mathbb{Z}_4$ above, $[\mathbb{Z}_n, +_n]$ only imperfectly simulates $[\mathbb{Z}, +]$. An integer "answer" in the computer is actually only good to within modulo n. Naturally, we expect that the use of something finite to simulate something infinite will not work perfectly, but perhaps there is another finite arithmetic that can do a better job than $[\mathbb{Z}_n, +_n]$. We shall see in Section 5.5 that this is not the case, and that $[\mathbb{Z}_n, +_n]$ for some n is the only possible choice.

ISOMORPHISMS

Now let's put a final condition on our homomorphism from S to T, and require it to be not only onto but also one-to-one.

5.70 Definition   Let $[S, \cdot]$ and $[T, +]$ be semigroups (monoids, groups). A mapping $f : S \to T$ is an **isomorphism** from $[S, \cdot]$ to $[T, +]$ if

  (a) f is a bijection
  (b) For all $x, y \in S$, $f(x \cdot y) = f(x) + f(y)$ □

5.71 Example   Let $M_2^0(\mathbb{Z})$ be the set of all 2×2 matrices of the form $\begin{bmatrix} 1 & z \\ 0 & 1 \end{bmatrix}$ where $z \in \mathbb{Z}$.

Then $M_2^0(\mathbb{Z})$ is a group under matrix multiplication. The function $f: M_2^0(\mathbb{Z}) \to \mathbb{Z}$ given by $f\left(\begin{bmatrix} 1 & z \\ 0 & 1 \end{bmatrix}\right) = z$ is an isomorphism from $[M_2^0(\mathbb{Z}), \cdot]$ to $[\mathbb{Z}, +]$.

Clearly, f is onto. To show that f is one-to-one, let $f\left(\begin{bmatrix} 1 & x \\ 0 & 1 \end{bmatrix}\right) = f\left(\begin{bmatrix} 1 & y \\ 0 & 1 \end{bmatrix}\right)$. Then $x = y$ and $\begin{bmatrix} 1 & x \\ 0 & 1 \end{bmatrix} = \begin{bmatrix} 1 & y \\ 0 & 1 \end{bmatrix}$. And f is a homomorphism because for any $\begin{bmatrix} 1 & x \\ 0 & 1 \end{bmatrix}$ and $\begin{bmatrix} 1 & y \\ 0 & 1 \end{bmatrix}$ in $M_2^0(\mathbb{Z})$, $f\left(\begin{bmatrix} 1 & x \\ 0 & 1 \end{bmatrix} \cdot \begin{bmatrix} 1 & y \\ 0 & 1 \end{bmatrix}\right) = f\left(\begin{bmatrix} 1 & x+y \\ 0 & 1 \end{bmatrix}\right) = x+y = f\left(\begin{bmatrix} 1 & x \\ 0 & 1 \end{bmatrix}\right) + f\left(\begin{bmatrix} 1 & y \\ 0 & 1 \end{bmatrix}\right)$. □

If f is an isomorphism from $[S, \cdot]$ to $[T, +]$, then f^{-1} exists and is a bijection. Further, f^{-1} is also a homomorphism, this time from T to S. To see this, let t_1 and t_2 belong to T and consider $f^{-1}(t_1 + t_2)$. Because $t_1, t_2 \in T$ and f is onto, $t_1 = f(s_1)$ and $t_2 = f(s_2)$ for some s_1 and s_2 in S. Thus,

$$
\begin{aligned}
f^{-1}(t_1 + t_2) &= f^{-1}(f(s_1) + f(s_2)) \\
&= f^{-1}(f(s_1 \cdot s_2)) \\
&= (f^{-1} \circ f)(s_1 \cdot s_2)) \\
&= s_1 \cdot s_2 \\
&= f^{-1}(t_1) \cdot f^{-1}(t_2)
\end{aligned}
$$

We can thus say simply that S and T are **isomorphic**, denoted by $S \simeq T$.

5.72 Practice   Let $5\mathbb{Z} = \{5z \mid z \in \mathbb{Z}\}$. Then $[5\mathbb{Z}, +]$ is a group. Show that $f: \mathbb{Z} \to 5\mathbb{Z}$ given by $f(x) = 5x$ is an isomorphism from $[\mathbb{Z}, +]$ to $[5\mathbb{Z}, +]$. (*Answer, page A-16*)

Checking whether a given function is an isomorphism from S to T, as in Practice 5.72, is not hard. Deciding whether S and T are isomorphic may be harder. To prove that they are isomorphic, we must produce a function. To prove that they are not isomorphic, we must show that no such function exists. Since we can't try all possible functions, we use ideas such as: there is no one-to-one correspondence between S and T; S is commutative but T is not; and so on.

The "Mirror Picture"

If S and T are two isomorphic algebraic structures, then we expect each to simulate the other. To illustrate, we know by Example 5.71 that $[M_2^0(\mathbb{Z}), \cdot]$ and $[\mathbb{Z}, +]$ are isomorphic groups, and we can calculate in either structure by hopping over to its mirror image, calculating there, and hopping back. To multiply $\begin{bmatrix} 1 & 7 \\ 0 & 1 \end{bmatrix}$ and $\begin{bmatrix} 1 & -3 \\ 0 & 1 \end{bmatrix}$ in $[M_2^0(\mathbb{Z}), \cdot]$, for example, we map to 7 and -3 in $[\mathbb{Z}, +]$, add there, getting 4, and map back to $\begin{bmatrix} 1 & 4 \\ 0 & 1 \end{bmatrix}$ in $M_2^0(\mathbb{Z})$. The product of $\begin{bmatrix} 1 & 7 \\ 0 & 1 \end{bmatrix}$ and $\begin{bmatrix} 1 & -3 \\ 0 & 1 \end{bmatrix}$ in $[M_2^0(\mathbb{Z}), \cdot]$ is $\begin{bmatrix} 1 & 4 \\ 0 & 1 \end{bmatrix}$. Conversely, to add $2 + 3$ in $[\mathbb{Z}, +]$, we go to $\begin{bmatrix} 1 & 2 \\ 0 & 1 \end{bmatrix}$ and $\begin{bmatrix} 1 & 3 \\ 0 & 1 \end{bmatrix}$ in $[M_2^0(\mathbb{Z}), \cdot]$. The result of multiplying $\begin{bmatrix} 1 & 2 \\ 0 & 1 \end{bmatrix}$ and $\begin{bmatrix} 1 & 3 \\ 0 & 1 \end{bmatrix}$ in $[M_2^0(\mathbb{Z}), \cdot]$ is $\begin{bmatrix} 1 & 5 \\ 0 & 1 \end{bmatrix}$, which then maps to 5, and $2 + 3 = 5$ in $[\mathbb{Z}, +]$. Figure 5.8a shows how $[\mathbb{Z}, +]$ simulates $(M_2^0(\mathbb{Z}), \cdot]$, and Figure 5.8b shows how $[M_2^0(\mathbb{Z}), \cdot]$ simulates $[\mathbb{Z}, +]$.

The mirror picture of Figure 5.8 shows us the central fact about two

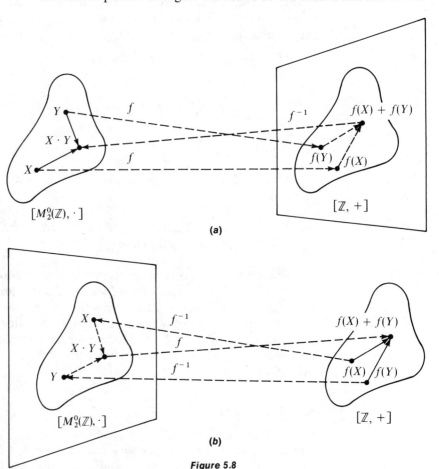

Figure 5.8

isomorphic structures, namely that each is merely a relabeling of the other. There can be no basic algebraic differences between the behavior of two isomorphic structures although the *names* of elements and operations may be different. If we ignore labels and concentrate on the underlying characteristics, then we make no distinction between isomorphic structures. Practice 5.73 shows that this classification by isomorphism is valid.

5.73 Practice

(a) Let $f:S \to T$ be an isomorphism from the semigroup $[S, \cdot]$ onto the semigroup $[T, +]$ and $g:T \to U$ be an isomorphism from $[T, +]$ onto the semigroup $[U, *]$. Show that $g \circ f$ is an isomorphism from S onto U.

(b) Let $\mathscr{S}$ be a collection of semigroups and define a binary relation ρ on $\mathscr{S}$ by $S \rho T \Leftrightarrow S \simeq T$. Show that ρ is an equivalence relation on $\mathscr{S}$.   (*Answer, page A-16*)

Classes of Isomorphic Groups

We will finish this section by looking at some equivalence classes of groups under isomorphism. Often we pick out one member of an equivalence class and note that it is the typical member of that class, and that all other groups in the class look just like it (with different names).

A result concerning the nature of very small groups follows immediately from Exercise 5 of Section 5.1.

5.74 Theorem

(a) Every group of order 2 is isomorphic to the group whose group table is

$\cdot$	1	a
1	1	a
a	a	1

(b) Every group of order 3 is isomorphic to the group whose group table is

$\cdot$	1	a	b
1	1	a	b
a	a	b	1
b	b	1	a

(c) Every group of order 4 is isomorphic to one of the two groups whose group tables are:

$\cdot$	1	a	b	c
1	1	a	b	c
a	a	1	c	b
b	b	c	1	a
c	c	b	a	1

$\cdot$	1	a	b	c
1	1	a	b	c
a	a	b	c	1
b	b	c	1	a
c	c	1	a	b

$\square$

We can also prove that any group is essentially a permutation group. Suppose $[G, \cdot]$ is a group. We want to establish an isomorphism from G to a permutation group; each element g of G must be associated with a permutation α_g on some set. In fact, the set will be G itself; for any $x \in G$, we define $\alpha_g(x)$ to be $g \cdot x$. We must show that $\{\alpha_g | g \in G\}$ forms a permutation group, and that this permutation group is isomorphic to G. First, we need to show that for any $g \in G$, α_g is indeed a permutation on G. From the definition $\alpha_g(x) = g \cdot x$, it is clear that $\alpha_g : G \to G$, but it must be shown that α_g is a bijection.

5.75 Practice   Show that α_g as defined above is a permutation on G.   *(Answer, page A-16)*

Now we consider $P = \{\alpha_g | g \in G\}$ and show that P is a group under function composition. P is nonempty because G is nonempty, and associativity always holds for function composition. We must show that P is closed, has an identity, and that each $\alpha_g \in P$ has an inverse in P. To show closure, let α_g and $\alpha_h \in P$. For any $x \in G$, $(\alpha_g \circ \alpha_h)(x) = \alpha_g(\alpha_h(x)) = \alpha_g(h \cdot x) = g(h \cdot x) = (g \cdot h)x$. Thus, $\alpha_g \circ \alpha_h = \alpha_{g \cdot h}$ and $\alpha_{g \cdot h} \in P$.

5.76 Practice   (a) Let 1 denote the identity of G. Show that α_1 is an identity for P under function composition.
(b) For $\alpha_g \in P$, $\alpha_{g-1} \in P$; show that $\alpha_{g-1} = (\alpha_g)^{-1}$.   *(Answer, page A-16)*

We know that $[P, \circ]$ is a permutation group, and it only remains to show that the function $f : G \to P$ given by $f(g) = \alpha_g$ is an isomorphism. Clearly, f is an onto function.

5.77 Practice   Show that $f : G \to P$ defined above is

(a) one-to-one
(b) a homomorphism   *(Answer, page A-16)*

We have now proved Theorem 5.78, first stated and proved by the English mathematician Cayley in the mid-1800s.

5.78 Theorem   Every group is isomorphic to a permutation group. $\square$

✓ CHECKLIST

Definitions

homomorphism (*p. 190*)   isomorphism (*p. 194*)
algebraic property (*p. 193*)   isomorphic structures (*p. 195*)
homomorphic image (*p. 193*)

Techniques

Test whether a given function from a structure S to a structure T is a homomorphism, an isomorphism, or neither.

Decide whether two structures are isomorphic.

Main Ideas

If f is a homomorphism from a structure S to a structure T, then the range $f(S)$ has the structure of S as well as the algebraic properties of S.

If S and T are isomorphic structures, they are identical except for relabeling, and each simulates the other.

The relation of isomorphism divides any collection of semigroups (monoids, or groups) into equivalence classes.

Except for isomorphisms, there is essentially only one group of order 2, one group of order 3, and two groups of order 4.

Every group is essentially a permutation group.

EXERCISES, SECTION 5.3

★ 1. In each case, decide whether the given function is a homomorphism from the group on the left to the one on the right. Are any of the homomorphisms also isomorphisms?

(a) $[\mathbb{Z}, +], [\mathbb{Z}, +]; f(x) = 2$

(b) $[\mathbb{R}, +], [\mathbb{R}, +]; f(x) = x + 1$

(c) $[\mathbb{Z} \times \mathbb{Z}, +], [\mathbb{Z}, +]; f(x, y)) = x + 2y$

(d) $[\mathbb{R}, +], [\mathbb{R}, +]; f(x) = |x|$

(e) $[\mathbb{R}^*, \cdot], [\mathbb{R}^*, \cdot];$ (where $\mathbb{R}^*$ denotes the set of nonzero real numbers); $f(x) = |x|$

(f) $[\mathbb{R}[x], +], [\mathbb{R}, +]; f(a_n x^n + \cdots + a_1 x + a_0) = a_n + \cdots + a_0$

(g) $[M_2^D(\mathbb{R}^+), \cdot], [\mathbb{R} \times \mathbb{R}, +]; f\left(\begin{bmatrix} a & 0 \\ 0 & b \end{bmatrix} \right) = (a, b)$

(h) $[S_3, \circ], [\mathbb{Z}_2, +_2]; f(\alpha) = \begin{cases} 1 & \text{if } \alpha \text{ is an even permutation} \\ 0 & \text{if } \alpha \text{ is an odd permutation} \end{cases}$

2. In each case, decide whether the given groups are isomorphic. If they are, produce an isomorphism function. If they are not, give a reason why they are not.
   (a) $[\mathbb{Z}, +], [12\mathbb{Z}, +]$ (where $12\mathbb{Z} = \{12z \mid z \in \mathbb{Z}\}$)
   (b) $[\mathbb{Z}_5, +_5], [5\mathbb{Z}, +]$
   (c) $[5\mathbb{Z}, +], [12\mathbb{Z}, +]$
   (d) $[S_3, \circ], [\mathbb{Z}_6, +_6]$
   (e) $[\{a_1 x + a_0 \mid a_1, a_0 \in \mathbb{R}\}, +], [\mathbb{C}, +]$
   (f) $[\mathbb{Z}_6, +_6], [S_6, \circ]$
   (g) $[\mathbb{Z}_2, +_2], [S_2, \circ]$

3. Let f be the function from the group $[\mathbb{Z}_{18}, +_{18}]$ to the group $[\mathbb{Z}_{12}, +_{12}]$ given by $f(x) = 4 \cdot_{12} x$. Then f can be shown to be a homomorphism.
   (a) The range $f(\mathbb{Z}_{18})$ of the function f is a subgroup of $[\mathbb{Z}_{12}, +_{12}]$. Find the elements of $f(\mathbb{Z}_{18})$.
   (b) Define an equivalence relation ρ on $\mathbb{Z}_{18}$ by $x \rho y \Leftrightarrow f(x) = f(y)$. Describe the resulting equivalence classes.
   (c) Use $[\mathbb{Z}_{18}, +_{18}]$ (and the classes of part (b)) to simulate the computation $4 +_{12} 8$ in $[f(\mathbb{Z}_{18}), +_{12}]$.
   (d) Use $[f(\mathbb{Z}_{18}), +_{12}]$ to simulate the computation $14 +_{18} 8$ in $[\mathbb{Z}_{18}, +_{18}]$ to within an equivalence class.

4. A group $[G, \cdot]$ is called **divisible** if for any $g \in G$ and any positive integer n, the equation $x^n = g$ has at least one solution in G.
   (a) Show that $[\mathbb{Q}, +]$ is a divisible group. (Note that the equation $x^n = g$ becomes $nx = g$ when the group operation is addition.)
   (b) Let f be a homomorphism from a divisible group $[G, \cdot]$ to a group $[H, \cdot]$ Show that the range $f(G)$ is a divisible group.

★ 5. Let $[S, \cdot]$ and $[T, +]$ be semigroups, and suppose that $t \in T$ is an indempotent element, that is, $t + t = t$. Let f be a homomorphism from S onto T. Prove that the set $U = \{s \in S \mid f(s) = t\}$ is a subsemigroup of S under $\cdot$.

6. (a) Let $[S, \cdot]$ be a semigroup. An isomorphism from S to S is called an **automorphism** on S. Let Aut(S) be the set of all automorphisms on S, and show that Aut(S) is a group under function composition.
   (b) For the group $[\mathbb{Z}_4, +_4]$, find the set of automorphisms and show its group table under $\circ$.

7. (a) Let $[G, +]$ be a commutative group, and let Hom(G) be the set of all homomorphisms from G into G. (A homomorphism from a group G into itself is called an **endomorphism** of G.) Define an operation of addition on Hom(G) as follows: for $f, g \in$ Hom(G), $(f + g)(x) = f(x) + g(x)$ for each $x \in G$. Show that $[\text{Hom}(G), +]$ is a commutative semigroup.
   (b) Let 0_G be the identity element of the group $[G, +]$ and define a function $i: G \to G$ by $i(x) = 0_G$ for all $x \in G$. Prove that $i \in$ Hom(G) and is an identity of $[\text{Hom}(G), +]$.

(c) For $f \in \text{Hom}(G)$, define a function $-f : G \to G$ by $(-f)(x) = -f(x)$ for each $x \in G$. Show that $-f \in \text{Hom}(G)$ and is the inverse of f in $[\text{Hom}(G), +]$, which will conclude the proof that $[\text{Hom}(G), +]$ is a commutative group.

8. Prove that a homomorphism f from a group G to a group H is an isomorphism if and only if there exists a homomorphism g, $g : H \to G$, such that $g \circ f = i_G$ and $f \circ g = i_H$, where i_G and i_H are the identity functions on G and H, respectively.

9. Let f be a homomorphism from a group G onto a group H. Show that f is an isomorphism if and only if the only element of G that is mapped to the identity of H is the identity of G.

★ 10. Let $[G, \cdot]$ and $[H, +]$ be groups.
(a) Show that the group $[G \times H, \cdot]$ is isomorphic to the group $[H \times G, \cdot]$
(b) Prove that the function $f : G \times H \to G$ defined by $f(x, y) = x$ is a homomorphism.

11. A bijection f is defined on a group $[G, \cdot]$ by the equation $f(x) = x^{-1}$. Prove that if f is an isomorphism, then $[G, \cdot]$ is a commutative group.

12. Prove that for any semigroup $[S, \cdot]$ there is a homomorphism from S onto a transformation semigroup.

13. Let $[G, \cdot]$ be a group and g a fixed element of G. Define $f : G \to G$ by $f(x) = g \cdot x \cdot g^{-1}$ for any $x \in G$. Prove that f is an isomorphism from G to G.

Section 5.4   *HOMOMORPHISM THEOREMS*

If we have some instance $[A, \cdot]$ of an algebraic structure, we could certainly simulate it by constructing some $[B, +]$ isomorphic to $[A, \cdot]$. (Remember the mirror picture.) But practically speaking, if we can "build" $[B, +]$, we could probably "build" $[A, \cdot]$ to begin with. Often we must settle for the imperfect simulation produced by a homomorphism from A onto B that is not an isomorphism. We have considered two examples of this situation, namely, the homomorphism example at the end of Section 3.2 and Example 5.69. (You might want to review these now.) In this section we will see why these examples work.

*DEFINING THE [S/f, *] STRUCTURE*

Let's assume that $[S, \cdot]$ and $[T, +]$ are semigroups and that there is a homomorphism f from S to T. We will also assume that f is not a one-to-one function (everything we say would still be true, but trivial, if f is one-to-

one). The range $f(S)$ is a subset of T, a semigroup under $+$, and the homo-morphic image of S under f. We recall from our examples that the key to the relationship between S and its homomorphic image $f(S)$ is in the equivalence classes of the domain S. We define a binary relation ρ on S by
$$x \rho y \Leftrightarrow f(x) = f(y).$$

5.79 Practice   Show that ρ as defined above is an equivalence relation on S. *(Answer, page A-16)*

Thus, ρ partitions S into equivalence classes. For $x \in S$, $[x]$ denotes the equivalence class to which x belongs. The image of x under f is some element in $f(S)$; the other members of $[x]$ are those members of S mapping under f to the same element. We denote the set of all distinct equivalence classes of S by S/f. A member of S/f will in general have more than one name.

5.80 Example   Let $f: \mathbb{N} \times \mathbb{N} \to \mathbb{N}$ be given by $f((x, y)) = x$. Then f is a homomorphism from $[\mathbb{N} \times \mathbb{N}, +]$ (use componentwise addition) to $[\mathbb{N}, +]$ because $f((x_1, y_1) + (x_2, y_2)) = f((x_1 + x_2, y_1 + y_2)) = x_1 + x_2 = f((x_1, y_1)) + f((x_2, y_2))$. Here f is also an onto function, and $f(\mathbb{N} \times \mathbb{N}) = \mathbb{N}$. One equivalence class of $\mathbb{N} \times \mathbb{N}$ is $[(5, 7)]$; this class is also named $[(5, 2)]$, $[(5, 8001)]$, and so on since f takes any of these ordered pairs to 5. In general, $[(x, y)] = \{(x, n) | n \in \mathbb{N}\}$.   □

5.81 Practice   Let $f: \mathbb{Z} \to \mathbb{Z}$ be given by $f(x) = x \cdot_3 1$. (Here $x \cdot_3 1$ denotes the *nonnegative* remainder when x is divided by 3.) Then f is a homomorphism from $[\mathbb{Z}, +]$ to $[\mathbb{Z}, +_3]$. Find the homomorphic image $f(\mathbb{Z})$, and show that $\mathbb{Z}/f = \{[16], [-12], [8]\}$.   *(Answer, page A-16)*

Operating in S/f

We now want to create an algebraic structure using the set S/f. To do so we must define a binary operation $*$ on the classes of S. Suppose that $[x]$ and $[y]$ are members of S/f. We want to define $[x] * [y]$ to be some unique member of S/f. A reasonable approach is to take x and y, representative members of the classes $[x]$ and $[y]$, respectively, form $x \cdot y$ in S, and then use the class $[x \cdot y]$ as the product $[x] * [y]$. Thus, $[x] * [y] = [x \cdot y]$. There is one problem with this definition of $*$, and it is a difficulty occurring whenever we operate on classes by choosing representatives of the classes. Since we want $*$ to be an operation on the *classes* $[x]$ and $[y]$, the resulting class $[x] * [y]$ must be independent of the particular representatives of $[x]$ and $[y]$ that we use to do the computation. This property must hold in order for $*$ to be a well-defined operation. In our previous two homomor-phism examples, we ignored this question, blithely saying "take any member of the class ..."; now we prove that the operation $*$ is well defined. Suppose then that we want to compute $[x] * [y]$ and that $[x] = [x']$ and $[y] = [y']$. We can therefore choose x' and y' as representatives of $[x]$ and $[y]$, and compute $[x' \cdot y']$. We want to show that $[x' \cdot y'] = [x \cdot y]$, which will be true if $f(x' \cdot y') = f(x \cdot y)$. Because f is a homomorphism, $f(x' \cdot y') = f(x') + f(y')$ and $f(x) + f(y) = f(x \cdot y)$. And because $[x] = [x']$ and $[y] = [y']$, it follows that $f(x) = f(x')$ and $f(y) = f(y')$. Thus,

$$f(x' \cdot y') = f(x') + f(y') = f(x) + f(y) = f(x \cdot y)$$

and $[x' \cdot y'] = [x \cdot y]$.

Not only do we have a binary operation on S/f, but $[S/f, *]$ is a semigroup.

5.82 Practice   Prove that $*$ is associative on S/f.   *(Answer, page A-16)*

5.83 Example   Consider the semigroup $[(\mathbb{N} \times \mathbb{N})/f, *]$ where f is defined as in Example 5.80. In this semigroup, $[(3, 8)] * [(7, 14)] = [(3, 8) + (7, 14)] = [(10, 22)]$. However, it is also correct to write $[(3, 8)] * [(7, 14)] = [(10, 6)]$. $\square$

5.84 Practice   Which of the following computations in $[(\mathbb{N} \times \mathbb{N})/f, *]$ are correct?

      (a) $[(0, 4)] * [(6, 2)] = [(0, 6)]$
      (b) $[(4, 10)] * [(4, 11)] = [(8, 10)]$
      (c) $[(15, 0)] * [(2, 12)] = [(17, 21)]$ *(Answer, page A-16)*

ISOMORPHISM FROM S/f TO f(S)

Look again at Practice 5.81. Here we can name the elements of $\mathbb{Z}/f$ as $[0]$, $[1]$, and $[2]$. In the semigroup $[\mathbb{Z}/f, *]$, the computation $[1] * [2] = [3] = [0]$ is correct. This equation strongly reminds us of arithmetic in the homomorphic image $[\mathbb{Z}_3, +_3]$. The similarity is no accident. We next show that for the general case where f is a homomorphism from semigroup $[S, \cdot]$ to semigroup $[T, +]$, the semigroup $[S/f, *]$ is isomorphic to the homomorphic image $[f(S), +]$. Since we are discussing four different semigroups here, perhaps Figure 5.9 can clarify the situation. Its solid lines denote the homo-

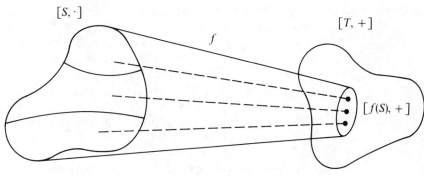

Figure 5.9

morphism f mapping S onto $f(S)$, and its dotted lines show the isomorphism we hope to establish between the classes in S/f and the elements of $f(S)$.

To show that $[S/f, *]$ is isomorphic to $[f(S), +]$, we need a bijection $g: S/f \to f(S)$. It should be easy to guess g because f maps all the members of a given class to the same member of $f(S)$.

5.85 Practice   Let $[x] \in S/f$. What would you suggest as a definition for $g([x])$?

(Answer, page A-16)

We define $g: S/f \to f(S)$ by $g([x]) = f(x)$. Notice that g is well defined in that if $[x] = [y]$, then $f(x) = f(y)$ and $g([y]) = f(y) = f(x) = g([x])$. Now we will show that g is a bijection. To show that g is onto, let $t \in f(S)$ where $t = f(x)$ for some x in S. Then $[x] \in S/f$, and $g([x]) = f(x) = t$. To show that g is one-to-one, suppose that $g([x]) = g([y])$. Then $f(x) = f(y)$, and $[x] = [y]$.

Finally, we can show that g is a homomorphism by proving that $g([x] * [y]) = g[x] + g[y]$.

5.86 Practice   Show that $g([x] * [y]) = g[x] + g[y]$. (Hint: use the definition of $*$ and the fact that f is a homomorphism from S to T.)   *(Answer, page A-16)*

We have now proved the following theorem.

5.87 Theorem   A Homomorphism Theorem for Semigroups
If f is a homomorphism from a semigroup S to a semigroup T, then the homomorphic image $f(S)$ is isomorphic to the semigroup S/f.   □

Shortly, we will see similar results for monoids and groups. These results explain the "imperfect" simulations done in our previous examples. If there is a homomorphism f from a structure $[A, \cdot]$ *onto* a structure $[B, +]$, then B is the homomorphic image $f(A)$ and is isomorphic to A/f. In this case, A (actually A/f) simulates B, but B can only simulate A/f, not A itself.

5.88 Example   Consider again the function $f: \mathbb{N} \times \mathbb{N} \to \mathbb{N}$ given by $f((x, y)) = x$, a homomorphism from $[\mathbb{N} \times \mathbb{N}, +]$ onto $[\mathbb{N}, +]$. A typical member of $(\mathbb{N} \times \mathbb{N})/f$ is $[(x, y)] = \{(x, n) | n \in \mathbb{N}\}$. The isomorphism $g: (\mathbb{N} \times \mathbb{N})/f \to N$ associates $[(x, y)]$ with $f((x, y)) = x$. To compute $2 + 5$ in $[\mathbb{N}, +]$, we look at the associated classes and compute $[(2, 1)] * [(5, 1)] = [(2, 1) + (5, 1)] = [(7, 2)]$, and then $g[(7, 2)] = f((7, 2)) = 7$. However, to simulate the computation $(3, 4) + (5, 6)$ in $\mathbb{N} \times \mathbb{N}$ by using $[\mathbb{N}, +]$, we find $f(3, 4) + f(5, 6) = 3 + 5 = 8$, and we know only that the answer is some member of $\{(8, n) | n \in \mathbb{N}\}$.   □

We want to extend Theorem 5.87 to the monoid and group cases. Thus, suppose that f is a homomorphism from a monoid $[S, \cdot]$ with identity i_S to a monoid $[T, +]$. Then the semigroup $[S/f, *]$ is also a monoid with identity $[i_S]$.

5.89 Practice Show that $[i_S]$ is an identity element for $[S/f, *]$, that is, show that for $[x] \in S/f$, $[x] * [i_S] = [i_S] * [x] = [x]$. *(Answer, page A-16)*

We already know that the homomorphic image of a monoid is a monoid. Thus, Theorem 5.87 becomes the following:

5.90 Theorem A Homomorphism Theorem for Monoids
If f is a homomorphism from a monoid S to a monoid T, then the homomorphic image $f(S)$ is isomorphic to the monoid S/f. □

And finally, let f be a homomorphism from a group $[S, \cdot]$ with identity i_S to a group $[T, +]$. Then $[S/f, *]$ is a group. We must show that each $[x] \in S/f$ has an inverse with respect to the identity $[i_S]$.

5.91 Practice For $[x] \in S/f$, find an inverse element in S/f. *(Answer, page A-16)*

Therefore, we arrive at Theorem 5.92.

5.92 Theorem A Homomorphism Theorem for Groups
If f is a homomorphism from a group S to a group T, then the homomorphic image $f(S)$ is isomorphic to the group S/f. □

5.93 Example Let $f : \mathbb{Z} \to \mathbb{Z}_8$ be defined by $f(x) = x \cdot_8 2$. Then f is a homomorphism from the group $[\mathbb{Z}, +]$ into the group $[\mathbb{Z}_8, +_8]$. The homomorphic image $f(\mathbb{Z})$ is $\{0, 2, 4, 6\}$. The distinct elements of $\mathbb{Z}/f$ are the classes $[0]$, $[1]$, $[2]$, and $[3]$ where

$$[0] = \{0, 4, 8, \ldots, -4, -8, \ldots\}$$
$$[1] = \{1, 5, 9, \ldots, -3, -7, \ldots\}$$
$$[2] = \{2, 6, 10, \ldots, -2, -6, \ldots\}$$
$$[3] = \{3, 7, 11, \ldots, -1, -5, \ldots\}$$

The isomorphism g from $\mathbb{Z}/f$ to $f(\mathbb{Z})$ is given by

$$[0] \to 0$$
$$[1] \to 2$$
$$[2] \to 4$$
$$[3] \to 6$$

□

5.94   *Practice*   For the above example, compute
(a) $g([2] * [3])$
(b) $g([2]) +_8 g([3])$.   (*Answer, page A-16*)

In Chapter 7 we will see a homomorphism theorem for the finite-state machine structure. Meanwhile, in the next section, we will expand on the situation for groups. When f is a homomorphism from a group $[S, \cdot]$ to a group $[T, +]$, we will see a new way to describe the members of S/f. We will also discover a sort of converse to the Homomorphism Theorem for Groups.

✔ CHECKLIST

Techniques

Given a homomorphism f from a structure $[S, \cdot]$ to a structure $[T, +]$, find $f(S)$, find the distinct members of S/f, compute in $[S/f, *]$, find the isomorphism from S/f to $f(S)$, and use S/f to simulate $f(S)$.

Main Ideas

Homomorphism Theorem for semigroups, monoids, groups: If f is a homomorphism from a semigroup (monoid, group) S to a semigroup (monoid, group) T, then the homomorphic image $f(S)$ is isomorphic to S/f.

EXERCISES, SECTION 5.4

★1. Let $\mathbb{N}^2[x] = \{a_2x^2 + a_1x + a_0 | a_0, a_1, a_2 \in \mathbb{N}\}$ and let $+$ denote polynomial addition. Then $[\mathbb{N}^2[x], +]$ is a monoid.
   (a) Prove that $f : \mathbb{N}^2[x] \to \mathbb{N}$ defined by $f(a_2x^2 + a_1x + a_0) = a_2$ is a homomorphism from $[\mathbb{N}^2[x], +]$ onto $[\mathbb{N}, +]$.
   (b) Which of the following are correct computations in $[\mathbb{N}^2[x]/f, *]$?
      (1) $[5x^2 + 3x + 2] * [3x^2 + 2x + 4] = [8x^2 + 3x + 2]$
      (2) $[3x + 7] * [x + 1] = [3x^2 + 10x + 7]$
      (3) $[x^2] * [2x] = [x^2 + 7]$

2. For each pair of groups S and T and the given homomorphism $f : S \to T$, find $f(S)$, find the distinct members of S/f, and indicate the isomorphism g from S/f to $f(S)$.
   ★(a) $[\mathbb{Z}_{18}, +_{18}], [\mathbb{Z}_{12}, +_{12}]; f(x) = 4 \cdot_{12} x$ (see Exercise 3 of Section 5.3)
   (b) $[\mathbb{Z}, +], [\mathbb{Z}_7, +_7]; f(x) = x \cdot_7 1$
   (c) $[\mathbb{Z}_{12}, +_{12}], [\mathbb{Z}_{18}, +_{18}]; f(x) = 6 \cdot_{18} x$

   (d) $[M_2^0(\mathbb{Q}), \cdot]$ where $M_2^0(\mathbb{Q}) = \left\{ \begin{bmatrix} 1 & q \\ 0 & 1 \end{bmatrix} \middle| q \in \mathbb{Q} \right\}$ and $\cdot$ is matrix multiplication, $[\mathbb{Q} \times \mathbb{Q}, +]$ where $+$ denotes componentwise addition;

$$f\left(\begin{bmatrix} 1 & q \\ 0 & 1 \end{bmatrix}\right) = (0, q).$$

3. Each problem below refers to the corresponding problem in Exercise 2.
   ★ (a) Compute $g([1] * [2])$ and $g([1]) +_{12} g([2])$
   (b) Compute $g([3] * [3])$ and $g([3]) +_7 g([3])$
   (c) Compute $g([2] * [2])$ and $g([2]) +_{18} g([2])$

   (d) Compute $g\left(\begin{bmatrix} 1 & 5 \\ 0 & 1 \end{bmatrix} * \begin{bmatrix} 1 & -4 \\ 0 & 1 \end{bmatrix}\right)$

   and $g\left(\begin{bmatrix} 1 & 5 \\ 0 & 1 \end{bmatrix}\right) + g\left(\begin{bmatrix} 1 & -4 \\ 0 & 1 \end{bmatrix}\right)$

4. Suppose you have a "machine" or "black box" that is a modulo 8 adder, that is, it performs arithmetic in $[\mathbb{Z}_8, +_8]$. Such a machine has 8 inputs and 8 outputs, denoted by $0, 1, 2, \ldots, 7$. You would like to use this machine to do arithmetic in $[\mathbb{Z}_4, +_4]$. The numbers 0, 1, 2, 3 can be fed into the $\mathbb{Z}_8$ machine. Define a function from the outputs of the $\mathbb{Z}_8$ machine to $\{0, 1, 2, 3\}$ allowing this machine to simulate a modulo 4 adder.

5. Let f be a homomorphism from a group $[S, \cdot]$ with identity i_S to a group $[T, +]$ with identity i_T. Then $f(i_S) = i_T$. Show that $[i_S]$ is a subgroup of S.

Section 5.5   QUOTIENT GROUPS

In this section we will concentrate on further refinement of the Homomorphism Theorem for Groups (Theorem 5.92). Because you have now had lots of practice in keeping track of the various operations encountered in homomorphisms from one group to another, we will simply use $\cdot$ to denote an arbitrary group operation. Of course in any specific group, the abstract $\cdot$ must be interpreted correctly.

Suppose then that $[G, \cdot]$ and $[H, \cdot]$ are groups, and that f is a homomorphism from G to H. We know that G/f is a group structure, and we will first find a nice way to describe the equivalence classes making up this group. The description involves the use of the equivalence class $[i_G]$, which is the identity of the group G/f.

THE KERNEL

5.95 Definition   For f a homomorphism from a group G with identity i_G to a group H with identity i_H, the set $[i_G] = \{x \in G \mid f(x) = i_H\}$ is called the **kernel** K of f.   □

5.96 Example

(a) Let $\mathbb{R}^* = \mathbb{R} - \{0\}$, and $\mathbb{R}^+$ be the set of all positive real numbers. Then the function f defined by $f(x) = |x|$ is a homomorphism from the group $[\mathbb{R}^*, \cdot]$ to the group $[\mathbb{R}^+, \cdot]$. The kernel K of f is $\{x \in \mathbb{R}^* \mid f(x) = 1\}$; therefore, $K = \{1, -1\}$.

(b) The unction f defined by $f(x) = x \cdot_8 2$ is a homomorphism from $[\mathbb{Z}, +]$ to $[\mathbb{Z}_8, +_8]$. Here $K = \{x \in \mathbb{Z} \mid f(x) = x \cdot_8 2 = 0\} = \{0, 4, 8, \ldots, -4, -8, \ldots\} = 4\mathbb{Z}$. □

5.97 Practice

(a) The function f defined by $f(x) = x \cdot_3 1$ is a homomorphism from $[\mathbb{Z}, +]$ to $[\mathbb{Z}_3, +_3]$. Find the kernel K.

(b) The function f defined by $f(x) = x \cdot_8 4$ is a homomorphism from $[\mathbb{Z}_{12}, +_{12}]$ to $[\mathbb{Z}_8, +_8]$. Find the kernel K. *(Answer, page A-16)*

Because i_G is always an element of K, K is a nonempty subset of G. Furthermore, K is a subgroup of $[G, \cdot]$.

5.98 Practice

Let $x, y \in K$ and show that $x \cdot y^{-1} \in K$, thus proving that K is a subgroup of $[G, \cdot]$ (see Exercise 13(a) of Section 5.2). *(Answer, page A-17)*

Although K is the only member of G/f that is a subgroup of G (because it is the only equivalence class containing i_G), all other members of G/f are related to K. Thus, let $[x]$ be a member of G/f and let $y \in [x]$. We can then write

$$f(y) = f(x)$$

$$f(y) = f(x) \cdot i_H$$

$$(f(x))^{-1} \cdot f(y) = i_H$$

$$f(x^{-1}) \cdot f(y) = i_H$$

$$f(x^{-1} \cdot y) = i_H$$

and $x^{-1} \cdot y \in K$. Let $x^{-1} \cdot y = k$ where $k \in K$. Then $y = xk$. If we let $xK = \{xk \mid k \in K\}$, we have shown that every member y of $[x]$ is a member of xK, or $[x] \subseteq xK$. But we can also show that $xK \subseteq [x]$. To prove this, let $xk \in xK$. Then $f(xk) = f(x) \cdot f(k) = f(x) \cdot i_H$ (since $k \in K$) $= f(x)$, and $xk \in [x]$. Therefore, we know that $[x] = xK$. Any equivalence class in G/f consists of a representative of the class multiplied by all the various members of the kernel. Henceforth, we will use xK to denote the class $[x]$, and G/K to denote the collection G/f of classes.

5.99 Example

(a) In the homomorphism of Example 5.96(a), $K = \{1, -1\}$. A typical member of $\mathbb{R}^*/K$ is $3K = \{3, -3\}$. As before, an equivalence class can have more than one name; here, $3K = -3K$.

(b) In the homomorphism of Example 5.96(b), $K = 4\mathbb{Z}$. The distinct members of $\mathbb{Z}/4\mathbb{Z}$ are (using additive notation this time)

$$0 + K = 0 + 4\mathbb{Z} = \{0, 4, 8, \ldots, -4, -8, \ldots\}$$
$$1 + K = 1 + 4\mathbb{Z} = \{1, 5, 9, \ldots, -3, -7, \ldots\}$$
$$2 + K = 2 + 4\mathbb{Z} = \{2, 6, 10, \ldots, -2, -6, \ldots\}$$
$$3 + K = 3 + 4\mathbb{Z} = \{3, 7, 11, \ldots, -1, -5, \ldots\} \qquad \square$$

5.100 Practice   (a) For Practice 5.97a, name the distinct members of $\mathbb{Z}/3\mathbb{Z}$ in terms of $K = 3\mathbb{Z}$.

(b) For Practice 5.97b, name the distinct members of $\mathbb{Z}_{12}/K$ in terms of K.   *(Answer, page A-17)*

We now know that for any $[x]$ in G/K, $[x] = xK = \{xk \,|\, k \in K\}$. A similar argument shows that $[x] = Kx = \{kx \,|\, k \in K\}$. Thus, for any $x \in G$, $xK = Kx$. (Note that this is set equality, and not a claim that $xk = kx$ for $k \in K$.)

5.101 Practice   Show that $[x] = Kx$.   *(Answer, page A-17)*

Multiplication Formula

The notation xK for members of the set G/K is useful in finding the product of two members of the *group* G/K. For $[x] = xK$, $[y] = yK$, we know that multiplication in the group G/K is defined by $[x] \cdot [y] = [x \cdot y]$, but this equation leads to $xK \cdot yK = xyK$, which gives us an easy formula for multiplying in G/K. (By Practice 5.101, the formula $Kx \cdot Ky = Kxy$ is also valid.)

5.102 Example   (a) In $\mathbb{R}^*/K$, the product of $3K$ and $(-5)K$ is $(3(-5))K = -15K = 15K$.

(b) In $\mathbb{Z}/4\mathbb{Z}$, the sum of $2 + 4\mathbb{Z}$ and $3 + 4\mathbb{Z}$ is $(2 + 3) + 4\mathbb{Z} = 5 + 4\mathbb{Z} = 1 + 4\mathbb{Z}$. $\square$

5.103 Practice   In $\mathbb{Z}/3\mathbb{Z}$,

(a) Find the sum of $7 + 3\mathbb{Z}$ and $5 + 3\mathbb{Z}$.

(b) Find the additive inverse of $4 + 3\mathbb{Z}$.   *(Answer, page A-17)*

QUOTIENT GROUPS

In summary, we know that for f a homomorphism from group G to group H, the kernel K of f is a subgroup of G; for every $x \in G$, $xK = Kx$; and, finally, $G/K = \{xK \,|\, x \in G\}$ is a group under the operation $xK \cdot yK = xyK$ and is isomorphic to the image group $f(G)$.

Now let's forget about homomorphisms (for the moment), and see how much of our construction for G/K still holds when, instead of K, we use an arbitrary subgroup of G. Thus we will ask, for S an arbitrary subgroup of G, the following questions:

1. Do the sets $xS = \{xs \mid s \in S\}$ for the various x's in G form a partition of G?
2. If so, will the multiplication formula $xS \cdot yS = xyS$ be a well-defined binary operation on these sets?
3. If so, will the sets form a group under this operation?

Only 2 above will cause us any difficulty. First, we need some terminology.

5.104 Definition   For S a subgroup of a group G, sets of the form $xS = \{xs \mid s \in S\}$ for $x \in G$ are called **left cosets** of S in G; sets of the form $Sx = \{sx \mid s \in S\}$ for $x \in G$ are called **right cosets** of S in G.   □

5.105 Example   The set $S = \{1, 5, 8, 12\}$ forms a subgroup of $[\mathbb{Z}_{13}^*, \cdot_{13}]$. The left coset $3S$ equals $\{3 \cdot_{13} 1, 3 \cdot_{13} 5, 3 \cdot_{13} 8, 3 \cdot_{13} 12\} = \{3, 2, 11, 10\}$.   □

5.106 Practice   (a) The set $S = \{0, 4, 8\}$ forms a subgroup of $[\mathbb{Z}_{12}, +_{12}]$. Find the members of the left coset $6 + S$ (note the additive notation).
(b) The set $S = \{i, (2, 3)\}$ forms a subgroup of the group $[S_3, \circ]$. Find the elements of the left cosets $(1, 2, 3)S$ and $(1, 3, 2)S$. Find the elements of the right cosets $S(1, 2, 3)$ and $S(1, 3, 2)$.

(Answer, page A-17)

Coset Partitions

Question (1) above asks whether the collection of distinct left cosets of S in G partitions G. This partitioning will certainly occur if these cosets are the equivalence classes of some equivalence relation ρ on G. We define a relation ρ in such a way that if ρ is an equivalence relation, the left cosets will be its equivalence classes, and then show that ρ is in fact an equivalence relation.

Thus, for S a subgroup of G, we define a binary relation ρ on G by $x \rho y \Leftrightarrow x$ and y are members of the same left coset of S in G. Because $x = x \cdot i_G$, $x \in xS$; to say that y is a member of the same left coset is to say that $y = xs$ for some $s \in S$. Therefore, we can redefine ρ as $x \rho y \Leftrightarrow y = xs$ for some $s \in S$.

5.107 Practice   Show that ρ as defined above is an equivalence relation on G.

(Answer, page A-17)

We conclude that the distinct left cosets of S in G partition G. A similar argument would show that the right cosets partition G (but perhaps differently).

Normal Subgroups and Coset Multiplication

If we propose treating the left cosets of S in G as objects and defining an operation · on the left cosets by $xS \cdot yS = xyS$, Question (2) asks whether such an operation is well defined. Recall that every time we define an operation on classes by using representatives of the classes, we must check to see that the answer is independent of the representatives used.

5.108 Example Consider the left cosets $(1, 2, 3)S$ and $(1, 3, 2)S$ where $S = \{i, (2, 3)\}$ in $[S_3, \circ]$. Suppose we define coset multiplication by $(1, 2, 3)S \cdot (1, 3, 2)S = (1, 2, 3)$ $(1, 3, 2)S = S$. Then (see Practice 5.106b) $(1, 2) \in (1, 2, 3)S$ and $(1, 3) \in (1, 3, 2)S$, but $(1, 2)(1, 3) = (1, 3, 2) \notin S$. The natural rule for coset multiplication does not work! □

Why isn't coset multiplication well defined in Example 5.108? What property does the kernel K of some homomorphism have that the subgroup S of S_3 lacks? For K the kernel of a homomorphism, any left coset xK is equal to the right coset Kx. By Practice 5.106b, $(1, 2, 3)S \notin S(1, 2, 3)$ and $(1, 3, 2)S \notin S(1, 3, 2)$. Equality of left and right cosets turns out to be exactly what we need to make our coset multiplication well defined.

5.109 Definition Let S be a subgroup of a group G. S is a **normal subgroup** of G if for all $x \in G$, $xS = Sx$. □

First we will show that normality is a sufficient condition for coset multiplication to be well defined. Suppose, then, that S is a normal subgroup of a group $[G, \cdot]$ and that coset multiplication is to be defined by the rule $xS \cdot yS = xyS$. We must show that for $x_1 \in xS$ and $y_1 \in yS$, the product $x_1 y_1 \in xyS$. We know that $x_1 = xs_1$ and $y_1 = ys_2$ for some $s_1, s_2 \in S$. The product $x_1 y_1$ is $(xs_1)(ys_2) = x(s_1 y)s_2$. The element $s_1 y$ is a member of the right coset Sy, which, by normality, equals the left coset yS. Thus, $s_1 y \in yS$ and $s_1 y = ys_3$ for some $s_3 \in S$. Continuing our computation,

$$x_1 y_1 = (xs_1)(ys_2) = x(s_1 y)s_2 = x(ys_3)s_2 = xy(s_3 s_2) \in xyS.$$

Coset multiplication is well defined when S is normal.

Conversely, when coset multiplication $xS \cdot yS = xyS$ is well defined, the subgroup S must be normal. To prove this, we assume the operation to be well defined, and then show that $xS = Sx$ for any $x \in G$. We will show set inclusion in each direction. First, let $sx \in Sx$. We need (temporarily) an arbitrary element $y \in G$ in order to consider the product $yS \cdot xS = yxS$. Because $ys \in yS$, $xs \in xS$, and coset multiplication is well defined, we must have $(ys)(xs) \in yxS$. Thus, $ys \cdot xs = yxs_1$ for some $s_1 \in S$. Multiplying both sides of this equation by y^{-1} on the left and s^{-1} on the right, we get $sx = xs_1 s^{-1} \in xS$, which proves that $Sx \subseteq xS$. Now for the opposite inclusion. Let $xs \in xS$. Then $x^{-1} \in G$ and, by the inclusion we have already proved, $Sx^{-1} \subseteq x^{-1}S$, so $sx^{-1} = x^{-1}s_1$ for some $s_1 \in S$. Multiplying both sides of this equation on the left and right by x, we get $xs = s_1x \in Sx$, proving that $xS \subseteq Sx$. Therefore, $xS = Sx$ and S is normal. The answer to Question (2) can be summarized as follows:

5.110 Theorem Let S be a subgroup of a group G. Coset multiplication $xS \cdot yS = xyS$ is well defined if and only if S is a normal subgroup. □

If S is normal, left and right cosets are the same and we could also write the formula for coset multiplication in the form $Sx \cdot Sy = Sxy$.

The Group Structure

Given that we have a normal subgroup and that as a result coset multiplication is well defined, the answer to Question (3) is easy: the set of cosets under this operation forms a group.

5.111 Practice Show that for S a normal subgroup of a group G, the collection of (left) cosets of S in G forms a group under coset multiplication. *(Answer, page A-17)*

5.112 Definition For S a normal subgroup of a group G, the group of cosets of S in G under coset multiplication is called the **quotient group (factor group) of G modulo S**, denoted by G/S. □

Of course, G/K is a quotient group where K is the kernel of a homomorphism. To find other normal subgroups, and therefore other examples of quotient groups, let's develop a test for normality. Suppose that S is a subgroup of G with the property that for any $x \in G$ and $s \in S$, $x^{-1}sx \in S$. This property insures that S is normal. To prove this, let $sx \in Sx$. Then by our hypothesis, $x^{-1}sx \in S$, so $x^{-1}sx = s_1$ where $s_1 \in S$, and $sx = xs_1 \in xS$. Thus, $Sx \subseteq xS$. Now let $xs \in xS$. Then by our hypothesis, $(x^{-1})^{-1}sx^{-1} = xsx^{-1} \in S$, so $xsx^{-1} = s_1$ where $s_1 \in S$, and $xs = s_1x \in Sx$. Thus, $xS \subseteq Sx$ and $xS = Sx$, so that S is normal.

5.113 Theorem If S is a subgroup of G such that for any $x \in G$ and $s \in S$, $x^{-1}sx \in S$, then S is normal in G. □

5.114 Example (a) Any subgroup S of a commutative group G is normal because for $x \in G$, $s \in S$, $x^{-1}sx = sx^{-1}x = s \in S$.

(b) A_n is normal in S_n, $n > 1$. For $\alpha \in S_n$, if α is even then so is α^{-1}, and if α is odd, α^{-1} is too. For $\beta \in A_n$, β is even. Thus, $\alpha^{-1}\beta\alpha$ is even. □

FUNDAMENTAL HOMOMORPHISM THEOREM

Recall that a quotient group G/S can be constructed without mention of a homomorphism function; all that is required is that S be a normal subgroup of G. But we can show that for any quotient group G/S, there is a related homomorphism.

5.115 Practice   Let S be a normal subgroup of the group $[G, \cdot]$ and define a function $f: G \to G/S$ by $f(x) = xS$ (thus f maps each element of G to its coset). Show that f is a homomorphism from G onto G/S.   (*Answer, page A-17*)

To tie things together, we will restate the Homomorphism Theorem for Groups (Theorem 5.92) in the terminology we have developed in this section, and also include Practice 5.115 which is, essentially, the converse of this theorem.

5.116 Theorem   Fundamental Homomorphism Theorem for Groups
If f is a homomorphism from a group $[G, \cdot]$ to a group $[H, \cdot]$, then the kernel K of f is a normal subgroup of G, and the homomorphic image $f(G)$ is isomorphic to the quotient group G/K. Conversely, any quotient group of G is a homomorphic image of G.   ☐

Theorem 5.116 says that every homomorphic image of a group G is (to within an isomorphism) a quotient group of G, and every quotient group of G is a homomorphic image of G. Homomorphic images and quotient groups of a given group are practically the same things, or at least there is a one-to-one correspondence between them.

Computer Arithmetic

Now to settle the issue about computer arithmetic raised after Example 5.69. We want to simulate, as best we can, $[\mathbb{Z}, +]$ by a finite structure, call it $[H, +]$. Because $[\mathbb{Z}, +]$ is a group, $[H, +]$ should be a group. To preserve the operation of $[\mathbb{Z}, +]$ in $[H, +]$, we need a homomorphism from $[\mathbb{Z}, +]$ onto $[H, +]$. Thus, H is a homomorphic image of $\mathbb{Z}$, and, by the Fundamental Theorem (5.116), it is essentially a quotient group of $\mathbb{Z}$. What are the possible quotient groups of $\mathbb{Z}$? They must be groups of the form $\mathbb{Z}/S$ where S is a normal subgroup of $[\mathbb{Z}, +]$. Because $[\mathbb{Z}, +]$ is commutative, normality presents no problem, so we concentrate on possible subgroups of $[\mathbb{Z}, +]$. By Theorem 5.58, any subgroup S must be $n\mathbb{Z}$ for some n. Thus, H is isomorphic to $\mathbb{Z}/n\mathbb{Z}$. If $n = 0$, then $n\mathbb{Z} = \{0\}$, and the distinct members of $\mathbb{Z}/\{0\}$ are $0 + \{0\} = 0, 1 + \{0\} = 1, -1 + \{0\} = -1, \ldots$. Hence, $\mathbb{Z}/\{0\}$ is infinite although we require H to be finite. Therefore, $n > 0$, and the distinct members of $\mathbb{Z}/n\mathbb{Z}$ are $0 + n\mathbb{Z}, 1 + n\mathbb{Z}, \ldots, (n-1) + n\mathbb{Z}$. The function $h: \mathbb{Z}/n\mathbb{Z} \to \mathbb{Z}_n$ given by $h(k + n\mathbb{Z}) = k$ is clearly an isomorphism from $[\mathbb{Z}/n\mathbb{Z}, +]$ to $[\mathbb{Z}_n, +_n]$. Therefore, $H \simeq \mathbb{Z}/n\mathbb{Z} \simeq \mathbb{Z}_n$, or $H \simeq \mathbb{Z}_n$. We conclude that to simulate $[\mathbb{Z}, +]$ by a finite structure, we must use $[\mathbb{Z}_n, +_n]$ for some n (or an isomorphic copy of $[\mathbb{Z}_n, +_n]$).

Conversely, $\mathbb{Z}/n\mathbb{Z}$ is a quotient group of $[\mathbb{Z}, +]$ for any positive integer

n and, by the converse half of the Fundamental Theorem, is a homomorphic image of $[\mathbb{Z}, +]$. Therefore, for any positive integer n, $\mathbb{Z}/n\mathbb{Z}$ (which again is essentially $[\mathbb{Z}_n, +_n]$) can be used to (imperfectly) simulate $[\mathbb{Z}, +]$.

LAGRANGE'S THEOREM

The final result of this chapter is a "counting" theorem about finite groups. But first, suppose that G is any group, not necessarily finite, and that S is a subgroup of G. The set of left cosets of S in G forms a partition of G, and so does the set of right cosets of S in G. Left and right cosets coincide if and only if S is normal in G. But regardless of whether S is normal, we can find a one-to-one correspondence between the elements of any left coset xS and the elements of S itself. The obvious mapping to try is $xs \to s$.

5.117   Practice   Let S be a subgroup of a group G, and let $x \in G$. Show that the function $f : xS \to S$ given by $f(xs) = s$ is a bijection.   *(Answer, page A-17)*

If S is finite, say $|S| = k$, then by Practice 5.117, every left coset of S in G has k elements. A similar argument shows that every right coset of S in G also has k elements. Now suppose that G itself is finite, say $|G| = n$. The left cosets of S in G partition G; the n elements of G are distributed among the left cosets, and each left coset has k elements. Thus, we must have an equation of the form $n = km$ where m is the number of left cosets of S in G, giving us Lagrange's Theorem, first proved about 1771.

5.118   Theorem   Lagrange's Theorem:
The order of a subgroup of a finite group divides the order of the group.   □

Each right coset of S in G also has k elements, so the equation $n = km$ also holds for the partition of G into right cosets where there are m right cosets. Thus, the number of left cosets of S in G equals the number of right cosets of S in G. (Naturally, this statement is trivial if S is normal in G.)

5.119   Definition   Let S be a subgroup of a finite group G. The number of left cosets (right cosets) of S in G is called the **index** of S in G, denoted by $[G:S]$. $[G:S]$ divides $|G|$.   □

Note that if S is normal in G, then $[G:S] = |G/S|$.

Lagrange's Theorem helps us narrow down the possibilities for subgroups of a finite group. If $|G| = 12$, for example, we would not look for any subgroups of order 7 since 7 does not divide 12. Also, we would know that G could not be used to simulate any group H of order 7. Such a simulation would involve a homomorphism from G onto H, and, by the Fundamental Homomorphism Theorem for Groups, $G/K \simeq H$, so $|G/K| = 7$ and again

7 does not divide 12.

Once more, let $|G| = 12$. The fact that 6 divides 12 does not imply the existence of a subgroup of G of order 6. In fact, A_4 is a group of order $4!/2 = 12$, but it can be shown that A_4 has no subgroups of order 6. The converse to Lagrange's Theorem does not always hold. There are certain cases where the converse can be shown to be true—for example, in finite commutative groups (note that A_4 is not commutative).

5.120   Practice   (a) From Lagrange's Theorem, what are the possible orders of subgroups of the group $G = \mathbb{Z}_2 \times \mathbb{Z}_3 \times \mathbb{Z}_4$?
   (b) For each number in the answer to (a), will G have a subgroup of this order?   (*Answer, page A-17*)

✔ *CHECKLIST*

Definitions

kernel of a homomorphism   normal subgroups (*p. 211*)
   (*p. 207*)   quotient group (*p. 212*)
left and right cosets (*p. 210*)   index of a subgroup (*p. 214*)

Techniques

Find the kernel of a homomorphism from a group G to a group H.

Given any normal subgroup S of G, find the distinct cosets of S in G and perform arithmetic in the quotient group G/S.

Main Ideas

If S is a normal subgroup of a group G, then the (left) cosets of S in G form a group G/S under the operation $xS \cdot yS = xyS$. Normality is necessary and sufficient for this operation to be well defined.

The kernel K of a homomorphism f from a group G to a group H is a normal subgroup of G, and $G/K \simeq f(G)$.

Quotient groups of a group and homomorphic images of that group are essentially the same.

Computer simulation of $[\mathbb{Z}, +]$ must be done with $[\mathbb{Z}_n, +_n]$ for some n, but n can be any positive integer.

The order of a subgroup of a finite group divides the order of the group.

EXERCISES, SECTION 5.5

1. Let $8\mathbb{Z} = \{8z \mid z \in \mathbb{Z}\}$. $8\mathbb{Z}$ is a normal subgroup of the commutative group $[\mathbb{Z}, +]$.

(a) List the distinct cosets that are the elements of the quotient group $\mathbb{Z}/8\mathbb{Z}$ (name them and indicate their members).

(b) What is the sum $(3 + 8\mathbb{Z}) + (6 + 8\mathbb{Z})$ in $\mathbb{Z}/8\mathbb{Z}$?

★ 2. Let $S = \{(0, 0), (1, 2), (2, 0), (0, 2), (1, 0), (2, 2)\}$.

Then S is a normal subgroup of the commutative group $[\mathbb{Z}_3 \times \mathbb{Z}_4, +]$.

(a) List the distinct cosets that are the elements of the quotient group $(\mathbb{Z}_3 \times \mathbb{Z}_4)/S$ (name them and indicate their members).

(b) What is the sum of $(1, 1) + S$ and $(1, 1) + S$ in $(\mathbb{Z}_3 \times \mathbb{Z}_4)/S$?

★ 3. Let $S = \{1, 2, 4\}$. Then S is a normal subgroup of the commutative group $[\mathbb{Z}_7^*, \cdot_7]$ where $\mathbb{Z}_7^* = \mathbb{Z}_7 - \{0\}$.

(a) List the distinct cosets that are the elements of the quotient group $\mathbb{Z}_7^*/S$.

(b) What is the product of $3S$ and $4S$ in $\mathbb{Z}_7^*/S$?

4. In each case below, two groups G and H are listed and a homomorphism $f: G \to H$ is given. Describe the kernel K and describe the quotient group G/K.

(a) $[\mathbb{Z}, +], [\mathbb{Z}_{24}, +_{24}]; f(x) = 3 \cdot_{24} x$

(b) $[\mathbb{Z}_{12}, +_{12}], [\mathbb{Z}_8, +_8]; f(x) = 2 \cdot_8 x$

(c) $[\mathbb{R}^* \times \mathbb{R}^*, \cdot], [\mathbb{R}^*, \cdot]$ where $\mathbb{R}^* = \mathbb{R} - \{0\}; f(x, y) = xy$

(d) $[\mathbb{R}[x], +], [\mathbb{R}, +]; f(a_n x^n + \cdots + a_0) = a_0$

(e) $[M_2(\mathbb{Q}), +], [\mathbb{Q}, +]; f\left(\begin{bmatrix} a & b \\ c & d \end{bmatrix}\right) = a - d$

5. A function $f: \mathbb{Z} \times \mathbb{Z} \to \mathbb{Z}$ is defined by $f((x, y)) = x + y$.

(a) Show that f is a homomorphism from $[\mathbb{Z} \times \mathbb{Z}, +]$ onto $[\mathbb{Z}, +]$.

(b) Describe the kernel K of f.

(c) Which of the following elements belong to the coset $(2, 3) + K$? $(4, 1), (3, 4), (1, 4), (5, 0)$

(d) Write the function that makes $(\mathbb{Z} \times \mathbb{Z})/K \simeq \mathbb{Z}$.

★ 6. A function $f: \mathbb{Q} \times \mathbb{Q} \times \mathbb{Q} \to \mathbb{Q} \times \mathbb{Q}$ is defined by $f((x, y, z)) = (y, y)$.

(a) Find $f(\mathbb{Q} \times \mathbb{Q} \times \mathbb{Q})$ and show that f is a homomorphism from $[\mathbb{Q} \times \mathbb{Q} \times \mathbb{Q}, +]$ onto $[f(\mathbb{Q} \times \mathbb{Q} \times \mathbb{Q}), +]$.

(b) Describe the kernel K of f.

(c) Which of the following elements belong to the coset $(3, 4, 5) + K$? $(2, 3, 4), (4, 4, 4), (4, 5, 6), (5, 4, 3)$

(d) Is the computation $((-7, 2, 6) + K) + ((5, 1, 9) + K) = (10, 3, 2) + K$ in $(\mathbb{Q} \times \mathbb{Q} \times \mathbb{Q})/K$ correct?

(e) Write the function that makes $(\mathbb{Q} \times \mathbb{Q} \times \mathbb{Q})/K \simeq f(\mathbb{Q} \times \mathbb{Q} \times \mathbb{Q})$.

7. A function $f: \mathbb{Q}^* \to \mathbb{Q}^+$ is defined by $f(x) = x^2$.

(a) Find $f(\mathbb{Q}^*)$ and show that f is a homomorphism from $[\mathbb{Q}^*, \cdot]$ onto $[f(\mathbb{Q}^*), \cdot]$.

(b) Describe the kernel K of f.

(c) List all the elements of the product $(3K)(7K)$ in $\mathbb{Q}^*/K$.

(d) Write the function that makes $\mathbb{Q}^*/K \simeq f(\mathbb{Q}^*)$.

8. Let F denote the set of all functions $f: \mathbb{R} \to \mathbb{R}$. For $f, g \in F$, define addition of functions as follows: for $x \in \mathbb{R}$, $(f + g)(x) = f(x) + g(x)$.
   (a) Prove that $[F, +]$ is a group.
   (b) Let $a \in \mathbb{R}$. Define a function $\alpha: F \to \mathbb{R}$ by $\alpha(f) = f(a)$. Prove that α is a homomorphism from $[F, +]$ onto $[\mathbb{R}, +]$.
   (c) Describe the kernel K of α.
   (d) Let $g + K$ denote a member of F/K. For any member h of $g + K$, what is the value of $h(a)$?
   (e) Write the function that makes $F/K \simeq \mathbb{R}$.

★ 9. Let f be a homomorphism from a group G onto a group H. Show that f is an isomorphism if and only if the kernel of f is $\{i_G\}$.

★ 10. For any group $[G, \cdot]$, the **center** of the group is $A = \{x \in G \mid x \cdot g = g \cdot x$ for all $g \in G\}$. $[A, \cdot]$ is a subgroup of $[G, \cdot]$ (see Exercise 10 of Section 5.2). Show that A is a normal subgroup.

11. For $n > 1$, find the order of the quotient group S_n/A_n.

12. Let n and k be fixed positive integers with k a divisor of n. Show that $[\mathbb{Z}_n, +_n]$ has a quotient group of order k.

13. (a) Prove that a quotient group of a commutative group is commutative.
    (b) A group $[G, \cdot]$ is **divisible** if for any $g \in G$ and any positive integer n, the equation $x^n = g$ has at least one solution in G. Prove that a quotient group of a divisible group is divisible.

★ 14. Prove the converse of Theorem 5.113: if S is a normal subgroup of a group G, then for any $x \in G$ and $s \in S$, $x^{-1}sx \in S$.

15. (a) Let S and T be normal subgroups of a group G. Show that $S \cap T$ is a normal subgroup of G. (Hint: use Exercise 14 above.)
    (b) Let S_1 and S_2 be normal subgroups of the groups G_1 and G_2, respectively. Show that $S_1 \times S_2$ is a normal subgroup of the group $G_1 \times G_2$.

16. (a) Let H be a subgroup of a group G. Show that $g^{-1}Hg$ is a subgroup of G for any $g \in G$.
    (b) Let H be a finite subgroup of a group G and suppose that H is the only subgroup of G with order $|H|$. Show that H is normal in G.

17. Let f be a homomorphism from a group G to a group H, and let S be a normal subgroup of G. Show that $f(S)$ is a normal subgroup of $f(G)$.

18. Let S and T be normal subgroups of a group G with $T \subseteq S$. Show that T is normal in S and that S/T is a normal subgroup of G/T.

19. Let S and T be normal subgroups of a group G with $T \subseteq S$.
    (a) Show that if $xT = yT$ in G/T, then $xS = yS$ in G/S.
    (b) Let $f: G/T \to G/S$ be defined by $f(xT) = xS$. By part (a), f is a well-defined function. Show that f is a homomorphism.
    (c) Prove that the quotient group S/T is the kernel of f.
    (d) Prove that $G/T/S/T \simeq G/S$. (Note the suggestion of cancellation here.)

20. Let G be a group and for each $g \in G$, define a function $f_g : G \to G$ by $f_g(x) = gxg^{-1}$.
    (a) Prove that each f_g is an isomorphism from G onto G.
    (b) Let $F = \{f_g \mid g \in G\}$. Show that F is a group under function composition.
    (c) Let A be the center of the group G, $A = \{g \in G \mid xg = gx$ for all $x \in G\}$. Prove that $G/A \simeq F$.

21. Let S be a subgroup of a finite group $[G, \cdot]$ with $[G:S] = 2$. Show that S is a normal subgroup of G.

★ 22. Let S and T be subgroups of a finite group $[G, \cdot]$ with $T \subseteq S$. Show that $[G:T] = [G:S] \cdot [S:T]$.

23. We know that for any finite group G and subgroup S, the number of left cosets of S in G equals the number of right cosets of S in G. Thus, if L is the set of all left cosets of S in G and R is the set of all right cosets of S in G, then there is a bijection from L onto R. Show that such a bijection exists even if G is infinite.

For Further Reading

Out of the many books on abstract or modern algebra, it is difficult to select a few. [1] is closest in spirit to much of the presentation in this chapter. Other very readable introductory texts are [2], [3], and [4]. Many algebra books, such as [5], now have an applied slant.

1. Joseph E. Kuczkowski and Judith L. Gersting, *Abstract Algebra, A First Look.* New York: Marcel Dekker, 1977.

2. John B. Fraleigh, *A First Course in Abstract Algebra*, 2nd ed. Reading Mass.: Addison-Wesley, 1976.

3. Neal H. McCoy, *Introduction to Modern Algebra*, 3rd ed. Boston: Allyn & Bacon, 1975.

4. Larry J. Goldstein, *Abstract Algebra: A First Course.* Englewood Cliffs. N.J.: Prentice-Hall, 1973.

5. William J. Gilbert, *Modern Algebra with Applications.* New York: Wiley, 1976.

Chapter 6

CODING THEORY

Errors can be introduced into data transmission through a variety of means such as hardware failure, interference, or the general random "glitches" to which sensitive and complex electronic equipment is subject. One can protect against errors by coding the information to be sent and decoding it at the other end in such a way as to maximize the probability of correcting, or at least detecting, such errors. Successful coding schemes often rely heavily on algebraic structures. This chapter introduces group codes, but barely scratches the surface of the work that has been done in coding theory. More sophisticated coding methods involve more complex algebraic structures.

Section 6.1   **ENCODING**

At the mention of coding information, most people probably think of double agents scribbling mysterious symbols on a piece of paper to be passed in a folded newspaper to someone sitting on a park bench (at least that's how its always done in the movies). Coding and decoding information for the purpose of insuring its secrecy is known as **cryptology** and involves much interesting mathematics. Exercises 1–3 at the end of this section discuss some coding and decoding techniques suitable for cryptology. There are, however, many other types of codes where secrecy is not the main object. What is your Zip code? What is your area code? What is your Social Security number? Zip codes are used by the postal service to represent geographic localities in numerical form. The telephone company uses area codes to represent clusters of switching networks in numerical form. And many organizations, both government and private, use your Social Security number to represent *you* in numerical form!

Coding, in general, is simply a translation of information from one form to another, more convenient, form. Some other familiar examples are the Universal Product Code—the series of black vertical lines containing product information found on frozen vegetable packages, and so on—and the code a computer uses (ASCII, for example or EBCDIC) to convert alphabetic characters into binary form for computer representation.

GENERAL IDEAS ABOUT INFORMATION CODING

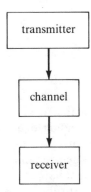

transmitter

channel

receiver

Figure 6.1

In this chapter we will be interested in codes intended to protect the information being transmitted. By protection we do not mean secrecy, but protection against corruption in the information itself. If we consider the communication model shown in Figure 6.1, the channel medium could be air (to transmit radio waves or voices or statellite signals), wires (to transmit electronic signals), magnetic tape (to transmit sequences of binary digits), and so on. Any channel, however, is subject to disturbances that can corrupt the information being transmitted so that what is received is not what was sent. The disturbances can be due to hardware failures, random interference **(noise)**, or deterioration, since we may consider transmission to take place over time as well as over space.

Let's play the children's game of "telephone," which involves voice transmission. Suppose the transmitted message is *the black hat* but the received message (in the case of voice transmission we might say the perceived message) is *the black cat*. Because the received message makes sense and *could* have been the transmitted message, there is no way to detect that an error has occurred. An alternative is to encode the message to be transmitted by repeating it. Thus, the code word for *the black hat* would be *the black hat the black hat*, and the code word for *the black cat* would be *the black cat the black cat*. A received message of *the black hat the black cat* would alert the receiver that an error has occurred in transmission. However, the received message is equally close to either of the two code words, so there would be no way to guess what the correct code word was. Notice that two errors could still go undetected.

Now let's make our code words three copies of the message: *the black hat the black hat the black hat* and *the black cat the black cat the black cat*. A received message of *the black hat the black cat the black hat* would signal that either one or two errors has occurred. If we assume it more likely that one error has occurred, we decode the message to the closest code word, *the black hat the black hat the black hat*. This process is called **maximum likelihood decoding** and gives us the correct code word for a received message where no more than one error has occurred. Since we can detect up to two errors and correct the effects of one error, we have designed a **double-error detecting, single-error correcting code**.

Out of this silly example we get three useful ideas: maximum likelihood decoding, redundancy in coding, and distance between code words. These ideas are related, and we will look at each again. We can at least note from this example that encoding messages with a certain amount of redundancy, although it makes the code words longer than the original messages and thus introduces more opportunities for errors to take place, also increases the capability of detecting and perhaps correcting such errors. In addition, we will want our code words to be far enough apart, whatever that may mean, to recognize when a received word is closer to one code word than to any other code word.

BINARY CODES

We will assume that any information we want to transmit can be represented as a sequence of binary digits, that is, 0's and 1's. We will also assume that errors occurring during transmission are "independent," so that a given disturbance affects only one binary digit and not the surrounding digits. Our final assumption is that it is just as likely for a 0 to be corrupted to a 1 as it is for a 1 to be corrupted to a 0 (although in actuality 1's get corrupted more frequently than 0's), and that the probability of either event happening is relatively small, say .01. For a sequence of three binary digits, the probability of correct transmission of all three digits is $(.99)(.99)(.99) = (.99)^3$. The probability of a single digit being in error (which can occur in any of three places) is $.01(.99)(.99) + (.99)(.01)(.99) + (.99)(.99)(.01) = 3(.01)(.99)^2$. Similarly, the probability of two digits being in error is $= 3(.01)^2(.99)$ and the probability of errors in all three digits is $(.01)^3$. Maximum likelihood decoding simply assumes the most probable situation—that the smallest possible number of errors has occurred. A single-error correcting code, for example, does not always decode correctly, but it will do so in the most likely cases, namely, in any case where only one error has occurred (or no errors have occurred).

Parity checks

A simple redundancy in coding binary words involves adding a final digit to the end of the word, called a **parity bit digit**. Thus, code words for m-tuple messages become $(m + 1)$-tuples. In an **even parity check**, the last digit is chosen so as to make the total number of 1's in the $(m + 1)$-tuple code word an even number. (In an **odd parity check**, the last digit is chosen so as to make the total number of 1's an odd number.) For example, in an even parity check code, the code word for 1011 is 10111. Any single error in transmission can be detected because it produces an odd number of 1's in the received word. It is impossible to tell which digit is in error, however, so this code has no correcting capabilities. Also, any even number of errors in transmission cannot be detected, and any odd number of errors is indistinguishable from a single error. This code is single-error detecting. Such codes are used in storing data on computer auxiliary memory media such as magnetic tapes or disk cartridges.

The codes we will consider in this chapter are generalizations of the even parity-check code in which the m-tuple message becomes the first m components of an n-tuple code word, and the additional $n - m$ binary digits are all special sorts of parity checks. This technique is often used to detect and correct errors in the main memory of a computer where the computer word length is m bits. For a machine advertising an m-bit word length, the hardware itself incorporates the check digits and the actual word length stored is n. Before we consider the general case, we need a definition of distance between code words. (Richard W. Hamming pioneered the study of error-detecting and error-correcting codes in 1950).

Distance

6.1 Definition   Let X and Y be binary n-tuples. The **Hamming distance** between X and Y, $H(X, Y)$, is the number of components in which X and Y differ.   □

6.2 Practice   For $X = 01011$ and $Y = 11001$, what is $H(X, Y)$? (Note that we'll often leave out parentheses and commas when describing our n-tuples.)

(Answer, page A-17)

6.3 Practice   Show that Hamming distance is a **metric** on the set of binary n-tuples; that is, show that for all binary n-tuples X, Y, and Z,
   (a) $H(X, Y) \geq 0$
   (b) $H(X, Y) = 0$ if and only if $X = Y$
   (c) $H(X, Y) = H(Y, X)$
   (d) $H(X, Z) \leq H(X, Y) + H(Y, Z)$ *(Answer, page A-17)*

Now we will consider that we have a code, that is, a collection of code words all of which are binary n-tuples.

6.4 Definition   The **minimum distance** of a code is the minimum Hamming distance between all possible pairs of distinct code words.   □

Each error that occurs in the transmission of a code word adds one unit to the Hamming distance between that code word and the received word. According to our maximum-likelihood decoding rule, we will decode a received word as the closest code word in terms of Hamming distance. Thus, it is not surprising that the error-detecting and error-correcting capabilities of our code are directly related to distance. Let's consider this further.

Suppose we picture the code words as specific binary n-tuples distinguished from the set S of all binary n-tuples—as in Figure 6.2. Suppose also that the minimum distance of the code is at least $d + 1$. Then any time a code word is corrupted by d or fewer errors, it will be changed to an n-tuple that is not another code word, and the occurrence of errors can be detected. Conversely, if any combination of d or fewer errors can be detected, code words must be at least $d + 1$ apart.

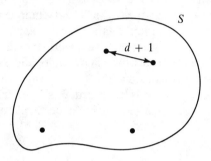

Figure 6.2

6.5   Theorem   A code is d-error detecting (can detect all combinations of d or fewer errors) if and only if its minimum distance is at least $d + 1$.   □

Now suppose that the minimum distance of the code is at least $2d + 1$. Then any time a code word X is corrupted by d or fewer errors, the received word X' will be such that $H(X, X') \leq d$ but for any other code word Y, $H(X', Y) \geq d + 1$; X' will be correctly decoded as X. Conversely, to correct any received word with d or fewer errors, the minimum distance of the code must be at least $2d + 1$ so that neighborhoods of radius d around code words do not intersect (see Figure 6.3).

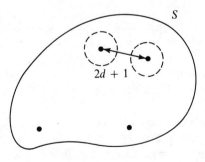

Figure 6.3

6.6   Theorem   A code is d-error correcting (can correct all combinations of d or fewer errors) if and only if its minimum distance is at least $2d + 1$.   □

6.7 Example   Suppose a code has a minimum distance of 6. It can then detect any combination of ≤ 5 errors. It can correct any combination of ≤ 2 errors. If code words X and Y are such that $H(X, Y) = 6$, then there will be a word X' produced by 4 errors on X that will be incorrectly decoded as Y; there will be a word X'' produced by 3 errors on X that can be arbitrarily decoded correctly as X or incorrectly as Y. This code is double-error correcting, 5-error detecting.   □

Theorems 6.5 and 6.6 show that, as we would expect, it requires a stronger condition to correct errors than to merely detect them.

GROUP CODES

Suppose once again that our code words are binary n-tuples. The set of all binary n-tuples can be expressed as $\mathbb{Z}_2^n$, and this set forms a group under componentwise addition modulo 2; we'll denote this group by $[\mathbb{Z}_2^n, +_2]$.

6.8 Practice   In the group $[\mathbb{Z}_2^5, +_2]$, what is

(a) $(01101) +_2 (11011)$?

(b) $-(10110)$?   *(Answer, page A-17)*

The set of code words is a subset of $\mathbb{Z}_2^n$, and we want these code words to be sufficiently widely scattered in $\mathbb{Z}_2^n$ so that the minimum distance is large enough to allow some error correction. The minimum distance of the code, as we shall see, is easy to compute if the code words form a subgroup of $[\mathbb{Z}_2^n, +_2]$. In this case we have a **group code**. The n-tuple of all 0's is the group identity, denoted by 0.

6.9 Definition   The **weight** of a code word X in $\mathbb{Z}_2^n$, $W(X)$, is the number of 1's it contains.
                                                                                          □

6.10 Theorem   The minimum distance of a group code is the minimum weight of all the nonzero code words.

Proof:
Let d be the minimum distance of a group code; then there are distinct code words X and Y with $H(X, Y) = d$. Because we have a group code, closure holds and $X +_2 Y = Z$ is a code word. $Z \neq 0$ because X and Y are distinct. Z will have 1's in exactly those components where X and Y differ, so $W(Z) = H(X, Y) = d$. Thus, the minimum weight of the code is $\leq d$. If the minimum weight is $<d$, let W be a nonzero code word with $W(W)$ the minimum weight. Then $H(W, 0) = W(W) < d$ (remember 0 is a code word), which contradicts the fact that d is the minimum distance of the code. Therefore, the minimum distance equals the minimum weight.   □

6.11 *Example*   The set $\{00000, 01111, 10101, 11010\}$ is a group code in $\mathbb{Z}_2^5$. The minimum distance of the code is 3. $\square$

6.12 *Practice*   Verify that closure holds for the code words of Example 6.11

(Answer, page A-17)

Generating Group Codes

How can we produce subgroups of $\mathbb{Z}_2^n$ to use as code words, and how can we control the minimum distance of the code? Again, some algebraic ideas come to our rescue. Let H be any $n \times r$ binary matrix with $r < n$. If $X \in \mathbb{Z}_2^n$, we can perform the matrix multiplication $X \cdot H$, where all multiplications and additions are done modulo 2. The result of this multiplication is a binary r-tuple.

6.13 *Example*   Let $r = 3, n = 5$, and

$$H = \begin{bmatrix} 1 & 1 & 0 \\ 0 & 1 & 1 \\ 1 & 0 & 1 \\ 1 & 0 & 0 \\ 1 & 1 & 0 \end{bmatrix}$$

Then

$$(11010) \begin{bmatrix} 1 & 1 & 0 \\ 0 & 1 & 1 \\ 1 & 0 & 1 \\ 1 & 0 & 0 \\ 1 & 1 & 0 \end{bmatrix} = (001)$$

$\square$

We can think of multiplication by H as a mapping from the group $[\mathbb{Z}_2^n, +_2]$ to the group $[\mathbb{Z}_2^r, +_2]$. Furthermore, this mapping is a group homomorphism (see Exercise 6, this section.) Now let's consider exactly those members X of $\mathbb{Z}_2^n$ such that $X \cdot H = 0_r$, the zero of the group $[\mathbb{Z}_2^r, +_2]$. This set is the kernel of the group homomorphism and is therefore a subgroup of $[\mathbb{Z}_2^n, +_2]$. We take this set to be the set of code words. Then we can easily determine the minimum weight (minimum distance) of the code simply by looking at H. If H has d distinct rows that add to 0_r in $[\mathbb{Z}_2^r, +_2]$, say $i_1, \ldots, i_d$, we can choose an X in $\mathbb{Z}_2^n$ having 1's exactly in the $i_1, \ldots, i_d$ components. Then $X \cdot H = 0_r$ so that X is a code word, and $W(X) = d$. On the other hand, if X is a code word with $W(X) = d$ and X has 1's exactly in components $i_1, \ldots, i_d$, then the equation $X \cdot H = 0_r$ forces rows $i_1, \ldots, i_d$ of H to sum to 0_r. Therefore, the minimum weight of the code equals the minimum number of distinct rows of H that add to 0_r. In particular, to produce a single-error correcting code, we must have distance at least 3, so we would have to choose an H with no row consisting of all 0's, and no two rows that are alike (these would add to 0_r).

6.14 Example   The code words of Example 6.11 were generated using the matrix H where

$$H = \begin{bmatrix} 1 & 0 & 1 \\ 1 & 1 & 1 \\ 1 & 0 & 0 \\ 0 & 1 & 0 \\ 0 & 0 & 1 \end{bmatrix}$$

H has no row of all 0's and no two rows that are alike, but rows 1, 2, and 4 add to $(0, 0, 0)$. Again, we see that the minimum distance of this code is 3.
□

6.15 Practice   For each code word X of Example 6.11, verify that $X \cdot H = 0_3$ where H is given above.   *(Answer, page A-17)*

In computing the product $X \cdot H$, we multiply elements of X by corresponding elements of the columns of H, and then sum. For each column of H, the pattern of 1's in the column determines which components of X contribute to the sum. If the sum is to be 0 (as is true when $X \cdot H = 0_r$), then those selected components of X must sum to 0, and, therefore, must consist of an even number of 1's. Thus, for a code word X, each column of H performs an even parity check on selected components of X. H is called a **parity-check matrix**.

Canonical Parity-Check Matrix

Given an $n \times r$ parity-check matrix H, we have not yet said anything about the *size* of the code H generates (how many code words we can have). We can best address this question by assuming that our parity-check matrix has the **canonical form**

$$H = \left[\frac{B}{I_r} \right]$$

where I_r is the $r \times r$ identity matrix and B is an arbitrary $(n - r) \times r$ binary matrix. The matrix of Example 6.14 is in canonical form. The I_r portion of H has the effect that each column of H selects a distinct component from among the last r components of X when the multiplication $X \cdot H$ is performed. Each of the last r components of X therefore controls the even parity check for one of the r multiplications that are done. For X to be a code word, the first $n - r$ components can be arbitrary, but the final r components will then be determined. The maximum number of code words is the maximum number of ways to select binary $(n - r)$-tuples, or 2^{n-r}. Let $m = n - r$. We can code all members of $\mathbb{Z}_2^m$ as code words in $\mathbb{Z}_2^n$ by leaving the first m components alone and then choosing the last r components so that the even parity check works for each column of H. This procedure helps us encode members of $\mathbb{Z}_2^m$ in $\mathbb{Z}_2^n$; such a code is called an (n, m) code. The first m components of a code word are the **information digits**, and the last r components are the **check digits**.

6.16 Example   The matrix H of Example 6.14 is

$$H = \begin{bmatrix} 1 & 0 & 1 \\ 1 & 1 & 1 \\ 1 & 0 & 0 \\ 0 & 1 & 0 \\ 0 & 0 & 1 \end{bmatrix}$$

This matrix is in canonical form, where $n = 5$, $r = 3$, and $m = n - r = 2$. H can thus generate $2^2 = 4$ code words. The 4 members of $\mathbb{Z}_2^2$ are 00, 01, 10, and 11. Each can be coded as a member of $\mathbb{Z}_2^5$ by keeping the first two digits and adding the appropriate check digits. To code 10, for instance, we have

$$(10\ C_1\ C_2\ C_3) \begin{bmatrix} 1 & 0 & 1 \\ 1 & 1 & 1 \\ 1 & 0 & 0 \\ 0 & 1 & 0 \\ 0 & 0 & 1 \end{bmatrix} = (000)$$

and, thus, $C_1 = 1$, $C_2 = 0$, $C_3 = 1$. We encode 10 as 10101.   □

6.17 Practice   Use the encoding procedure and the matrix H above to code 00, 01, and 11 in $\mathbb{Z}_2^5$. Compare the results with Example 6.11.   (*Answer, page A-17*)

HAMMING CODES

For a given $n \times r$ canonical parity-check matrix H, we now know how to encode all of $\mathbb{Z}_2^m = \mathbb{Z}_2^{n-r}$ as a subgroup of $\mathbb{Z}_2^n$, and how to determine from H the minimum distance of the resulting code. Now let's turn the problem around. Suppose we want to encode $\mathbb{Z}_2^m$ as, say, a single-error correcting code. How big will the code words have to be (what is n), or, equivalently, how many check digits must be added (what is r)? Once we know the dimensions of a parity-check matrix H, how can we find a canonical H that generates the code?

If the code is to be single-error correcting, no two rows of H can be alike. Since each row has r elements, there can be no more than 2^r rows; however, we cannot have the zero row, so the number of possible rows is $\leq 2^r - 1$. Thus, $n \leq 2^r - 1$ and $m = n - r \leq 2^r - r - 1$. This inequality says that to code the set of all binary m-tuples in a single-error correcting code requires r check digits where r is such that $m \leq 2^r - r - 1$. Then H can be taken to be any matrix of the form $\left[\dfrac{B}{I_r} \right]$ where B is an $m \times r$ binary matrix with at least two 1's in each row (and no rows alike). If r and m are such that $m = 2^r - r - 1$, then the code is said to be a **perfect code**. In this case, the rows of the matrix B consist of all the $2^r - r - 1$ r-tuples with at least two 1's, and the code generated by H is called a **Hamming code**. We have shown the following result.

6.18 Theorem   For any number $n = 2^r - 1$ there exists a perfect, single-error correcting Hamming code of length $2^r - 1$ and size $2^{2^r - r - 1}$.   □

6.19 Example   The code of Example 6.16 is a (5, 2) code; $n = 5$, $m = 2$, $r = 3$. The code is single-error correcting and satisfies the relation $m \leq 2^r - r - 1$, but it is not a perfect code. The perfect code which has 3 check digits is a (7, 4) code. A matrix generating such a code is

$$H = \left[\begin{array}{ccc} 1 & 1 & 0 \\ 1 & 0 & 1 \\ 0 & 1 & 1 \\ 1 & 1 & 1 \\ \hline 1 & 0 & 0 \\ 0 & 1 & 0 \\ 0 & 0 & 1 \end{array} \right]$$

□

6.20 Practice   The H given in Example 6.19 can encode $\mathbb{Z}_2^4$ in $\mathbb{Z}_2^7$. Give the list of 16 code words generated by the encoding procedure.   (*Answer, page A-18*)

In this section we have primarily been concerned with defining "good" codes and developing an encoding procedure. The next section concerns itself with the corresponding decoding procedure.

✔ CHECKLIST

Definitions

maximum likelihood
  decoding (*p. 221*)
parity bit digit (*p. 222*)
even parity check (*p. 222*)
odd parity check (*p. 222*)
Hamming distance (*p. 223*)
minimum distance of a code
  (*p. 223*)
group code (*p. 225*)

weight of a code word (*p. 225*)
parity-check matrix (*p. 227*)
canonical parity-check
  matrix (*p. 227*)
information digits (*p. 228*)
check digits (*p. 228*)
perfect code (*p. 229*)
Hamming code (*p. 229*)

Techniques

Given a canonical $n \times r$ parity-check matrix, find m such that $\mathbb{Z}_2^m$ can be encoded and produce the code words for $\mathbb{Z}_2^m$.

Given m such that $\mathbb{Z}_2^m$ is to be encoded as a single-error correcting code, find a canonical parity-check matrix to generate the code.

Main Ideas

The error-detecting and error-correcting capabilities of a binary code are functions of the minimum distance of the code.

In a group code, the minimum distance is the minimum weight of the nonzero code words.

A parity-check matrix H can be used to generate a group code, in which case the minimum distance of the code can be determined from H.

A canonical $n \times r$ parity-check matrix provides an easy procedure to encode $\mathbb{Z}_2^m$ in $\mathbb{Z}_2^n$, where $m = n - r$.

For any number $n = 2^r - 1$, there exists a perfect single-error correcting Hamming code of length n and size $2^{2^r - r - 1}$.

EXERCISES, SECTION 6.1

★ 1. A simple code for transmitting secret messages is a shift cypher or a Caesar cypher, so-called because Julius Caesar used it to transmit military information. An integer s is chosen as the amount of shift, $1 \leq s \leq 26$, and each letter in the message is encoded as the letter s units farther along in the alphabet, with the last s letters of the alphabet shifted in a cycle to the first s letters. For example, if $s = 4$, then A is encoded as E, B is encoded as F, W is encoded as A, and so on, as in

Figure 6.4. Decoding a message requires knowledge of s. Thus, for $s = 4$, the message KIVSRMQS is easily decoded as GERONIMO.

A	B	C	D	E	F	G	H	I	J	K	L	M	N	O	P	Q	R	S	T	U	V	W	X	Y	Z
↓	↓	↓	↓	↓	↓	↓	↓	↓	↓	↓	↓	↓	↓	↓	↓	↓	↓	↓	↓	↓	↓	↓	↓	↓	↓
E	F	G	H	I	J	K	L	M	N	O	P	Q	R	S	T	U	V	W	X	Y	Z	A	B	C	D

Figure 6.4

(a) The Centurian who was supposed to deliver "s" was killed enroute, but you have received the message

$\qquad$ MXXSMGXUEPUHUPQP

in a Caesar cypher. Find the value of s and decode the message.

(b) Deciphering codes where letters of the alphabet are coded by a one-to-one function to letters of the alphabet can sometimes be made easier by making a frequency count of the letters in the received text. In general, E is the most frequently used letter of the English alphabet, T is the next most frequently used, and so on. Thus, a long passage of code in which K is the most frequent letter suggests trying to decode K as E. Frequency counts of combinations of letters can be used also. If the following paragraph were coded with such a function, what would make it difficult to decode?

$\qquad$ Unusual paragraphs of this sort go far toward driving a spy crazy. If a paragraph such as this should turn up in your mailbox, how would you go about solving it to find out what your contact was actually saying? Or is it only proof that your mailbox is too dusty?

(c) Decode the received word HAL under a Caesar cypher where $s = 25$.

★ 2. Let us assume that we have a one-to-one function f mapping letters of the alphabet onto $\mathbb{Z}_{26}$, the integers from 0 to 25. We will use the simplest such mapping where $f(A) = 1, f(B) = 2, \ldots, f(Y) = 25, f(Z) = 0$. (The code we are about to describe can of course be made harder to decipher by a trickier function.) We extend f to strings of letters "homomorphically" by defining $f(x_1 \cdots x_n) = f(x_1), \cdots, f(x_n)$. The message to be sent is broken down into blocks of k units in length; f is applied to each block X to generate a member X' of $\mathbb{Z}_{26}^k$. The key to the encoding is a $k \times k$ matrix M with entries in $\mathbb{Z}_{26}$ having a multiplicative inverse M^{-1}. The block X is coded as $f^{-1}(X'M)$ where all arithmetic is done in $\mathbb{Z}_{26}$. The decoding process consists of applying the function f, multiplying by M^{-1}, and applying f^{-1}. Thus,

$$f^{-1}([f(f^{-1}(X'M))]M^{-1}) = f^{-1}([X'M]M^{-1}) = f^{-1}(X') = X$$

For example, if $M = \begin{bmatrix} 3 & 5 \\ 2 & 3 \end{bmatrix}$, then $M^{-1} = \begin{bmatrix} 23 & 5 \\ 2 & 23 \end{bmatrix}$. If X is the

block IF, then $X' = (9, 6)$, and the code word is $f^{-1}\left((9, 6) \begin{bmatrix} 3 & 5 \\ 2 & 3 \end{bmatrix} \right) =$

$f^{-1}(13, 11) = $ MK. The decoding process (which requires knowledge of both M and f) translates MK back to $(13, 11)$ and then applies M^{-1}:

$$(13, 11) \begin{bmatrix} 23 & 5 \\ 2 & 23 \end{bmatrix} = (9, 6)$$

Finally, $f^{-1}(9, 6) = $ IF.

(a) Given a matrix encoding where the code matrix M is $\begin{bmatrix} 3 & 2 \\ 7 & 5 \end{bmatrix}$ (M has previously been delivered to you by a trusted agent), find M^{-1} and decode the message MXOSHI.

(b) Errors introduced in transmission of code words with matrix encoding are not independent. If the code word MXOSHI from part (a) is corrupted to MXOTHI, what word does the decoding process produce?

3. Standard cryptologic codes have the disadvantage that for efficient decoding a key to the decoding process must be separately passed to the receiver (the shift factor s of Exercise 1 and the matrix M of Exercise 2, for example). A coding process has been developed (see "A Method for Obtaining Digital Signatures and Public-key Cryptosystems," R. L. Rivest, A. Shamir, and L. Adleman, *Communications of the ACM* 21, no. 2 (February 1978): 120–126) that allows any participant P in a communications network to make public to the network his encoding scheme. Doing so, however, does not disclose the key to the decoding process. Any member S of the network wishing to send a secret communication to P may do so by encoding it with P's known coding scheme, and only P will be able to decode it. Secure communication of a decoding key is not necessary. (The method has the further advantage that S can encode the message to P in such a way as to include S's "signature," proof that this particular message indeed came from S.)

   The method works as follows. Two large prime numbers p and q are chosen at random and $p \cdot q = n$ is computed. Letting $m = (p - 1)(q - 1)$, a large random number y is chosen such that the greatest common divisor of y and m is 1. This step guarantees the existence of an integer x, $0 < x < m$, such that $x \cdot_m y = 1$. Efficient computer algorithms exist for producing p, q, y, and x. The message to be coded is translated into a string of integers in $\mathbb{Z}_{26}$, and the resulting string treated as a single number T, $0 \le T \le n - 1$, or as a sequence of such numbers. The coding process consists of computing $T^x \cdot_n 1$.

   The coding scheme is made public by announcing n and x. To decode $T^x \cdot_n 1$, compute $(T^x \cdot_n 1)^y \cdot_n 1$. It can be shown that this computation produces T as a result. The key to the decoding, y, is not made public. Both the coding and decoding processes can be efficiently done by computer. To crack the code, however, requires finding the prime factors p and q of n, which, at present, cannot be done efficiently. If n is, say, a 200-digit number, it is estimated that finding its prime factors using a high-speed computer and the best current algorithms could require 3 billion years! Thus, such a code is currently unbreakable in practice, if not in theory.

As an example, we'll use this coding and decoding scheme with trivially small numbers. Let $p = 3$, $q = 5$. Then $n = 15$, $m = 8$. We choose $y = 11$.

(a) Compute x.

(b) Represent the letter C as a 3 and find the code for 3.

(c) Decode your answer to part (b) to retrieve the 3.

4. A simple code for transmission of numeric data is to add a single check digit d obtained as follows: if the string of digits $x_1 x_2 x_3 \cdots x_n$ is to be sent, $0 \le x_i \le 9$, then the check digit d consists of the units digit of the sum

$$x_1 + 2x_2 + x_3 + 2x_4 + \cdots + x_n^*, \quad \text{where} \quad x_n^* = \begin{cases} x_n & \text{if } n \text{ is odd} \\ 2x_n & \text{if } n \text{ is even} \end{cases}$$

The code word is then $x_1 x_2 \cdots x_n d$. This code can detect many digit transposition errors (a common error introduced by human copying of data).

★(a) Determine whether an error has occurred if the received word is 21347.

★(b) What will the code word be for the digits 15247? What will the received word be if the digits 4 and 7 are transposed? What will the received word be if the digits 5 and 2 are transposed, as well as the digits 4 and 7? Which of these errors can be detected?

(c) Write a computer program using this scheme to detect errors in received strings of up to 20 digits.

5. Let's define a function $A(n, d)$ giving the size (number of code words) of the largest code of length n (code words are binary n-tuples) with minimum distance d. (Here, we are considering arbitrary codes, not necessarily group codes.) There is no known equation to express this function, but some of its values can be computed. Find

(a) $A(n, n)$

(b) $A(n, 1)$

(c) $A(n, 2)$, $n \ge 2$

6. Let H be an $n \times r$ binary matrix mapping $\mathbb{Z}_2^n \to \mathbb{Z}_2^r$ by the operation $X \cdot H$ for $X \in \mathbb{Z}_2^n$. Show that this operation is a homomorphism from the group $[\mathbb{Z}_2^n, +_2]$ to the group $[\mathbb{Z}_2^r, +_2]$ by showing that for $X, Y \in \mathbb{Z}_2^n$, $(X +_2 Y)H = X \cdot H +_2 Y \cdot H$.

7. Consider the canonical parity-check matrix

$$H = \begin{bmatrix} 1 & 1 & 1 \\ 0 & 1 & 1 \\ 1 & 0 & 1 \\ 1 & 0 & 0 \\ 0 & 1 & 0 \\ 0 & 0 & 1 \end{bmatrix}$$

(a) Show that the code generated by H is single-error correcting.
(b) Write the set of binary m-tuples H encodes, and write the code word for each.

8. Consider the canonical parity-check matrix

$$H = \begin{bmatrix} 1 & 1 & 0 & 1 \\ 0 & 1 & 1 & 1 \\ 0 & 1 & 0 & 1 \\ 1 & 0 & 0 & 1 \\ 1 & 1 & 0 & 0 \\ 1 & 0 & 0 & 0 \\ 0 & 1 & 0 & 0 \\ 0 & 0 & 1 & 0 \\ 0 & 0 & 0 & 1 \end{bmatrix}$$

(a) Show that the code generated by H is single-error correcting.
(b) Write the set of binary m-tuples H encodes, and write the code word for each.
(c) Is this a perfect code?

9. Write a computer program that, given a canonical $n \times r$ parity-check matrix H for a single-error correcting code with $r \leq 4$ and $n \leq 2^r - 1$, will write the set of binary m-tuples H encodes and the code word for each.

★ 10. Which of the following are perfect codes? Which are single-error correcting?
(a) (5, 3)
(b) (12, 7)
(c) (15, 11)

★ 11. For each computer word length given, what is the minimum number of check digits required to code the set of computer words as a single-error correcting code?
(a) 32 (IBM)
(b) 36 (DEC 10)
(c) 60 (CDC 6600)

12. Give an example of a canonical parity-check matrix that will generate a single-error correcting code for the set of words in $\mathbb{Z}_2^6$.

13. Give a canonical parity-check matrix for a single-error correcting $(15, 11)$ code. Is this a perfect code?

14. Let H be an $n \times r$ matrix for a perfect, single-error correcting Hamming code, $r \geq 3$. Let H' be the $n \times (r + 1)$ matrix obtained by attaching a column of $1's$ to H. Show that H' generates a code of minimum distance 4.

Section 6.2   **DECODING**

GENERAL DECODING

In this section we put ourselves at the receiving end of the communications channel and attempt to decode received words. As before, we assume that our code words are binary n-tuples. Regardless of the encoding procedure used to create the code words, a direct decoding process involves comparing the received n-tuple X with all the code words. Then we decode X as the closest code word in terms of Hamming distance. There may not be a unique closest code word, and even if there is, our process may cause us to decode incorrectly, depending on the number of errors incurred in transmission and the minimum distance of the code (recall Theorem 6.6). If no errors have occurred, then X is itself a code word and will be correctly decoded to itself. The difficulty with this decoding process is that if the code is of size 2^m, we have to do 2^m comparisons in order to find the code word closest to the received word X. We must have an array of all 2^m code words against which to compare. For $m = 32$, for example (see Exercise 11 of the previous section), 2^{32} is an extremely large number and this approach is not practical.

DECODING A GROUP CODE

Suppose, however, that our code is a group code generated by an $n \times r$ canonical parity-check matrix H. Then we can devise a decoding procedure ultimately requiring storage of only 2^r elements. In a single-error correcting code, if $m = 32$, then r can be 6, and $2^6 = 64$ is a reasonable number of pieces of information to keep available. Let's see how the decoding procedure works.

We recall that the set $\mathscr{C}$ of code words is the kernel of the homomorphism H induces from $[\mathbb{Z}_2^n, +_2]$ to $[\mathbb{Z}_2^r, +_2]$. $\mathscr{C}$ is thus a normal subgroup of $[\mathbb{Z}_2^n, +_2]$, and $\mathbb{Z}_2^n$ can be partitioned into cosets of the form $X + \mathscr{C}$,

$X \in \mathbb{Z}_2^n$. $\mathscr{C}$ has 2^m elements, and the number of cosets (by Lagrange's Theorem) is $2^n/2^m = 2^{n-m} = 2^r$. The cosets provide the key to the decoding. Suppose a word $X \in \mathbb{Z}_2^n$ is received. X belongs to the coset $X + \mathscr{C}$. Each element E_i of this coset is an n-tuple of the form $X + C_i$ where $C_i \in \mathscr{C}$. Because E_i and C_i are in $\mathbb{Z}_2^n$, they have the property that $-E_i = E_i$ and $-C_i = C_i$. The equation

$$E_i = X + C_i$$

can be written as

$$X = C_i + E_i \qquad (1)$$

or

$$X + E_i = C_i \qquad (2)$$

From (1) we see that 1's in E_i occur in exactly those components where X and C_i differ. Thus, the weight of E_i equals the distance between X and C_i. The code word C_i closest to X is the one for which the corresponding E_i has minimum weight. To decode X, we look for the element in the coset of X having minimum weight and, according to (2), add that element to X. The result is the code word to which we decode X. The coset element having minimum weight is called the **coset leader**, and it may not be unique. If two n-tuples of minimum weight occur in the same coset, one is arbitrarily chosen as the coset leader, which simply means that no word in this particular coset is sufficiently close to a code word to allow accurate decoding. Remember that in a d-error correcting code, the decoding procedure can be applied to any received n-tuple, but it will only "correctly" correct those with $\leq d$ errors.

To summarize this decoding procedure, when X is received, we must find the coset to which X belongs, and add that coset's leader to X.

6.21 Example   Consider the code of Example 6.11. Here $n = 5$ and $\mathscr{C} = \{00000, 01111, 10101, 11010\}$. Suppose the 5-tuple $X = 11011$ is received. By inspection, we see that the closest code word is 11010, and we would decode X as 11010. Let's use the decoding procedure. X belongs to the coset $X + \mathscr{C}$. The members of this coset are:

$$11011 + 00000 = 11011$$
$$11011 + 01111 = 10100$$
$$11011 + 10101 = 01110$$
$$11011 + 11010 = 00001$$

The coset leader is 00001. Adding this to X, we get

$$11011 + 00001 = 11010$$

and we decode X to 11010.   □

SYNDROME

Given a received X, if we generate all the members of the coset $X + \mathscr{C}$ by adding the code words to X, as in Example 6.21, we must still have the 2^m code words on hand, so we have not gained anything. Suppose, however, that we have somehow arrived at a list of the 2^r coset leaders. We can then identify the coset leader corresponding to X by recalling some work from Chapter 5. There we learned that if f is a homomorphism from a group S to a group T with kernel K, the cosets $s + K$ are equivalence classes of S where elements are equivalent if and only if they map to the same place under f. Translating this result to our group code, the parity-check matrix H provides a homomorphism from $\mathbb{Z}_2^n$ to $\mathbb{Z}_2^r$ with kernel $\mathscr{C}$, and elements of $\mathbb{Z}_2^n$ are in the same coset if and only if they map to the same place under this homomorphism. Thus, X and Y are in the same coset of $\mathscr{C}$ in $\mathbb{Z}_2^n$ if and only if $X \cdot H = Y \cdot H$.

6.22 Definition   In a binary group code generated by the $n \times r$ parity-check matrix H, for any $X \in \mathbb{Z}_2^n$, the r-tuple $X \cdot H$ is the **syndrome** of X. $\qquad\square$

6.23 Theorem   Let H be an $n \times r$ parity-check matrix generating a group code $\mathscr{C}$. Then, for $X, Y \in \mathbb{Z}_2^n$, X and Y are in the same coset of $\mathscr{C}$ in $\mathbb{Z}_2^n$ if and only if X and Y have the same syndrome. $\qquad\square$

Theorem 6.23 can also be proved directly (see Exercise 1, this section).

6.24 Practice   In Example 6.21 we found the four members of one coset. The parity-check matrix that generated the code for this example is

$$H = \begin{bmatrix} 1 & 0 & 1 \\ 1 & 1 & 1 \\ 1 & 0 & 0 \\ 0 & 1 & 0 \\ 0 & 0 & 1 \end{bmatrix}$$

Compute the syndrome for each member of the coset.   *(Answer, page A-18)*

6.25 Example   Again considering Example 6.21, suppose we know that 00000, 00001, and 00010 are coset leaders, and we receive the word $X = 01101$. To decode X, we compute its syndrome

$$XH = (01101) = \begin{bmatrix} 1 & 0 & 1 \\ 1 & 1 & 1 \\ 1 & 0 & 0 \\ 0 & 1 & 0 \\ 0 & 0 & 1 \end{bmatrix} = (010)$$

and the syndrome of each coset leader

$$(00000)H = (000)$$

$$(00001)H = (001)$$

$$(00010)H = (010)$$

We conclude that X and the coset leader (00010) belong to the same coset, and we decode X as $01101 + 00010 = 01111$. $\square$

FINDING COSET LEADERS

Now, to decode we only need to have available the 2^r coset leaders (plus the $n \times r$ encoding matrix). But how can the coset leaders be determined? There's no good answer to this general problem. In theory, of course, we can simply examine each member of $\mathbb{Z}_2^n$ in turn, compute its syndrome, and, for each distinct syndrome, save the new member of $\mathbb{Z}_2^n$ if its weight is less than the previous member of $\mathbb{Z}_2^n$ that had that syndrome. However, this process involves searching through 2^n n-tuples, which can be an unacceptably large set. In certain cases, we can narrow down to a reasonable size the set of n-tuples qualifying as possible coset leaders.

For example, suppose our code is a perfect, single-error correcting code. (A more general case is considered in Exercise 2 at the end of this section.) Then $n = 2^r - 1$, and the rows of H are binary representations of the integers $1, 2, \ldots, 2^r - 1$. The code word 0_n is the coset leader corresponding to the syndrome 0_r. Any other syndrome is a binary r-tuple representing a digit d, $1 \le d \le 2^r - 1$. Let row i_d be the row of H representing d; then the n-tuple with 1 in component i_d and 0's elsewhere is the coset leader for this syndrome.

On the other hand, each n-tuple with a weight of 1 has a unique syndrome. Thus the set of coset leaders for the $2^r = n + 1$ cosets consists precisely of 0_n (the coset leader for the coset consisting of code words) plus the n n-tuples with a weight of 1. In a perfect, single-error correcting code, there are no arbitrary choices of coset leaders between two possibilities both of minimum weight. Every received word is within one unit of a unique code word.

Applied to Figure 6.3, this result means that if the "spheres" of the figure have radius 1, they should cover the whole space. And, in fact, there are 2^m spheres, each with $n + 1$ elements, and $2^m(n + 1) = 2^m \cdot 2^r = 2^{m+r} = 2^n$, which is the size of the whole space. The sphere partition of the set $\mathbb{Z}_2^n$ is "orthogonal" to the coset partition in that the intersection of any block from one partition with any block from the other contains just one element of $\mathbb{Z}_2^n$.

6.26 Example   The perfect $(7, 4)$ code of Example 6.19 has the coset leaders and corresponding syndromes of Figure 6.5. A received word of 1101101 is decoded by computing its syndrome

$$(1101101) \cdot H = (001)$$

Coset leaders	Syndromes
0000000	000
0000001	001
0000010	010
0000100	100
0001000	111
0010000	011
0100000	101
1000000	110

Figure 6.5

The received word is then decoded to

$$1101101 + 0000001 = 1101100 \qquad \square$$

6.27 Practice In the perfect (7, 4) code, how would the received word 1000100 be decoded?

(*Answer, page A-18*)

6.28 Example The (5, 2) code of Example 6.14 is not a perfect code (nor does it satisfy the requirements of Exercise 2). Since n is small, we can find its coset leaders by "brute force." Figure 6.6 shows the 8 cosets of $\mathbb{Z}_2^5$, together with the coset leader and syndrome for each. Note that in two cosets, there was an arbitrary choice of coset leader between two candidates.

Coset leaders				Syndromes
1. 00000	01111	10101	11010	000
2. 00001	01110	10100	11011	001
3. 00010	01101	11000	10111	010
4. 00011	01100	11001	10110	011
5. 00100	11110	01011	10001	100
6. 10000	00101	01010	11111	101
7. 00110	01001	10011	11100	110
8. 01000	00111	10010	11101	111

Figure 6.6

A received word of 10101 will be decoded as 10101 (this is a code word). A received word of 11000 will be decoded as 11010. A received word of 10011 will be decoded as 10101, or because its coset leader has weight 2, it can be flagged to indicate that at least two errors have occurred and that decoding cannot be done with certainty. $\qquad \square$

6.29 Example   The matrix of Example 6.14 allows encoding of $\mathbb{Z}_2^2$ in $\mathbb{Z}_2^5$. Using Exercise 2 at the end of this section, we choose a different canonical parity-check matrix.

$$H^* = \begin{bmatrix} 1 & 0 & 1 \\ 0 & 1 & 1 \\ 1 & 0 & 0 \\ 0 & 1 & 0 \\ 0 & 0 & 1 \end{bmatrix}$$

The set of code words is now {00000, 01011, 10101, 11110}. The coset leaders (except for the last one) and syndromes are given in Figure 6.7. We do not

Coset leaders	Syndromes
00000	000
00001	001
00010	010
00100	011
01000	100
10000	101
00110	110
———	111

Figure 6.7

have to search through all of $\mathbb{Z}_2^5$ to find the coset leaders. For the coset with syndrome 110, for example, it is clear from the form of H^* that we can add rows 3 and 4, so we put 1's in these components and 0's elsewhere. It is also clear from the form of H^* that no 5-tuple with weight 1 can produce this syndrome, so 00110 is the coset leader (11000 will also work). (Because our original H for this problem was quite small, this sort of procedure could also have been used in Example 6.28. The advantage to having H organized as suggested in Exercise 2 is that coset leaders can be found more systematically, which is what we are looking for in the case of large n.)   □

6.30 Practice   What is the coset leader for the syndrome 111?   *(Answer, page A-18)*

Even easier decoding (at the expense of a slight complication in encoding) occurs if we use a noncanonical parity-check matrix as discussed in Exercise 3, this section.

✔ **CHECKLIST**

Definitions

coset leader (*p. 236*)
syndrome (*p. 237*)

Techniques

Given an $n \times r$ parity-check matrix for a group code, classify the elements of $\mathbb{Z}_2^n$ into cosets with respect to the set of code words, identify coset leaders, and decode received words.

Main Ideas

For a group code $\mathscr{C}$ generated by an $n \times r$ parity-check matrix H, each word X in $\mathbb{Z}_2^n$ is decoded by using its syndrome to locate the coset of $\mathscr{C}$ in $\mathbb{Z}_2^n$ to which it belongs, and adding the coset leader to X. The form of H may make identification of the coset leaders easier.

EXERCISES, SECTION 6.2

1. Let H be an $n \times r$ parity-check matrix generating a group code $\mathscr{C}$. Show that X and Y are in the same coset of $\mathscr{C}$ in $\mathbb{Z}_2^n$ if and only if $X \cdot H = Y \cdot H$.

2. Let H be an $n \times r$ parity-check matrix for a group code where $2^{r-1} \leq n < 2^r$ and the rows of H are binary representations of the integers $1, \ldots, n$. Prove that no coset leader has a weight greater than 2.

★ 3. Let H be an $n \times r$ parity-check matrix where the ith row is the binary representation of the integer i, $1 \leq i \leq n$. Then H generates a single-error correcting code. (Because H is not in canonical form, the encoding process is not quite the same.) Prove that any received word X in which a single error has occurred can be correctly decoded by knowing only the syndrome of X.

4. Consider the (6, 3) code of Exercise 7, Section 6.1.
   (a) Partition $\mathbb{Z}_2^6$ into cosets, indicating coset leaders and syndromes as in Figure 6.6.
   (b) How would each of the following words be decoded? Flag any for which you know that more than one error has occurred.

   010011
   110110
   011010
   110010

5. Consider the (6, 3) code of Exercise 7, Section 6.1. The H given there encodes all of $\mathbb{Z}_2^3$ in $\mathbb{Z}_2^6$.
   (a) Using Exercise 3 above, write a new matrix H^* encoding $\mathbb{Z}_2^3$ and allowing for easy decoding.
   (b) Write the new code word for each member of $\mathbb{Z}_2^3$.
   (c) If the word $X = 011101$ is received, decode it without generating a list of coset leaders.

6. Consider the (9, 5) code of Exercise 8, Section 6.1. Try to make up a table of coset leaders and syndromes such as Figure 6.7 without generating all the members of each coset.

★ 7. (a) Give a canonical parity-check matrix H for a perfect single-error correcting (15, 11) code (see Exercise 12 of Section 6.1).

(b) Decode the received word

011000010111001

8. (a) Write a canonical parity-check matrix for a single-error correcting (8, 4) code.

(b) Write the table of coset leaders and syndromes.

(c) Decode the following received words. Flag any for which you know that more than one error has occurred.

01110011
10010000
00111011

9. Given an $n \times r$ canonical parity-check matrix for a single-error correcting (n, m) code with $r \le 3$ and $n \le 2^r - 1$, write a computer program to decode received binary n-tuples.

FOR FURTHER READING

General references on coding theory (given roughly in order of increasing depth of coverage) are [1], [2], and [3]. Error detecting and correcting more from an engineering than from a mathematical viewpoint are discussed in [4]. Reference [5], although historical rather than mathematical, should be read by anyone with an interest in cryptology.

1. W. Wesley Peterson and E. J. Weldon, *Error-Correcting Codes*, 2nd ed. Cambridge, Mass.: MIT Press, 1972.

2. Elwyn R. Berlekamp, *Algebraic Coding Theory*. New York: McGraw-Hill, 1968.

3. F. J. MacWilliams and N. J. A. Sloane, *The Theory of Error-Correcting Codes, Part I* and *Part II*. New York: North-Holland, 1977.

4. John Wakerly, *Error Detecting Codes, Self-Checking Circuits and Applications*. New York: North-Holland, 1978.

5. David Kahn, *The Codebreakers*. New York: Macmillan, 1967.

Chapter 7

FINITE-STATE MACHINES

The algebraic structures of Chapter 5 were basically attempts to model, or simulate, arithmetic. In this chapter we consider an attempt to simulate much more general computation. In fact, a finite-state machine, the structure introduced in this chapter, simulates computational devices such as modern digital computers. We will also learn how one finite-state machine can simulate another. Then we will see how adequate this structure is as a model of computation in the most general sense by characterizing its capabilities as a "recognizer."

Section 7.1   MACHINES—SIMULATION I

Sy Burr stops at the vending machine for a cup of coffee before heading for his computer science class. Figure 7.1 shows what appears on the front of the vending machine. Sy inserts his money (this turns the machine on), presses the button for coffee, and then the button for sugar. Little does he realize that he is operating a "sequential network," or a "finite-state machine." Sy's insertion of the right amount of money and pressing buttons constitute input into the machine. This input is contributed at discrete (distinct or separated) moments of time, which we will denote by t_0, t_1, t_2, and so on. The machine has certain responses to these inputs. The response to a given input may occur at any time after that input and before the next input, but to synchronize the operation, we will only look at the machine at the fixed times, or clock pulses, t_0, t_1, t_2, and so on. Thus, the responses to an input at time t_i will appear at time t_{i+1}.

The machine's responses to inputs are of two types. One type of response is the visible output of the machine, which in this case is a drink of some type, or nothing. A less visible response is the state of the machine. Thus, after Sy inserts his money, the machine goes from an "off" state to a "waiting" state. After Sy presses the button for coffee, the machine goes from a "waiting" state to a "coffee" state. The output that occurs with all of these states is "nothing." After Sy presses the sugar button, the machine goes to a "coffee–sugar" state and there is output of coffee with sugar. Finally, the machine turns off to be ready to function for another customer. Figure 7.2 describes the action of the machine for Sy's input.

```
┌─────────────────────────────────────────────┐
│  Coffee            ○    Coin ⊟               │
│  Tea               ○                          │
│  Black             ○                          │
│  Cream             ○                          │
│  Sugar             ○                          │
│  Cream and Sugar   ○                          │
└─────────────────────────────────────────────┘
```

Figure 7.1

Time	t_0	t_1	t_2	t_3	t_4
Input	coin	coffee button	sugar button	—	—
State	off	waiting	coffee	coffee–sugar	off
Output	nothing	nothing	nothing	coffee with sugar	nothing

Figure 7.2

As we noted, the machine's reactions to an input occur at the next clock pulse. The output of coffee with sugar is a direct result of the machine being in the "coffee–sugar" state. However, reaching the "coffee–sugar" state at t_3 is a function of both the input at t_2 *and* the state at t_2. In turn, the state at t_2 is a function of both input and state at t_1. Essentially, the "coffee" state allows the machine to remember the "coffee" input and lets that input have an effect on the machine's behavior for longer than just one clock pulse.

Let's try to abstract some of the important features of the vending machine operation.

1. Operations of the machine are **synchronized**, or so we may assume, by discrete clock pulses.
2. The machine proceeds in a **deterministic** fashion; that is, its actions

in response to a given sequence of inputs are completely predictable. Randomness, probability, magic play no part here.

3. The machine responds to **inputs**.

4. There is a **finite number of states** the machine can attain. At any given moment, the machine is in exactly one of these states. Which state it will be in next is a function both of the present state and of the present input. The present state, however, depends upon the previous state and input, and so forth back to the initial operation. Thus, the state of the machine at any moment serves as a form of memory of past inputs.

5. The machine is capable of **output**. The nature of the output is a function of the present state of the machine, meaning that it also depends upon past inputs.

DEFINITION

We have noted the essential aspects of the vending machine and looked at them in a more abstract and out-of-context way. This process is, of course, the building of a mathematical model describing the vending machine. (Recall the discussion of Simulation I in Section 3.1.)

Our model is called a finite-state machine (the formal definition appears in 7.1). It is a useful model only if it describes a wide variety of situations, and, indeed, it does. Even the modern digital computer is a finite-state machine. Its operations are synchronized by very rapid clock pulses; it operates in a deterministic fashion and is capable of responding to inputs. A computer is composed of a large number of bistable ("on–off") elements. If there are n such elements, there are altogether 2^n on–off configurations in which the computer can be. These configurations are the states of the computer, and this number is finite although very large. The present state of the computer (the present memory configuration) reflects its history of past inputs. Finally, the output at any moment depends upon the present state of the machine.

7.1 Definition $M = [S, I, O, f_s, f_o]$ is a **finite-state machine** if S is a finite set of states, I is a finite set of input symbols (the **input alphabet**), O is a finite set of output symbols (the **output alphabet**), and f_s and f_o are functions where $f_s : S \times I \to S$ and $f_o : S \to O$. The machine is always initialized to begin in a fixed starting state s_o.   □

The function f_s is the next-state function. It maps a (state, input) pair to a state. Thus, the state at clock pulse t_{i+1}, state (t_{i+1}), is obtained by applying the next-state function to the state at time t_i and the input at time t_i:

$$\text{state } (t_{i+1}) = f_s(\text{state } (t_i), \text{input } (t_i))$$

The function f_o is the output function. When f_o is applied to a state at time t_i, we get the output at time t_i:

$$\text{output } (t_i) = f_o(\text{state } (t_i))$$

Notice that the effect of applying function f_o is available instantly, but the effect of applying function f_s is not available until the next clock pulse.

EXAMPLES

To describe a particular finite-state machine, we have to define the three sets and two functions involved.

7.2 Example A finite-state machine M is described as follows: $S = \{s_0, s_1, s_2\}$, $I = \{0, 1\}$, $O = \{0, 1\}$. Because the two functions f_s and f_o act on finite domains, they can be defined by a **state table** as in Figure 7.3. The machine M begins in

	Next state		
	Present input		
Present state	0	1	Output
s_0	s_1	s_0	0
s_1	s_2	s_1	1
s_2	s_2	s_0	1

Figure 7.3

state s_0, which has an output of 0. If the first input symbol is a 0, the next state of the machine is then s_1, which has an output of 1. If the next input symbol is a 1, the machine stays in state s_1 with an output of 1. Continuing this procedure, an input sequence consisting of the characters 01101 (read left to right) would produce the following effect:

Time	t_0	t_1	t_2	t_3	t_4	t_5
Input	0	1	1	0	1	—
State	s_0	s_1	s_1	s_1	s_2	s_0
Output	0	1	1	1	1	0

The initial 0 of the output string is spurious—it merely reflects the starting state, not the result of any input.

In a similar way, the input sequence 1010 produces an output of 00111.

Another way to define the functions f_s and f_o (in fact all of M) is by a directed graph called a **state graph**. Each state of M with its corresponding output is the label of a node of the graph. The next-state function is given by directed arcs of the graph, each arc showing the input symbol(s) that produces that particular state change. The state graph for M appears in Figure 7.4.   □

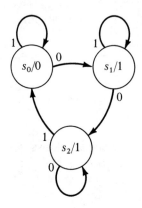

Figure 7.4

7.3 Practice   For the machine M of Example 7.2, what output sequence is produced by the input sequence 11001?   (*Answer, page A-18*)

7.4 Practice   A machine M is given by the state graph of Figure 7.5. Give the state table for M.   (*Answer, page A-18*)

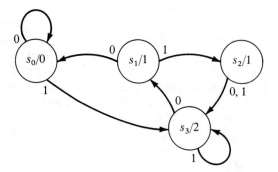

Figure 7.5

7.5 Practice   A machine M is described by the state table of Figure 7.6.

(a) Draw the state graph for M.

(b) What output corresponds to an input sequence of 2110?

                                            (*Answer, page A-18*)

Present state	Next state			Output
	Present input			
	0	1	2	
s_0	s_0	s_1	s_1	0
s_1	s_1	s_0	s_0	1

Figure 7.6

MORE USEFUL EXAMPLES

The machine of Example 7.2 is not particularly interesting. If finite-state machines model real-world computers, they should be able to do something.

Let's try to build a finite-state machine that will add two binary numbers. Machine input will consist of a sequence of pairs of binary digits, each of the form 00, 01, 10, or 11. These represent the digits of the two numbers to be added, read from right to left. Thus, the least significant digits of the numbers to be added are given first, and the output gives the least significant digits of the answer first. Recall the basic facts of binary addition:

$$
\begin{array}{cccc}
0 & 0 & 1 & 1 \\
0 & 1 & 0 & 1 \\
\hline
0 & 1 & 1 & 10
\end{array}
\quad \text{(here a carry to the next column takes place)}
$$

A moment's thought shows us that we might encounter four cases in any given column: (1) the output should be 0 with no carry; (2) the output should be 0 but there needs to be a carry to the next column; (3) the output should be 1 with no carry; and (4) the output should be 1 with a carry to the next column. We will let these cases be represented by the states s_0, s_1, s_2, s_3, respectively, of the machine; s_0, as always, is the starting state. We have already indicated the output for each state but we need to determine the next-state function. For example, suppose we are in state s_1 and the input is 11. The output for the present state is 0, but there is a carry, so in the next column we are adding $1 + 1 + 1$, which results in an output of 1 and a carry. The next state is s_3.

7.6 Practice

Referring to the binary adder under construction,

(a) What is the next state if the present state is s_2 and the input is 11?
(b) What is the next state if the present state is s_3 and the input is 10?

(Answer, page A-18)

After considering all possible cases, we have the complete state graph

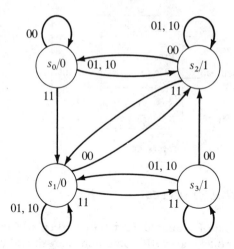

Figure 7.7

of Figure 7.7. The operation of this machine in adding the two numbers 01 and 11 can be traced as follows:

Time	t_0	t_1	t_2	t_3
Input	11	01	00	—
State	s_0	s_1	s_1	s_2
Output	0	0	0	1

The output is 100 when we ignore the initial 0, which does not reflect the action of any input. Converting this arithmetic to decimal form, we have computed $1 + 3 = 4$. Note the symmetry of this machine with respect to inputs of 10 and 01, reflecting that binary addition is commutative.

7.7 Practice   Compute the sum of 01110110 and 01010101 by using the binary adder machine of Figure 7.7.   (*Answer, page A-18*)

      We have already noted that a given input signal may affect the behavior of a finite-state machine for longer than just one clock pulse. The limited memory of past inputs that can be incorporated into the states of a machine allows us to use these machines as "recognizers." A machine can be built to recognize, say by producing an output of 1, when the input it has received matches a certain description. (This operation is essentially the one carried on by the lexical analyzer, or scanner, in a compiler. Input strings are broken down into recognizable substrings, and the substrings are then treated as units in the next stage of the compilation process. The scanner should be able to recognize, for example, a sequence of input symbols constituting a keyword or reserved word in the programming language.) We will discuss the capabilities of finite-state machines as recognizers more fully in Section 3 of this chapter. Here we will simply construct some examples.

7.8 Example   The machine described in Figure 7.8 is a parity-check machine. When the input received through time t_i contains an even number of 1's, then the output at time t_{i+1} is 1; otherwise, the output is 0.   □

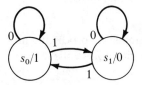

Figure 7.8

7.9 Example   Suppose we want to design a machine having an output of 1 exactly when the input string received to that point ends in 101. As a special case, an input sequence consisting of just 101 could be handled by progressing directly from state s_0 to states s_1, s_2, and s_3 with outputs of 0 except for s_3, which has an output of 1. This much of the design results in Figure 7.9a. From

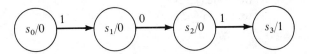

Figure 7.9 (a)

Figure 7.9a, we want to be in state s_2 whenever the input has been such that one more 1 should take us to s_3 (with an output of 1), meaning that we should be in s_2 whenever the two most recent input symbols were 10, regardless of what came before. In particular, a string of 1010 should put us in s_2; hence, the next-state function for s_3 with an input of 0 is s_2. Similarly, we can use s_1 to "remember" that the most recent input symbol received was 1, and that a 01 will take us to s_3. In particular, 1011 should put us in s_1; hence, the next-state function for s_3 with an input of 1 is s_1. The rest of the next-state function can be determined the same way; Figure 7.9b shows the complete state graph.

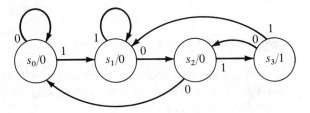

Figure 7.9(b)

Notice that the machine is in state s_2, for example, at the end of an input of 0110, at the end of an input of 011010, in fact, at the end of any input ending in 10; yet s_2 cannot distinguish between these inputs. *Each state of M represents a class of indistinguishable input histories, s_3 being the state representing all inputs ending in 101.*   □

7.10 Practice   Draw the state graph for a machine producing an output of 1 exactly when the input string received to that point ends in 00.   (*Answer, page A-19*)

✔ CHECKLIST

Definitions

finite-state machine (*p. 246*)
input alphabet (*p. 246*)
output alphabet (*p. 246*)
state table (*p. 247*)
state graph (*p. 247*)

Techniques

Compute the output string for a given finite-state machine and a given input string.

Draw a state graph from a state table and vice versa.

Construct a finite-state machine to act as a recognizer for a certain type of input.

Main Ideas

Finite-state machines with their synchronous, deterministic mode of operation and the limited memory capabilities available through the finite number of states.

EXERCISES, SECTION 7.1

★ 1. For each input sequence and machine given, compute the corresponding output sequence (starting state is always s_0).
(a) Input sequence 011011010.

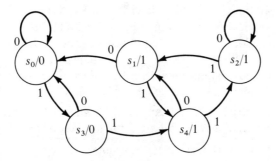

Figure 7.10

(b) Input sequence *abccaab*.

Present State	Next state			Output
	Present input			
	a	*b*	*c*	
s_0	s_2	s_0	s_3	*a*
s_1	s_0	s_2	s_3	*b*
s_2	s_2	s_0	s_1	*a*
s_3	s_1	s_2	s_0	*c*

Figure 7.11

(c) Input sequence 0100110.

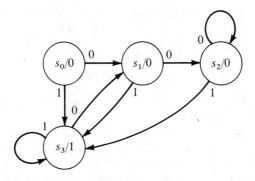

Figure 7.12

★ 2. (a) For the machine described in Exercise 1(a) find all input sequences yielding an output sequence of 0011110.
   (b) For the machine described in Exercise 1(b) find all input sequences yielding an output sequence of *abaaca*.
   (c) For the machine described in Exercise 1(c) what will be the output for an input sequence $a_1 a_2 a_3 a_4 a_5$ where $a_i \in \{0, 1\}$, $1 \leq i \leq 5$?

3. Construct a finite-state machine that will compute $x + 1$ where x is the input given in binary form, least significant digit first.

4. (a) Construct a "delay machine" having input and output alphabet $\{0, 1\}$ and that, for any input sequence $a_1 a_2 a_3 a_4 \ldots$, produces an output sequence of $00a_1 a_2 a_3 a_4 \ldots$.
   (b) Explain (intuitively) why it would not be possible to build a finite-state machine that for any input sequence $a_1 a_2 a_3 a_4 \ldots$ would produce the output sequence $0a_1 0a_2 0a_3 0a_4 0 \ldots$.

★ 5. You have an account at First National Usury Trust and a card to operate their bank machine. Once you have entered your card with your name, the bank machine will allow you to process a transaction only if you enter your correct code number, which is 417. Draw a state graph for a finite-state machine designed to recognize this code number. The output alphabet should have 3 symbols: "bingo" (correct code), "wait" (correct code so far), and "dead" (incorrect code). The input alphabet is $\{0, 1, 2, \ldots, 9\}$. To simplify notation, you may designate an arc by I-$\{3\}$, for example, meaning that the machine will take this path for an input symbol that is any digit except 3.

6. Construct finite-state machines that act as recognizers for the input described by producing an output of 1 exactly when the input received to that point matches the description. (The input and output alphabet in each case is $\{0, 1\}$.)
   (a) Set of all strings beginning with 000.
   (b) Set of all strings where the second input is 0 and the fourth input is 1.
   (c) Set of all strings consisting of any number (including none) of 01 pairs or consisting of two 1's followed by any number (including none) of 0's.
   (d) Set of all strings ending in 110.
   (e) Set of all strings containing 00.

7. A paragraph of English text is to be scanned and the number of words beginning with "con" counted. Design a finite-state machine that will output a 1 each time such a word is encountered. The output alphabet is $\{0, 1\}$. The input alphabet is the 26 letters of the English alphabet, a finite number of punctuation symbols (period, comma, etc.), and a special character β for blank. To simplify your description, you may use I-$\{m\}$, for example, to denote any input symbol not equal to m.

8. (a) In a certain computer language (BASIC, for example), any decimal number N must be presented in one of the following forms:

$$sd^* \quad \text{or} \quad sd^*.d^* \quad \text{or} \quad d^* \quad \text{or} \quad d^*.d^* \qquad (1)$$

where s denotes the sign (i.e., $s \in \{+, -\}$), d is a digit (i.e., $d \in \{0, 1, 2, \ldots, 9\}$), and d^* denotes a string of digits where the string may be of any length, including length zero (the empty string). Thus, the following would be examples of valid decimal numbers:

$$+2.74 \qquad -.58 \qquad 129 \qquad +$$

Design a finite-state machine to recognize valid decimal numbers by producing an output of 1. The input symbols are $+$, $-$, ., or any digit. To simplify notation, you may use "d" to represent an input of any digit.

(b) Modify the machine of part (a) to recognize any sequence of decimal numbers (as defined in part (a)) separated by commas. For example, such a machine would recognize

$$+2.74, -.58, 129, +$$

The input alphabet should be the same as for the machine of part (a) with the addition of the symbol c for comma.

(c) In Pascal, a decimal number must be presented in a form similar to that for BASIC except that any occurrence of a decimal point must have at least one digit before it and after it. Write an expression similar to expression (1) in part (a) to describe the valid form for a decimal number. How would you modify the machine of part (a) to recognize such a number?

★ 9. Let M be a finite-state machine with n states. The input alphabet is $\{0\}$. Show that for any input sequence that is long enough, M's output must eventually be periodic. What is the maximum number of inputs that can occur before periodic output begins? What is the maximum length of a period?

10. Write a computer program that, given the state table description of a finite-state machine with no more than 50 states and no more than 5 input symbols, will write the output string for any given input string.

Section 7.2   *MORPHISMS—SIMULATION II*

Once we have talked about a structure to simulate diverse phenomena (Simulation I), we also consider what it means for one instance of a structure to simulate another instance of that structure (Simulation II). (Perhaps we can build a simpler machine to do a given task.) As we saw in the algebraic structures of Chapter 5, if A and B are two instances of a structure, then simulation is directly tied to homomorphism. For finite-state machines we can reasonably define notions both of homomorphism and of simulation, but this time simulation will be a weaker idea than homomorphism. The difference is that, unlike our previous structures, we can view the behavior of a finite-state machine both internally and externally. External behavior considers the machine as a box where we see only input and output; internal behavior is concerned not only with input and output, but also with state changes of the machine. The definition of a finite-state machine certainly takes into account internal behavior, so this view is the one we will take first. For simplicity, we will assume that all the machines under consideration have the same input alphabet and the same output alphabet, usually $I = O = \{0, 1\}$.

HOMOMORPHISM

7.11   Definition   Let $M = [S, I, O, f_s, f_o]$ and $M' = [S', I, O, f'_s, f'_o]$ be two finite-state machines. A **homomorphism** from M to M' is a function $g: S \to S'$ such that for any $i \in I$ and $s \in S$,

$$g(f_s(s, i)) = f'_s(g(s), i) \qquad (1)$$

and

$$f_o(s) = f'_o(g(s)) \qquad (2)$$

$\square$

For a homomorphism from M to M', we thus associate with each state of M a corresponding state in M'. Equation (2) says that a state and its corresponding state have the same output. Equation (1) says that for any input symbol, a state and its corresponding state will proceed under the appropriate next-state functions to corresponding states. We can illustrate Equation (1) by the commutative diagram of Figure 7.13 (where $g: S \times I \to S' \times I$ means that (s, i) maps to $(g(s), i)$). If we think of applying the next-state function as an "operation," then this diagram says that in a machine

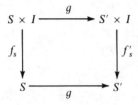

Figure 7.13

homomorphism, just as in any other homomorphism, you can operate and then map, or you can map and then operate—the end result will be the same.

7.12 Example   Figure 7.14 shows two machines, M and M', defined by state graphs. The function $g:\{0, 1, 2\} \to \{a, b\}$ given by

$$g:0 \to a$$
$$1 \to b$$
$$2 \to b$$

is a homomorphism from M onto M'. Clearly, a state in M and its corresponding state in M' have the same output, and it only requires six cases to test that for any input symbol, a state in M and its corresponding state in M' move to corresponding states. For example, Figure 7.15 traces the general commutative diagram starting with state 1, input 1 in M.   □

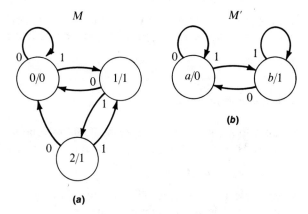

(a)

(b)

Figure 7.14

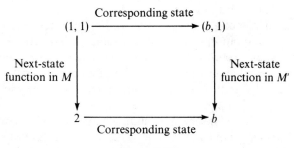

Figure 7.15

7.13 Practice   For the homomorphism of Example 7.12, draw the instance of the commutative diagram that begins with state 2, input 0 in M.   *(Answer, page A-19)*

In any finite-state machine M, we can consider $f_o(s)$ to represent the output beginning at state s and processing the empty input string λ, which is a special case of the following situation. If we begin at state s and process a nonempty input string α, that is, trace the operation of M through the various states determined by f_s, there will be a corresponding output string. We will also use f_o to designate this more general output function, so that now $f_o: S \times I^* \to O^*$ where I^* denotes the set of all strings over I and O^* denotes the set of all strings over O.

7.14 Practice   Let $M = [S, I, O, f_s, f_o]$ and $M' = [S', I, O, f'_s, f'_o]$ be finite-state machines with a homomorphism g from M to M'. The definition of a homomorphism says that for any state s in S, $f_o(s, \lambda) = f'_o(g(s), \lambda)$. Show that for any $s \in S$ and $\alpha \in I^*$,

$$f_o(s, \alpha) = f'_o(g(s), \alpha)$$

In other words, show that a state in M and its corresponding state in M' will produce the same output strings for the same input strings.

(Answer, page A-19)

A Homomorphism Theorem

Suppose that g is a homomorphism from machine $M = [S, I, O, f_s, f_o]$ *onto* machine $M' = [S', I, O, f'_s, f'_o]$. We define a binary relation ρ on S by

$$s_m \, \rho \, s_n \Leftrightarrow g(s_m) = g(s_n)$$

States in M are therefore related when they have the same corresponding states in M'. Clearly, ρ is an equivalence relation on S and will partition S into distinct equivalence classes. We want to treat the classes themselves as states of a new machine, which we will denote by M/g. The new machine will have input alphabet I and output alphabet O, but we must be able to define an output function and a next-state function. Any state of M/g is an equivalence class $[s]$ of states of M. All members of $[s]$ correspond under the homomorphism g to the same state in M'. From the definition of a homomorphism, a state and its corresponding state have the same output. Thus, all states in $[s]$ have the same output associated with them, and we will define this common output symbol to be the output associated with state $[s]$ of M/g.

To define the next-state function for M/g, consider state $[s_m]$ and an input symbol i. In machine M, the next state for s_m under i is given by $f_s(s_m, i)$. We can define the next state for $[s_m]$ under i to be the class $[f_s(s_m, i)]$. We are thus operating on a class by operating on a representative of the class, and we must be sure that such an operation is well defined—that it is independent of whatever representative of the class is chosen. Thus, suppose that $s_n \in [s_m]$. We then have

$$g(s_m) = g(s_n)$$

Also, by the definition of homomorphism,

$$g(f_s(s_m, i)) = f'_s(g(s_m), i)$$

and

$$g(f_s(s_n, i)) = f'_s(g(s_n), i)$$

Putting these facts together, we can write

$$g(f_s(s_m, i)) = f'_s(g(s_m), i) = f'_s(g(s_n), i) = g(f_s(s_n, i))$$

This equation says that the next states in M for s_m and s_n under i map under g to the same state in M', so they belong to the same equivalence class in M. Therefore, we can define the next state of a class under i to be the class of the next state of a representative, and the choice of representative does not matter.

7.15 Example   In Example 7.12, g is an onto homomorphism from M to M'. The states of M/g are $[0]$ and $[1]$ where $[0] = \{0\}$ and $[1] = \{1, 2\}$. The state table for M/g appears in Figure 7.16.   □

Present state	Next state Present input 0	1	Output
[0]	[0]	[1]	0
[1]	[0]	[1]	1

Figure 7.16

If the situation for machines continues to parallel what happened with algebraic structures in Section 5.4, we would expect that for g a homomorphism from M onto M', the homomorphic image M' is isomorphic to M/g. This is indeed the case, but we first make sure we have the definition well in mind.

7.16 Practice   Complete the definition. An **isomorphism** from machine $M = [S, I, O, f_s, f_o]$ to machine $M' = [S', I, O, f'_s, f'_o]$ is a function $g : S \to S'$ such that g is a homomorphism and a _____.   *(Answer, page A-19)*

If g is an isomorphism from M to M', then the number of states in S equals the number of states in S'. The inverse function g^{-1} exists. Furthermore, g^{-1} is an isomorphism from M' to M (see the next Practice problem), so we can speak of M and M' as being isomorphic.

7.17 Practice   If g is an isomorphism from M to M', show that g^{-1} is an isomorphism from M' to M.   (*Answer, page A-19*)

Isomorphic finite-state machines are structurally identical. Their states may have different names but their state graphs, if the nodes are relabeled, look exactly the same.

For g a homomorphism from M onto M', we want to define an isomorphism h from M/g to M'. Let $[s]$ be any state in M/g. Define h by $h([s]) = g(s)$. Any representative of $[s]$ will have the same image under g as s has, so h is well defined. We must show that h is a bijection and a homomorphism. To see that h is onto, let $s' \in S'$. Then $s' = g(s)$ for some $s \in S$, $[s]$ is a state in M/g, and $h[s] = s'$. To see that h is one-to-one, suppose $h([s_m]) = h([s_n])$. Then $g(s_m) = g(s_n)$, so s_m and s_n are in the same class and $[s_m] = [s_n]$. The output of $[s]$ in machine M/g is defined as the output of $g(s)$ in machine M'; thus, a state in M/g and its corresponding state (under h) in M' have the same output. Finally, the next state of $[s]$ in machine M/g under input i is defined as $[f_s(s, i)]$, and the next state of $h([s]) = g(s)$ under input i is $f'_s(g(s), i)$. And $h[f_s(s, i)] = g(f_s(s, i)) = f'_s(g(s), i)$. Thus, corresponding states (under h) proceed to corresponding states. The function h is therefore an isomorphism from M/g to M'. All of these facts can be summarized in the following theorem.

7.18 Theorem   A Homomorphism Theorem for Machines
Let g be a homomorphism from a finite-state machine M onto a finite-state machine M'. Then the homomorphic image M' is isomorphic to the machine M/g.   □

7.19 Practice   Refer to Examples 7.12 and 7.15.

(a) Draw the state graph for M/g.
(b) Define the isomorphism h from M/g to M'.   (*Answer, page A-19*)

Quotient Machines

There is a sort of converse to Theorem 7.18. Let $M = [S, I, O, f_s, f_o]$ be a finite-state machine. Whenever S can be partitioned so that a new machine can be defined using the blocks of the partition as states, the new machine is called a **quotient machine** of M. For a quotient machine to exist, all states in a given partition block must have the same output, and, for each input symbol, all must proceed under the next-state function f_s to states in the same partition block. The output function and next-state function for the quotient machine are then defined in the obvious way. Given that a quotient machine exists, it is always possible to define a machine homomorphism from M onto the quotient machine, namely the function g mapping a state s of M to the state $[s]$ in the quotient machine. The function g is clearly onto. Also a state in M and its corresponding state (under g) in the quotient machine have the same output and proceed, under any input symbol, to corresponding states. Hence, g is a homomorphism. When we combine this result with that stated in Theorem 7.18, we get the following theorem.

7.20 Theorem

Fundamental Homomorphism Theorem for Machines
Every homomorphic image of a finite-state machine M is essentially (up to isomorphism) a quotient machine of M, and every quotient machine of M is a homomorphic image of M.   □

Compare Theorem 7.20 with the Fundamental Homomorphism Theorem for Groups (Theorem 5.116).

7.21 Example

Figure 7.17 shows the state table for a machine M. We can construct a quotient machine by partitioning S into blocks $A = \{s_0, s_2\}$, $B = \{s_1\}$, and $C = \{s_3, s_4\}$. Note that all members of a given block have the same output and for a given input symbol, all proceed under the next-state function to states in the same block.   □

	Next state		
	Present input		
Present state	0	1	Output
s_0	s_2	s_0	0
s_1	s_3	s_0	1
s_2	s_0	s_0	0
s_3	s_3	s_1	1
s_4	s_4	s_1	1

Figure 7.17

7.22 Practice

Write the state table for the quotient machine of Example 7.21. Define the homomorphism g from M onto the quotient machine.   *(Answer, page A-20)*

SIMULATION

It is time that we turned our attention to the external view of the behavior of a finite-state machine where we care only about input strings and corresponding output strings. Our definition of simulation has to do with reflecting the external behavior of one machine in that of another.

7.23 Definition   Let $M = [S, I, O, f_s, f_o]$ and $M' = [S', I, O, f'_s, f'_o]$ be two finite-state machines. M **simulates** M' if there is a function $g: S' \to S$ such that for any string $\alpha \in I^*$ and $s' \in S'$,

$$f'_o(s', \alpha) = f_o(g(s'), \alpha) \qquad \qquad \square$$

Thus, M simulates M' if every state in M' has a corresponding state in M producing the same output string for any input string. In particular, if $\alpha = \lambda$, then $f'_o(s', \lambda) = f_o(g(s'), \lambda)$, so that a state and its corresponding state have the same output symbol.

7.24 Practice   For machines A and B described in Figure 7.18, show that A simulates B but that B does not simulate A.   (*Answer, page A-20*)

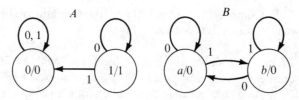

Figure 7.18

Simulation and Homomorphism

What is the connection between machine homomorphism and machine simulation? By Practice 7.14, if there is a homomorphism from M to M', then the homomorphism function is also a simulation function, and M' simulates M. If there is a homomorphism from M *onto* M', then every state s' in M' has at least one preimage in S; if a function is defined to map s' to one of its preimages, then this function is a simulation function, and M simulates M'.

7.25 Example   Consider the two machines of Example 7.12. The function

$$g: 0 \to a$$
$$1 \to b$$
$$2 \to b$$

guarantees that M' simulates M. Thus, for example, given any input string α, the output string produced by starting in state 2 of M and processing α is the same as that produced by starting in state b of M' and processing α. The function

$$h: a \to 0$$
$$b \to 1$$

guarantees that M simulates M'.

Here M is the more complicated machine. Because M' is a homomorphic image of M, we know that M' is isomorphic to a quotient machine of M. This means that, in some sense, a picture of M' is embedded in M, so it is not surprising that the more complex structure can simulate the simpler structure. What is surprising is that the simpler structure, M', also simulates the more complex structure, M. Note, however, that M' simulates only the external behavior of M. There is no homomorphism from M' to M, no way for M' to mimic the state behavior of M.   □

7.26   Definition   Machines M and M' are **equivalent** if each simulates the other.   □

It is easy to show (Exercise 5, this section) that the relation of equivalence is an equivalence relation on any collection of finite-state machines.

We have established that whenever there is a homomorphism from M onto M', then M and M' are equivalent. We have already mentioned, however, that simulation is a weaker notion than homomorphism. Although a homomorphism produces simulation, simulation can be present without homomorphism. In the following example two equivalent machines are given, each with the same number of states, but there is not a homomorphism from either one to the other.

7.27   Example A and B are two machines described by the state graphs of Figure 7.19. For any state in either A or B, the output upon processing any input string α

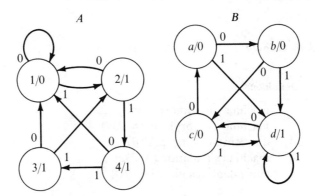

Figure 7.19

is, aside from the initial output symbol, α itself. Both A and B are "copy machines." Thus, A simulates B by the mapping

$a \to 1$

$b \to 1$

$c \to 1$

$d \to 2$

and B simulates A by the mapping

$$1 \to a$$
$$2 \to d$$
$$3 \to d$$
$$4 \to d$$

A and B are equivalent machines.

If a homomorphism g existed from A to B, it would have to map 1 to a, b, or c, and map 2, 3, and 4 to d. Suppose $g(1) = a$. Then consider the next states for both 1 and a under input 0. The next state for 1 under 0 is 1; the next state for a under 0 is b. But $g(1) \neq b$. Corresponding states do not proceed to corresponding states; so g is not a machine homomorphism. Similar difficulties occur if 1 maps to b or to c. Since these are all the possibilities, there is no homomorphism from A to B.

The proof that there is no homomorphism from B to A works the same way.   □

Although one machine can simulate another without involving a homomorphism, if we want to show that M simulates M', we will generally establish the existence of a homomorphism from M onto M'. Simulation is also established if we show the existence of a homomorphism from some submachine of M onto M'; not all the states of M need be used.

✔ CHECKLIST

Definitions

machine homomorphism (*p. 256*)
machine isomorphism (*p. 259*)
quotient machine (*p. 260*)
machine simulation (*p. 262*)
equivalent machines (*p. 263*)

Techniques

Verify that a given function is a machine homomorphism.

Main Ideas

Homomorphism from one machine to another (Simulation II—internal behavior).

The partition of the states of a machine into blocks that can sometimes serve as states of a quotient machine.

A homomorphic image of M is isomorphic to a quotient machine of M, and a quotient machine of M is a homomorphic image of M.

Simulation of one machine by another (Simulation II—external behavior).

Although machines can simulate each other with no homomorphism present, a homomorphism from one machine onto another means that each machine simulates the other.

EXERCISES, SECTION 7.2

1. Figure 7.20 shows two finite-state machines M and M'.

M

Present state	Next state		Output
	Present input		
	0	1	
1	3	4	0
2	3	1	0
3	2	3	1
4	2	3	1

(a)

M'

Present state	Next state		Output
	Present input		
	0	1	
a	c	c	0
b	c	a	0
c	b	c	1

(b)

Figure 7.20

The function g given by

$$1 \to a$$

$$2 \to b$$

$$3 \to c$$

$$4 \to c$$

is a homomorphism from M to M'. Verify that this is true by testing all possible cases.

★ 2. For each of the finite-state machine pairs M and M' of Figure 7.21, define a homomorphism g from M to M'.

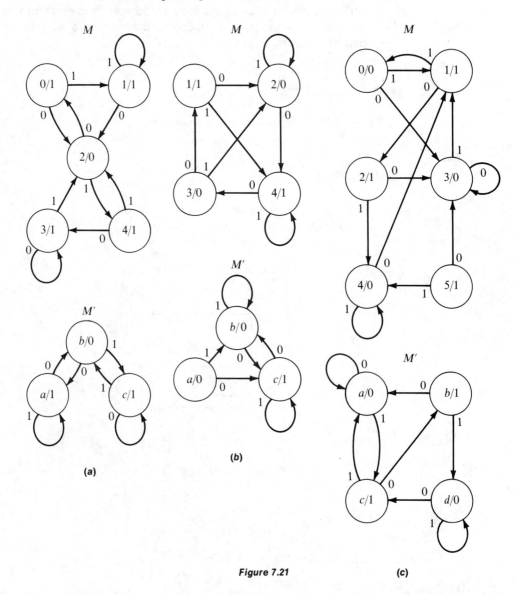

Figure 7.21

(a)   (b)   (c)

★ 3. For any onto homomorphisms of Exercise 2, draw the state graph of M/g and define the isomorphism from M/g to M'.

4. Partition the state set of each finite-state machine in Figure 7.22 so as to define a quotient machine. Write the state table for each quotient

machine. Define the homomorphism from the original machine onto the quotient machine.

	Next state		
	Present input		
Present state	0	1	Output
s_0	s_1	s_0	1
s_1	s_4	s_3	1
s_2	s_0	s_5	0
s_3	s_4	s_1	1
s_4	s_5	s_5	0
s_5	s_3	s_0	1

(a)

	Next state			
	Present input			
Present state	a	b	c	Output
s_0	s_2	s_1	s_4	0
s_1	s_2	s_0	s_4	1
s_2	s_3	s_4	s_2	0
s_3	s_3	s_1	s_3	0
s_4	s_3	s_0	s_1	1

(b)

Figure 7.22

5. Prove that the binary relation of equivalence on any collection of finite-state machines is an equivalence relation. Describe the corresponding equivalence classes.

6. Construct two finite-state machines M and M', different from any examples given in this section, so that there does not exist a homomorphism from M to M' but M' simulates M.

Section 7.3   *MACHINES AS RECOGNIZERS*

We have already seen some examples of finite-state machines as recognizers signalling with an output of 1 whenever an input string belonging to a particular set of possible input strings has been received. The machine of

Example 7.8, for instance, recognizes the set of all strings consisting of an even number of 1's. Now we want to see exactly what sets finite-state machines are capable of recognizing. Remember that recognition is possible because machine states can have a limited memory of past inputs. Even though the machine is finite, it is possible for a particular input signal to affect the behavior of a machine "forever." However, not every input signal can do so, and there will be some classes of inputs that require remembering so much information that no machine can detect them.

To avoid writing down outputs, we will designate those states of a finite-state machine with output of 1 as **final states**, and use a double circle to denote them in the state graph. Then we can give the following definition of recognition.

7.28   Definition   A finite-state machine M with input alphabet I **recognizes** a subset S of I^* if M, beginning in state s_0 and processing an input string α, ends in a final state if and only if $\alpha \in S$.   ☐

7.29   Practice   Describe the sets recognized by the machines of Figure 7.23.

(*Answer, page A-20*)

REGULAR SETS

We want a compact, symbolic way to describe sets such as those appearing in the answer to Practice 7.29. We will describe such sets by using "regular expressions"; each regular expression describes a particular set. First, we will define what regular expressions are; then we will see how a regular expression describes a set. We assume here that I is some finite set of symbols; later, I will be the input alphabet for a finite-state machine.

7.30   Definition,   Regular expression over I
Part I

(a) The symbol $\emptyset$ is a regular expression; the symbol λ is a regular expression.
(b) The symbol i for any $i \in I$ is a regular expression.
(c) If A and B are regular expressions, then (AB), $(A \vee B)$, and $(A)^*$ are regular expressions.   ☐

(Definition 7.30, Part I is another example of a recursive definition, specifying in (a) and (b) some particular objects as regular expressions and then in (c) giving rules for building regular expressions from already existing regular expressions.) Continuing our definition,

7.30   Definition,   Any set represented by a regular expression according to the conventions
Part II   described below is a **regular set**.

$\emptyset$ represents the empty set.
λ represents the set $\{\lambda\}$ containing the empty string.
i represents the set $\{i\}$.

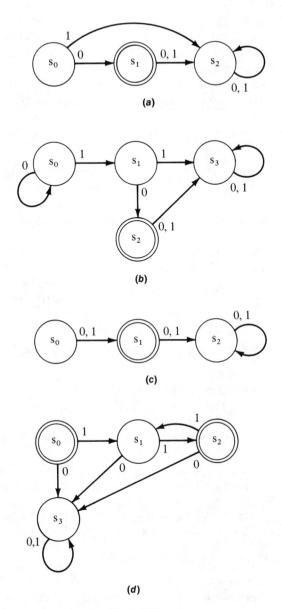

Figure 7.23

For regular expressions A and B,

(AB) represents the set of all elements of the form $\alpha\beta$ where α belongs to the set represented by A and B belongs to the set represented by B.

($A \vee B$) represents the union of A's set and B's set.

(A)* represents the concatenation of members of A's set.   □

We note that λ, the empty string, is a member of the set represented by A^*. In writing regular expressions, we eliminate parentheses when no ambiguity results.

We will be a little sloppy and say things like "The regular set $0^* \vee 10$" instead of "The set represented by the regular expression $0^* \vee 10$."

7.31 Example   Here we will give some regular expressions and then describe the set each one represents.

(a) $1^*0(01)^*$

(a′) Any number (including none) of 1's, followed by a single 0, followed by any number (including none) of 01 pairs.

(b) $0 \vee 1^*$

(b′) A single 0 or any number (including none) of 1's.

(c) $(0 \vee 1)^*$

(c′) Any string of 0's or 1's, including λ.

(d) $11((10)^*11)^*(00^*)$

(d′) A nonempty string of pairs of 1's interspersed with any number (including none) of 10 pairs, this string followed by at least one 0.   □

7.32 Practice   Which strings belong to the set described by the regular expression?

(a) 10100010; $(0^*10)^*$
(b) 011100; $(0 \vee (11)^*)^*$
(c) 000111100; $((011 \vee 11)^*(00)^*)^*$ *(Answer, page A-20)*

7.33 Practice   Write regular expressions for the sets recognized by the machines of Practice 7.29.   *(Answer, page A-20)*

A regular set may be described by more than one regular expression. For example, the set of all strings of 0's and 1's, which we already know from Example 7.31(c) to be described by $(0 \vee 1)^*$, is also described by the regular expression $((0 \vee 1^*)^* \vee (01)^*)^*$. We might, therefore, write the equation

$$(0 \vee 1)^* = ((0 \vee 1^*)^* \vee (01)^*)^*$$

Although we may be quite willing to accept this particular equation, how can we decide in general whether two regular expressions are equal, in the sense that they represent the same set? An efficient algorithm that will make this decision for any two regular expressions has not been found. Indeed, we will see in Chapter 9 why it seems likely that no such procedure will ever be found. (Exercise 8 at the end of Section 8.1 outlines a nonefficient procedure.)

KLEENE'S THEOREM

We have introduced regular sets because, as we will see, these are exactly the sets finite-state machines are capable of recognizing. Thus, any set recognized by some finite-state machine is regular, and conversely, any regular set can be recognized by some finite-state machine. This result was first proved by the American mathematician Stephen Kleene in 1956.

A Machine Recognizes a Regular Set

First, we will show that any set recognized by a finite-state machine is regular.

We have represented finite-state machines by directed graphs. Temporarily, we enlarge the set of machines to include structures whose graphs may not have a full complement of arrows, so that some states under a given input symbol may have no next state defined. If we call such structures Machines (with a capital M), then a (finite-state) machine is a special case of a Machine. Although we are ultimately interested in the set of strings taking a given machine from its starting state to any final state, we first consider the set of strings taking a Machine from any one state to another, not necessarily different, state. By using induction on the size of the Machine, we will prove that such a set is regular.

For the base step, assume that we have a Machine with only one state, s_0. Let $K = \{i_1, i_2, \ldots, i_k\}$ be the set of input symbols for which the next-state function on s_0 is defined. We want to find a regular expression for the set of all strings taking M from s_0 to s_0. Since there is nowhere else to go, any input string from K^* does this. Thus, the regular expression is $(i_1 \vee i_2 \vee \cdots \vee i_k)^*$. Note that the set includes λ, which certainly takes M from s_0 to s_0.

Now we assume that in any k-state Machine, the set of strings taking the Machine from any state s_m to any state s_n is regular. Finally, we let M be a Machine with $k + 1$ states, and we let s_m and s_n be states in M. We consider the two cases $s_m = s_n$ and $s_m \neq s_n$.

For the case $s_m = s_n$, we first consider nonempty strings taking M from s_m back to s_m for the first time. Such strings will be of two types:

1. A single input symbol $i \in I$.
2. A string of the form $i_p \alpha i_q$ where $i_p, i_q \in I$; i_p moves M from s_m to a different state s_{m1}; α is a string moving M from s_{m1} to some, not necessarily different, state s_{1m} but keeping it away from s_m; and then i_q takes M from s_{1m} back to s_m.

Let A be the set of all input strings taking M from s_{m1} to s_{1m} without going through s_m. If we disconnect s_m, the rest of the Machine is a k-state Machine, and A is regular by the induction hypothesis. For a fixed i_p and i_q, $i_p A i_q$ is thus a regular set. The set B of all strings of the form (2) is the union of a finite number of such sets (taking the union over the various i_p's and i_q's); hence, B is regular. And the set C of all strings described by (1) and (2) is the union of B with a finite number of single input symbols; C is also regular. Now C^* denotes the set of concatenations of members of C and describes the set of all input strings taking M from s_m to s_m; C^* is regular.

Now we need to handle the second case where $s_m \neq s_n$. Again, we first consider the set E of all strings moving M from s_m to s_n for the first time. Any such string is of the form αi where α takes M from s_m to some $s_{1m} \neq s_n$ but keeps it away from s_n, and i takes M from s_{1n} to s_n. Let D be the set of all input strings taking M from s_m to s_{1n} without going through s_n. If we disconnect s_n, the rest of the Machine is a k-state Machine, and D is regular by the induction hypothesis. For a fixed i, Di is thus a regular set. The set E consists of the union of a finite number of such sets (taking the union over the various i's); E is also regular. Now let F denote the set of all strings taking M from s_n to s_n; we know F is regular by the previous case. The regular set EF is the set of all input strings taking M from s_m to s_n.

We now have shown that the set of input strings taking a Machine M from any one state to any one state is regular. The set of strings taking a (finite-state) machine M from s_0 to any final state is the union of a finite number of such sets, so it is regular. If M has no final states, the empty set $\emptyset$ is the only set "recognized," and $\emptyset$ is also regular.

We have now proved the first half of Kleene's Theorem

7.34 Theorem   Any set recognized by a finite-state machine is regular.   □

Theorem 7.34 says that, given a finite-state machine M, there exists a regular expression describing the set of strings M recognizes. The proof of Theorem 7.34, however, does not tell us how to find such an expression easily.

Nondeterministic Finite-State Machines

The other half of the Kleene Theorem says that for any regular set, there is a finite-state machine recognizing it. To prove this result, we will introduce a new kind of machine, called a **nondeterministic finite-state machine**—defined as an ordinary finite-state machine except that for each state–input pair, the next state need not be uniquely determined and there is, in fact, a set of possible next states; this set could even be the empty set.

7.35 Example   Figure 7.24 shows the state table and the state graph for a nondeterministic machine M.   □

As a nondeterministic machine acts upon an input string α, the first input symbol processed leads M from the starting state to a set of possible next states. Each of these states, upon processing the second symbol, has a set of possible next states; the union of these sets is the set of possible states for M after processing two symbols of α. If we continue this procedure, we can find the set of possible states for M after processing α. If any of the states in this set is a final state of M, then we say that M **recognizes** α. The set of strings so recognized is the set recognized by M.

Present state	Next state Present input 0	1	Output
s_0	s_0, s_1	s_1, s_2	0
s_1	s_1	s_1	0
s_2	s_1, s_2	s_0	1

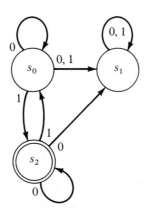

Figure 7.24

7.36 Example   The nondeterministic machine of Example 7.35 recognizes the set $0*1(0 \vee 10*1)*$. For any string α in this set, M has a possible sequence of moves that would result in M being in a final state at the end of processing α.   □

A nondeterministic machine M does not operate by choosing at each clock pulse some next state out of a set of possible next states. Rather, it operates like a parallel processor, keeping track at all times of all its possible configurations. In fact, we can simulate M's behavior by running in parallel a bunch of deterministic machines, each of which traces out a different possible sequence of moves for M. We can also simulate M's behavior by constructing a single big deterministic machine with enough states to represent all of M's possible configurations.

7.37 Theorem   For any nondeterministic machine M recognizing a set S, there is a deterministic machine M' also recognizing S.

Proof:
The states of M' are sets of states of M. If s_0 is the starting state of M, then $\{s_0\}$ is the starting state for M'. For each state $\{s_{i_1}, \ldots, s_{i_n}\}$ of M' and each input symbol i, we find the next state of M' by taking the union of the set

of next states for s_{i1} under i in M, s_{i2} under i in M, and so on. A state of M' is labeled a final state if and only if it contains a final state of M. □

7.38 Example Figure 7.25 shows the state table for the deterministic counterpart of the nondeterministic M of Example 7.35. For example, the next state of $\{s_1, s_2\}$ under 1 is $\{s_0, s_1\}$ because the set of next states in M for s_1 under 1 is $\{s_1\}$ and the set of next states in M for s_2 under 1 is $\{s_0\}$. From the state graph for M, Figure 7.26, we see that M indeed recognizes the set $0*1(0 \vee 10*1)*$, and we also see that states B and C are "unreachable" from the starting state A and could be eliminated. □

Present state	Next state		Output
	Present input		
	0	1	
$A = \{s_0\}$	$\{s_0, s_1\}$	$\{s_1, s_2\}$	0
$B = \{s_1\}$	$\{s_1\}$	$\{s_1\}$	0
$C = \{s_2\}$	$\{s_1, s_2\}$	$\{s_0\}$	1
$D = \{s_0, s_1\}$	$\{s_0, s_1\}$	$\{s_1, s_2\}$	0
$E = \{s_1, s_2\}$	$\{s_1, s_2\}$	$\{s_0, s_1\}$	1

Figure 7.25

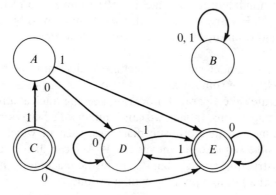

Figure 7.26

In Example 7.38, the number of states in the deterministic machine is close to the number of states in the original nondeterministic machine. This situation is unusual; if M has n states, M' could have as many as $2^n - 1$ states.

7.39 Practice Find a deterministic machine M' recognizing the same set as M.

(Answer, page A-20)

Present state	Next state		Output
	Present input		
	0	1	
s_0	s_0	s_1	0
s_1	s_1	s_0, s_1	1

A Regular Set Has A Machine Recognizer

Theorem 7.37 says that we gain no recognition capabilities by considering nondeterministic machines. Therefore, the proof of Kleene's Theorem will be complete if we can show that for any regular set, there is a nondeterministic finite-state machine recognizing it. We will prove that such a machine exists by showing how to construct it. Because the definition of a regular expression is inductive, we must construct our machine inductively. We let I be the set of symbols, and consider the various types of regular expressions.

(1) $\varnothing$ and λ. A trivial machine with a single, nonfinal state, as in Figure 7.27a, recognizes $\varnothing$. Figure 7.27b shows a deterministic machine that recognizes λ (a deterministic machine is a special case of a non-deterministic machine).

(2) $i \in I$. Figure 7.27c shows a deterministic machine that recognizes i.

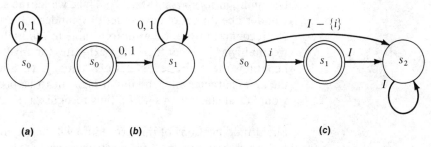

(a)   (b)   (c)

Figure 7.27

We now assume the inductive hypothesis that for regular expressions A and B there are nondeterministic recognizers M_A and M_B. To avoid mixups, we'll also assume the states in M_A and the states in M_B have different names.

(3) AB. The basic idea here is to connect the two machines M_A and M_B in series to create a machine M_{AB} recognizing AB. The set of states for M_{AB} is the union of the sets of states of M_A and M_B. The starting state for M_{AB} is the starting state of M_A, and the final states of M_{AB} are the final states of M_B. Whenever a state–input pair in M_A

could take M_A to a final state, we want to allow the possibility of jumping instead to the starting state of M_B, so that we can begin to process strings $\beta \in B$ in M_B. Hence, we modify the state table for M_A so that whenever the set of next states contains a final state in M_A, we add the starting state of M_B to that set. Then for any $\alpha\beta \in AB$, there is a sequence of moves taking M_{AB} from its starting state through the actions of M_A on α to the point of recognition, then transferring to perform the actions of M_B on β until β is recognized by M_B; hence, $\alpha\beta$ is recognized by M_{AB}.

(4) $A \vee B$. The basic idea here is to connect the two machines M_A and M_B in parallel to create a machine $M_{A \vee B}$ recognizing $A \vee B$. The states of $M_{A \vee B}$ are the states of M_A plus the states of M_B plus one additional state $\bar{s}$ designated as the starting state for $M_{A \vee B}$. The final states of $M_{A \vee B}$ are the final states of M_A plus the final states of M_B. When we process the first symbol i of a string γ, we want to allow the possibility of simulating either M_A's actions in processing i beginning in its starting state s_A or M_B's actions in processing i beginning in its starting state s_B. We define the set of next states for $\bar{s}$ under i to be the union of the set of next states of s_A under i and the set of next states of s_B under i. Thus, $M_{A \vee B}$ processes γ by simulating either M_A or M_B, recognizing γ if it is recognized by either M_A or M_B.

(5) A^*. M_{A^*} uses the set of states of M_A plus an additional starting state $\bar{s}$, which must be a final state in order to recognize λ. The final states of M_A are also final states of M_{A^*}. If i is the first symbol of a string γ, then M_{A^*} should simulate M_A's actions in processing i beginning in its starting state s_A. Thus, we let the set of next states of $\bar{s}$ under i be the set of next states of s_A under i. If an initial segment of γ is recognized by A, we want to be able to reinitialize at once. Hence, we modify M_A so that the set of next states for any final state and input j contains the set of next states of $\bar{s}$ under j. This modification allows the first character after the initial segment to be processed just as M_A would do it starting in s_A. M_{A^*} thus recognizes A^*.

Slight modifications of this procedure will be required to take care of troublesome cases involving λ. To construct a machine for 1^*0^*, for example, we would want to leave the starting state for the machine of 1^* as a final state, even though according to (3) only the final states of 0^* should remain final. Similarly, a machine for 1^*0 would call for a transfer on 0 from the starting state of the machine for 1^* to the final state of the machine for 0.

7.40 Example   Let's build the nondeterministic recognizer for the regular set $1 \vee 00^*$. First, we build the machine for 0^*. Beginning with the machine for 0, Figure 7.28a, we modify it as described in part (5) to get Figure 7.28b, the machine for 0^*. Figure 7.28c shows the machine for 00^*, and, finally, Figure 7.28d is a nondeterministic machine recognizing $1 \vee 00^*$.   □

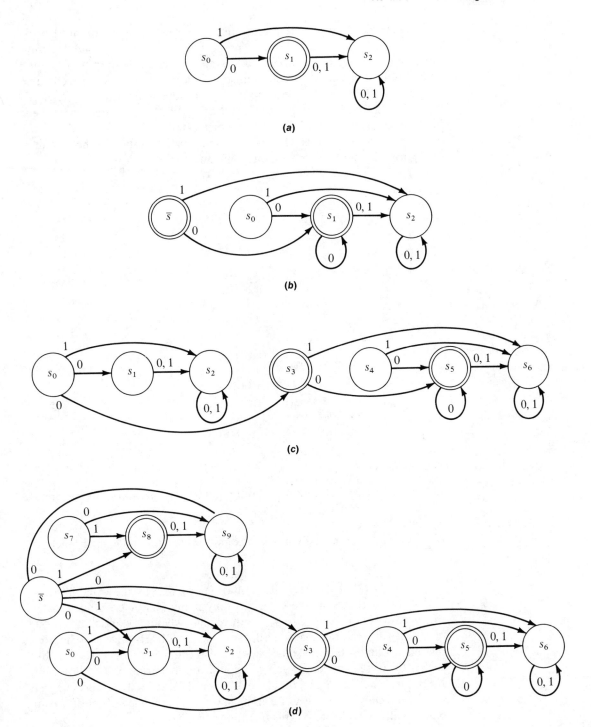

Figure 7.28

The procedure just described should be viewed as a canonical procedure, that is, it is completely general and always works. But for any particular case, we can probably come up with a much simpler machine. Thus, suppose we want a (deterministic) finite-state machine to recognize the regular set $1 \vee 00^*$. The canonical procedure is to produce the nondeterministic machine of Figure 7.28d, then use the procedure described in the proof of Theorem 7.37 to find the corresponding deterministic machine, which will be horrendous. However, it is not hard to think up the machine of Figure 7.29, a 4-state machine recognizing $1 \vee 00^*$.

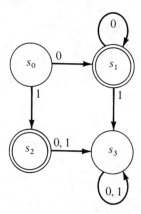

Figure 7.29

Final Comments

Let's summarize what we know about regular sets and finite-state machines.

7.41 Theorem   Kleene's Theorem
A set is regular if and only if it is recognized by some finite-state machine.

□

Theorem 7.41 outlines the limitations as well as the capabilities of finite-state machines, as there are certainly many sets that are not regular. For example, $S = \{0^n 1^n \mid n \geq 0\}$ is not regular where a^n stands for a string of n copies of a. (Notice that 0^*1^* does not do the job.) By Kleene's Theorem, there is no finite-state machine capable of recognizing S. Yet S seems like such a "nice" set, and surely you or I could count a string of 0's followed by 1's and see whether we had the same number of 0's as 1's. This lapse suggests some deficiency in our use of a finite-state machine as a model of a computational device. More on this possibility in Chapter 9.

✔ *CHECKLIST*

Definitions

final state (*p. 268*)
recognition (by deterministic machine) (*p. 268*)
regular expression (*p. 268*)
regular set (*p. 268*)
nondeterministic finite-state machine (*p. 272*)
recognition (by nondeterministic machine) (*p. 272*)

Techniques

Find a regular expression from the description of a regular set.

Decide whether a given string belongs to a given regular set.

Given a nondeterministic finite-state machine, construct a deterministic machine that recognizes the same set of input strings.

Given a regular set, construct the canonical nondeterministic machine that recognizes it.

Main Ideas

The class of nondeterministic finite-state machines cannot recognize any more sets than the class of finite-state machines.

The class of sets recognizable by finite-state machines is the class of all regular sets; hence, there are limitations to the recognition capabilities of finite-state machines.

EXERCISES, SECTION 7.3

1. Give a regular expression for the set recognized by each finite-state machine in Figure 7.30 (see page 280).

2. Give a regular expression for each set described below.
   (a) The set of all strings of 0's and 1's beginning with 0 and ending with 1.
   (b) The set of all strings of 0's and 1's having an odd number of 0's.
   ★ (c) $\{101, 1001, 10001, 100001, \ldots\}$

★ 3. Does the given string belong to the given regular set?
   (a) 01110111; $(1^*01)^*(11 \vee 0^*)$
   (b) 11100111; $((1^*0)^* \vee 0^*11)^*$
   (c) 011100101; $01^*10^*(11^*0)^*$
   (d) 1000011; $(10^* \vee 11)^*(0^*1)^*$

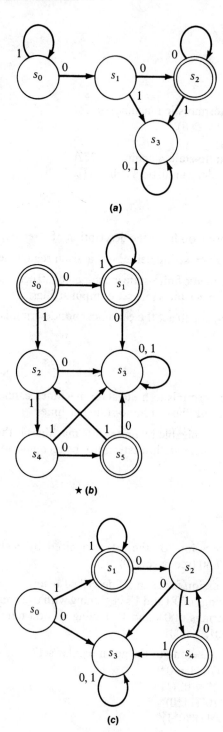

(a)

★ (b)

(c)

Figure 7.30

4. (a) Write a regular expression for the set of all alphanumeric strings beginning with a letter. (The set of legal identifiers in FORTRAN is a subset of this set.)

   (b) Write a regular expression for the set of all arithmetic expressions of the form addition or subtraction of two positive integers.

5. (a) Prove that if A is a regular set, then the set A^R consisting of the reverse of all strings in A is also regular.

   (b) For any string α, let α^R be the reverse string. Do you think the set $\{\alpha\alpha^R \mid \alpha \in I^*\}$ is regular?

6. For each machine in Figure 7.31, use the proof of Theorem 7.37 to find a deterministic machine recognizing the same set as the nondeterministic machine. Write a regular expression for the set.

	Next state		
	Present input		
Present state	0	1	Output
s_0	s_0, s_1	s_1	1
s_1	s_0	s_0, s_1	0

★ (a)

	Next state		
	Present input		
Present state	0	1	Output
s_0	s_1, s_2	s_0	1
s_1	s_0, s_1	s_2	0
s_2	s_2	s_2	0

(b)

	Next state		
	Present input		
Present state	0	1	Output
s_0	s_2	s_1, s_3	0
s_1	s_2	s_1	1
s_2	s_2	s_2	0
s_3	s_4	s_2	0
s_4	s_2	s_5	0
s_5	s_2	s_3	1

(c)

Figure 7.31

7. (a) Construct the canonical, nondeterministic recognizer for the regular set 1*.

   (b) Use the method of Theorem 7.37 to construct the deterministic equivalent of the nondeterministic machine of part (a). This new machine is the canonical recognizer for 1*.

   (c) By intuition, construct a deterministic machine to recognize 1*.

8. (a) Construct the canonical, nondeterministic recognizer for the regular set 1*0.

   (b) Contemplate (do not actually construct) the deterministic equivalent of the nondeterministic machine of part (a) constructed according to the method of Theorem 7.37.

   (c) By intuition, construct a deterministic machine to recognize 1*0.

9. (a) Construct the canonical, nondeterministic recognizer for the regular set $(0 \lor 01)*$.

   (b) Contemplate (do not actually construct) the deterministic equivalent of the nondeterministic machine of part (a) constructed according to the method of Theorem 7.37.

   (c) By intuition, construct a deterministic machine to recognize $(0 \lor 01)*$.

10. Prove that if A is a regular set whose symbols come from the alphabet I, then $I* - A$ is a regular set.

FOR FURTHER READING

Entry [1] is a very readable, general introduction to finite-state machines. A historical presentation of finite-state machines in terms of McCulloch–Pitts neurons is given in [2], which also has a good discussion of Kleene's Theorem. Kleene's Theorem is also covered thoroughly in [3]. Several equivalent definitions of finite-state machines are given in [4], and [5] discusses applications of finite-state machines to compilers.

1. Frederick C. Hennie, *Finite-State Models for Logical Machines.* New York: Wiley, 1968.

2. Marvin L. Minsky, *Computation: Finite and Infinite Machines.* Englewood Cliffs. N. J.: Prenctice-Hall, 1967.

3. Peter J. Denning, Jack B. Dennis, and Joseph E. Qualitz, *Machines, Languages and Computation.* Englewood Cliffs, N. J.: Prentice-Hall, 1978.

4. John E. Hopcroft and Jeffrey D. Ullman, *Introduction to Automata Theory, Languages, and Computation.* Reading, Mass.: Addison-Wesley, 1979.

5. Alfred V. Aho and Jeffrey D. Ullman, *Principles of Compiler Design.* Reading, Mass.: Addison-Wesley, 1977.

Chapter 8

MACHINE DESIGN AND CONSTRUCTION

Chapter 7 introduced finite-state machines and gave some indications of how they can be useful. In this chapter we deal with practical considerations in the actual physical implementation of a finite-state machine. A method to possibly reduce the number of states needed in the machine is developed. Also, we will see how any particular finite-state machine can be physically constructed from the simple logic gates of Chapter 4 plus one additional element, or how it can be built from a storehouse of larger stock machines.

Section 8.1 MACHINE MINIMIZATION

If we want to actually construct a finite-state machine M, the number of internal states is a factor in the cost of construction. If we can find a machine with the same external behavior as M but having fewer states, we will have a machine that can do the job we want done but will be more economical to build.

UNREACHABLE STATES

First, let's observe that we can remove any **unreachable states** of M.

8.1 Example   Let M be given by the state table of Figure 8.1. Although the state table

	Next state		
	Present input		
Present state	0	1	Output
s_0	s_1	s_3	0
s_1	s_3	s_0	0
s_2	s_1	s_3	1
s_3	s_0	s_1	1

Figure 8.1

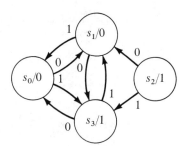

Figure 8.2

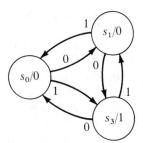

Figure 8.3

contains the same information as the state graph, the graph (Figure 8.2) shows us at a glance that state s_2 can never be reached from the starting state s_0. If we simply remove state s_2, we have the state graph (Figure 8.3) for a machine M' with one less state that behaves exactly like M, that is, gives the same output as M for any input string. Notice that M and M' are not equivalent. Although M clearly simulates M', M' does not simulate M because state s_2 of M has no corresponding state in M'. But because s_2 can never be reached under any possible input string, it does not matter that there is no corresponding state for it; machine equivalence is a stronger idea than we need here. $\square$

8.2 Practice   What state(s) are unreachable from s_0 in the machine of Figure 8.4? Try to get your answer directly from the state table.   *(Answer, page A-20)*

Present state	Next state Present input 0	1	Output
s_0	s_1	s_4	0
s_1	s_4	s_1	1
s_2	s_2	s_2	1
s_3	s_3	s_1	0
s_4	s_0	s_0	1

Figure 8.4

MINIMIZATION PROCEDURE

We will assume that all unreachable states have now been removed from M, and we will continue to look for a "reduced" machine M' with the same external behavior as M. This means that for any input string α, if we process α through M from the starting state s_0 and if we process α through M' from the starting state s_0', we want to produce identical output strings.

Because all states in M are now reachable, every state in M must have a corresponding state in M', one that will produce the same output string for all input strings. If there is some state s_i of M with no corresponding state in M', we could argue as follows: find an input string β taking M from s_0 to s_i. Such a string exists because s_i is reachable from s_0. Apply β to M' starting at s_0', and let M' be in state s_i' at the end of processing β. Because s_i' is not a corresponding state for s_i, there is some input string γ such that $f_o(s_i, \gamma) \neq f_o'(s_i', \gamma)$. The concatenation string $\beta\gamma$ is an input string producing different output when fed into machine M than when fed into machine M'. By this argument, it follows that M' must simulate M and conversely. In short, M and M' must be equivalent.

We know that machine equivalence is an equivalence relation on finite-state machines. Therefore, we want to find a minimal representative of $[M]$, one with the fewest states. The procedure for constructing this minimal machine involves equivalences of the states of M.

8.3 Definition   Two states s_i and s_j of M are **equivalent** if for any $\alpha \in I^*$, $f_o(s_i, \alpha) = f_o(s_j, \alpha)$.   □

Equivalent states of a machine produce identical output strings for any input string. Notice that we are now comparing states within a single machine and not between two machines as we were just a moment ago when talking about simulation.

8.4 Practice   Prove that state equivalence is an equivalence relation on the states of a machine.   *(Answer, page A-20)*

A Quotient Machine

For the time being, we will postpone the problem of how to identify equivalent states in a given machine M. Let's simply assume that we have somehow found which states are equivalent and have partitioned the states of M into the corresponding equivalence classes. These classes can serve as states for a quotient machine of M. There are two requirements for this to happen: (1) all states in the same class must have the same output; and (2) for each input symbol, all states in the same class must proceed under the next state function to states in the same class.

8.5 Practice   Show that conditions (1) and (2) are satisfied when M is partitioned into classes of equivalent states.   *(Answer, page A-20)*

We will use M^* to denote the quotient machine of M that we have found. By the Fundamental Homomorphism Theorem for Machines (Theorem 7.20), M^* is a homomorphic image of M. Therefore, M^* and M are equivalent machines. M^* belongs to $[M]$ and the number of states of M^* is less than or equal to the number of states of M. We still need to show that there is no member of $[M]$ with fewer states than M^*.

Let $M_1 \in [M]$. Then M_1 and M^* are equivalent. In particular, M_1 simulates M^*, and there is a mapping g from the states of M^* to the states of M_1 such that a state in M^* and the corresponding state in M_1 produce the same output string for any input string. The mapping g must be one-to-one. To see this, suppose s_i^* and s_j^* are two states of M^* with $g(s_i^*) = g(s_j^*)$. Because s_i^* and s_j^* have the same corresponding state in M_1, it also follows that s_i^* and s_j^* are themselves equivalent states in M^*. By the construction of M^*, $s_i^* = s_j^*$. Therefore, g is a one-to-one map from the states of M^* to the states of M_1, and hence, M_1 has at least as many states as M^*.

Equivalent States

The minimization problem for M thus boils down to finding the equivalent states of M. Perhaps we should note first that the obvious approach of directly trying to satisfy the definition of equivalent states will not work. Given two states s_i and s_j of M, we cannot actually compare the outputs corresponding to each possible input string. Fortunately, the problem is not as "infinite" as it first sounds; we can get by with identifying k-equivalent states.

8.6 Definition   Two states s_i and s_j of M are **k-equivalent** if for any $\alpha \in I^*$ where α has $\leq k$ symbols, $f_o(s_i, \alpha) = f_o(s_j, \alpha)$.   □

Clearly, k-equivalence is an equivalence relation on the states of M. It is possible to directly test two states of M for k-equivalence since we could actually produce the finite number of input strings having $\leq k$ symbols. But it turns out that we don't have to do this. We can begin by finding 0-equivalent states. These are states producing the same output for 0-length input strings, that is, states having the same associated output symbol. Thus, we can identify the 0-equivalence classes directly from the description of M.

8.7 Example   Let M be defined by the state table of Figure 8.5. The 0-equivalence classes of the states of M are

$$\{0, 2, 5\} \quad \text{and} \quad \{1, 3, 4, 6\}$$ □

	Next state		
	Present input		
Present state	0	1	Output
0	2	3	0
1	3	2	1
2	0	4	0
3	1	5	1
4	6	5	1
5	2	0	0
6	4	0	1

Figure 8.5

Our procedure to find k-equivalent states is a recursive one; we know how to find 0-equivalent states, and we will show how to find k-equivalent states once we have identified states that are $(k - 1)$-equivalent. Suppose, then, that we already know which states are $(k - 1)$-equivalent. If states s_i and s_j are k-equivalent, they must produce the same output strings for any input string of length $\leq k$, in particular, for any string of length $\leq k - 1$. Thus, s_i and s_j must at least be $(k - 1)$-equivalent. But also they must produce the same output strings for any k-length input string.

An arbitrary k-length input string consists of a single arbitrary input symbol followed by an arbitrary $(k - 1)$-length input string. Applying such a k-length string to states s_i and s_j (which themselves have the same output symbol), the single input symbol moves s_i and s_j to next states s_i' and s_j'; and s_i' and s_j' must produce identical output strings for the remaining, arbitrary $(k - 1)$-length string, which will surely happen if s_i' and s_j' are $(k - 1)$-

equivalent. Therefore, *to find k-equivalent states, look for* $(k - 1)$*-equivalent states whose next states under any input symbol are* $(k - 1)$*-equivalent.*

8.8 Example   Consider again the M of Example 8.7. We know the 0-equivalent states. To find 1-equivalent states, we look for 0-equivalent states with 0-equivalent next-states. For example, the states 3 and 4 are 0-equivalent; under the input symbol 0, they proceed to states 1 and 6, respectively, which are 0-equivalent states, and under the input symbol 1 they both proceed to 5, which of course is 0-equivalent to itself. Therefore, states 3 and 4 are 1-equivalent. But states 0 and 5, themselves 0-equivalent, proceed under the input symbol 1 to states 3 and 0, respectively, which are not 0-equivalent states. So states 0 and 5 are not 1-equivalent; the input string 1 will produce an output string of 01 from state 0 and 00 from state 5. The 1-equivalence classes for M are

$$\{0, 2\}, \{5\}, \{1, 3, 4, 6\}$$

To find 2-equivalent states, we look for 1-equivalent states with 1-equivalent next-states. Thus, states 1 and 3, although 1-equivalent, proceed under input 1 to states 2 and 5, respectively, which are not 1-equivalent states. Therefore, states 1 and 3 are not 2-equivalent. The 2-equivalence classes for M are

$$\{0, 2\}, \{5\}, \{1, 6\}, \{3, 4\}$$

The 3-equivalence classes for M are the same as the 2-equivalence classes.

$\square$

8.9 Definition   Given two partitions π_1 and π_2 of a set S, π_1 is a **refinement** of π_2 if each block of π_1 is a subset of a block of π_2. $\square$

In the Example above, each successive partition of the states of M into equivalence classes is a refinement of the previous partition. This refinement will always happen; k-equivalent states must also be $(k - 1)$-equivalent, so the blocks of the $(k - 1)$-partition can only be further subdivided. However, the subdivision process cannot continue indefinitely (at worst it can only go on until each partition block contains only one state); there will eventually be a point where $(k - 1)$-equivalent states and k-equivalent states coincide. (In Example 8.8, 2-equivalent and 3-equivalent states coincide.) Once this happens, all next-states for members of a partition block under any input symbol fall within a partition block. Thus, k-equivalent states are also $(k + 1)$-equivalent and $(k + 2)$-equivalent, and so on. Indeed, these states are equivalent.

Summary

The total procedure (called the Moore reduction algorithm) for finding equivalent states is to start with 0-equivalent states, then 1-equivalent states,

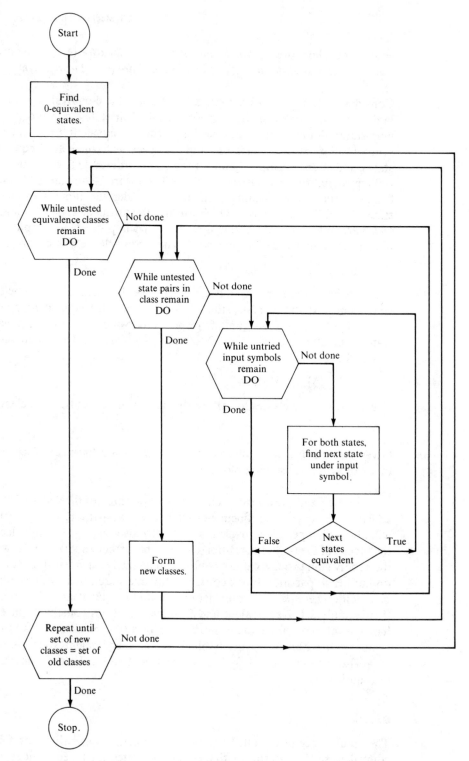

Figure 8.6

and so on until the partition no longer subdivides. Figure 8.6 gives a flow-chart outline of this procedure. You may have forgotten by now, but we wanted to use the partition into equivalent states to form a quotient machine of M that would be a minimal machine M^* equivalent to M.

8.10   Example   For M of Example 8.7 and 8.8, the quotient machine M^* will have states

$$A = \{0, 2\}$$
$$B = \{5\}$$
$$C = \{1, 6\}$$
$$D = \{3, 4\}$$

The state table for M^* is obtained from that for M (Figure 8.7). M^* (starting state A) is equivalent to M; it will reproduce M's output for any input string, but has four states instead of seven.   □

	Next state		
	Present input		
Present state	0	1	Output
A	A	D	0
B	A	A	0
C	D	A	1
D	C	B	1

Figure 8.7

8.11   Example   We will minimize M, where M is given by the state table of Figure 8.8.

	Next state		
	Present input		
Present state	0	1	Output
0	3	1	1
1	4	1	0
2	3	0	1
3	2	3	0
4	1	0	1

Figure 8.8

The 0-equivalence classes of M are

$$\{0, 2, 4\}, \{1, 3\}$$

The 1-equivalence classes of M are

$$\{0\}, \{2, 4\}, \{1, 3\}$$

No refinement is possible. Let

$$A = \{0\}$$
$$B = \{2, 4\}$$
$$C = \{1, 3\}.$$

The reduced machine is shown in Figure 8.9.   □

	Next state		
	Present input		
Present state	0	1	Output
A	C	C	1
B	C	A	1
C	B	C	0

Figure 8.9

8.12 Practice   Minimize each M of Figure 8.10.   *(Answer, page A-21)*

	Next state		
	Present input		
Present state	0	1	Output
0	2	1	1
1	2	0	1
2	4	3	0
3	2	3	1
4	0	1	0

(a)

	Next state		
	Present input		
Present state	0	1	Output
0	1	3	1
1	2	0	0
2	1	3	0
3	2	1	0

(b)

Figure 8.10

We will assume from here on that all finite-state machines have been minimized.

✔ **CHECKLIST**

Definitions

unreachable state (*p. 284*)   *k*-equivalent states (*p. 288*)
equivalent states (*p. 286*)   partition refinement (*p. 289*)

Technique

Minimize finite-state machines.

Main Ideas

Unreachable states can be removed from a machine.

Once unreachable states are removed, a machine with the same external behavior as M must be equivalent to M. A machine equivalent to M with the least possible number of states is the quotient machine of M formed from classes of equivalent states.

EXERCISES, SECTION 8.1

★ 1. Identify any unreachable states of M.

Present state	Next state Present input 0	1	Output
s_0	s_2	s_0	0
s_1	s_2	s_1	1
s_2	s_2	s_0	1

(a)

Present state	Next state Present input a	b	c	Output
s_0	s_1	s_0	s_3	0
s_1	s_1	s_3	s_0	1
s_2	s_3	s_2	s_1	0
s_3	s_1	s_1	s_0	0

(b)

Figure 8.11

★ 2. Minimize M.

Present state	Next state		Output
	Present input		
	0	1	
0	3	6	1
1	4	2	0
2	4	1	0
3	2	0	1
4	5	0	1
5	3	5	0
6	4	2	1

Figure 8.12

3. Minimize M.

Present state	Next state		Output
	Present input		
	0	1	
0	5	3	1
1	5	2	0
2	1	3	0
3	2	4	1
4	2	0	1
5	1	4	0

Figure 8.13

4. Minimize M.

Present state	Next state		Output
	Present input		
	0	1	
0	1	3	0
1	2	0	0
2	0	3	0
3	2	1	0

Figure 8.14

★ 5. Minimize M.

Present state	Next state			Output
	Present input			
	a	b	c	
0	1	4	0	1
1	4	2	3	0
2	3	4	2	1
3	4	0	1	0
4	1	0	2	0

Figure 8.15

6. Minimize M.

Present state	Next state		Output
	Present input		
	0	1	
0	1	3	1
1	2	0	0
2	4	3	1
3	0	1	1
4	2	4	0

Figure 8.16

7. Let M be a finite-state machine, and M^* be the minimized machine for M. Let M^{**} be the minimization of any machine equivalent to M. Prove that M^* and M^{**} are isomorphic, thus showing that the reduced machine for $[M]$ is essentially unique.

★ 8. Given any two regular expressions A and B, the following long, drawn-out procedure can be used to determine whether A and B represent the same set. We first use the algorithm of Section 7.3 (building the nondeterministic machine and then converting it to a deterministic machine) to build a finite-state machine M_A to recognize the regular set A, and a finite-state machine M_B to recognize the regular set B. We remove unreachable states from these machines (take each state s and backtrack to find the set of all states having access to s). Then $A = B$ if and only if M_A and M_B recognize the same set. But M_A and M_B recognize the same set if, for any string α, processing α through M_A from its starting state and processing

α through M_B from its starting state produces the same output strings, meaning that M_A and M_B are equivalent machines (see the discussion following Practice 8.2).

The next part of the procedure is to minimize M_A and M_B; denote the minimized machines by M_A^* and M_B^*. By Exercise 7 above, $A = B$ if and only if M_A^* and M_B^* are isomorphic. Because there are only a finite number of possible bijections, we can test each bijection to see whether it is an isomorphism by considering the behavior of each pair of corresponding states under each input symbol.

Try to decide intuitively which of the following equalities are true. If you have trouble, you can always resort to the above procedure!

(a) $0*0 = 00*$

(b) $0*1* = (0 \vee 1)*(01)*$

(c) $1*0 = 0 \vee 1*10$

(d) $1*0* = 1*10* \vee 1*00*$

Section 8.2   BUILDING MACHINES

SEQUENTIAL NETWORKS

We know that the output of a finite-state machine is a function of its present state, and that the present state of the machine is a function of past inputs. Thus, the states of a machine have certain memory capabilities. On the other hand, recall the logic networks of Chapter 4, which are combinations of AND gates, OR gates, and inverters, and where output is virtually instantaneous and a function only of the present input. Yet we will see that we can build any finite-state machine by using a logic network—provided we introduce one additional element. The new element must provide the memory missing from our previous logic networks.

Delay Elements

The element we need is a **delay element** (the simplest of a class of elements known as **flip-flops**). It has a single input that takes on values of 0 or 1, and it will output at time $t + 1$ the input signal it received at time t. Figure 8.17

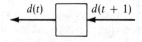

Figure 8.17

represents the delay element *at time t + 1*. (Note we have shown signals propagating from right to left as this is generally how we will see delay elements used.) The delay element has a memory that captures input for the duration of one clock pulse. Our old networks with no delay elements are also called **combinational networks**; when one or more delay elements are introduced, the network is known as a **sequential network**. Input sequences can be run through sequential networks (hence the name) provided that the clock pulse synchronizes input signals as well as delay elements.

8.13 **Example**   In Figure 8.18 the effect of the delay element is to feed the output from the terminal AND gate of the network back into the initial OR gate at the next clock pulse. Figure 8.19 shows the behavior of this network for a certain sequence of input values. The initial output of the delay element is assumed to be 0.   □

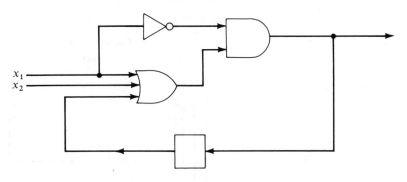

Figure 8.18

Time	0	1	2	3	4
x_1	1	0	0	1	0
x_2	1	1	0	0	0
Output	0	1	1	0	0

Figure 8.19

There are some general rules for the arrangements allowed in sequential networks just as there are in combinational networks. It is still the case that input or output lines cannot be tied together except by passing through an element, and that they can be split to serve as input to more than one element. Unlike combinational networks, loops (where the output of an element becomes part of its input) are allowed, provided that at least one delay element is incorporated into the loop. The delay element prevents the confusion resulting from an element trying to act upon its current output.

Networks For Finite-State Machines

We can build any finite-state machine by using a sequential network. A cluster of delay elements in the network functions as the states of the machine, retaining some memory of past inputs. The general structure of such a network is shown in Figure 8.20. It consists of two parts: (1) a combinational network (no delay elements); and (2) some loops containing all the delay elements.

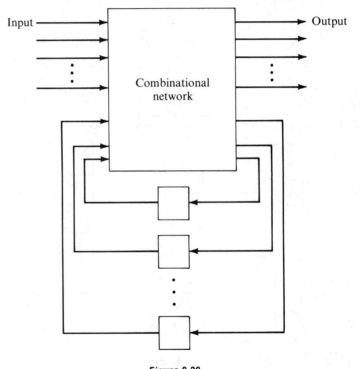

Figure 8.20

Now for some specifics of the construction. All of our logic elements are binary, and input and output values can only be 0 or 1. If the finite-state machine has an input alphabet other than $\{0, 1\}$, we must encode the input symbols in binary form, done by using n-tuples of binary numbers to represent the input symbols (making sure that n is big enough to encode all the symbols). The set of binary n-tuples has 2^n members, so if we have k

input symbols to encode, for example, we pick n just big enough so that $2^n \geq k$, or $n \geq \log_2 k$. Because $\log_2 k$ may not be an integer, we want n to be the smallest integer greater than or equal to $\log_2 k$. We represent this integer by $\lceil \log_2 k \rceil$ ($\lceil x \rceil$ denotes the least integer greater than or equal to x; this is the **ceiling function** of x). If the output alphabet of our machine is not $\{0, 1\}$, we do a similar encoding for output.

8.14 Example Suppose we want to build a sequential network for a finite-state machine with six input symbols. We compute $\lceil \log_2 6 \rceil = 3$. Our sequential network will have three input wires. Notice that there are eight combinations of 0's and 1's that can appear on these three wires. We will assign the six input symbols to six of these combinations, and two combinations will remain unused. □

If the finite-state machine we are building has q states, we compute $\lceil \log_2 q \rceil = p$. Our network will then have p delay elements; different combinations of 0's and 1's on the inputs (or outputs) of the delay elements will represent the different states.

The essence of the construction is to translate the state table of the finite-state machine into truth functions for the outputs and delay elements, and then construct the logic network for each truth function as we did in Chapter 4. Let's look at an example.

8.15 Example We will build a sequential network for the finite-state machine of Example 7.2. The input and output alphabet for this machine is $\{0, 1\}$, and the state table appears in Figure 8.21.

Present state	Next state — Present Input 0	1	Output
s_0	s_1	s_0	0
s_1	s_2	s_1	1
s_2	s_2	s_0	1

Figure 8.21

Because there are only two input symbols and two output symbols, there need only be one input line x (taking on values of 0 or 1) and one output line y for the network. There are three states, so we need $\lceil \log_2 3 \rceil = 2$ delay elements. We arbitrarily associate states with configurations of the inputs or outputs of the delay elements as in Figure 8.22.

	d_1	d_2
s_0	0	0
s_1	0	1
s_2	1	0

Figure 8.22

Now we use the information contained in the state table to write three truth functions. One truth function describes the behavior of the output $y(t)$; it is a function of the two variables $d_1(t)$ and $d_2(t)$, representing the present state. The other two truth functions describe the behavior of $d_1(t + 1)$ and $d_2(t + 1)$, representing the next state; these are functions of $x(t)$, $d_1(t)$ and $d_2(t)$, the present input and present state. Figure 8.23 shows these truth functions.

$x(t)$	$d_1(t)$	$d_2(t)$	$y(t)$	$d_1(t + 1)$	$d_2(t + 1)$
0	0	0	0	0	1
1	0	0	0	0	0
0	0	1	1	1	0
1	0	1	1	0	1
0	1	0	1	1	0
1	1	0	1	0	0

Figure 8.23

In constructing the third line of Figure 8.23, for example, $x(t) = 0$, $d_1(t) = 0$, and $d_2(t) = 1$, meaning that the present input is 0 and the present state is s_1. The output associated with state s_1 is 1, so $y(t) = 1$. The next state associated with input 0 and present state s_1 is s_2; so $d_1(t + 1) = 1$ and $d_2(t + 1) = 0$. Notice that there are some don't care conditions for these functions because the configuration $d_1(t) = 1, d_2(t) = 1$ does not occur.

The canonical sum-of-products form for each of these truth functions is

$$y(t) = d_1'd_2 + d_1d_2' \qquad (y \text{ is not a function of } x)$$

$$d_1(t + 1) = x'd_1'd_2 + x'd_1d_2'$$

$$d_2(t + 1) = x'd_1'd_2' + xd_1'd_2$$

The logic networks for these expressions go into the large box in Figure 8.20. Thus, Figure 8.24 is a wiring diagram for our finite-state machine. ☐

8.16 Practice   One possible behavior pattern for the finite-state machine of Example 7.2 is shown in Figure 8.25. Follow the wiring diagram of Figure 8.24 to see that it reproduces this behavior. Note that d_1 and d_2 are initially both 0 to correspond to the start state s_0.   (*Answer, page A-21*)

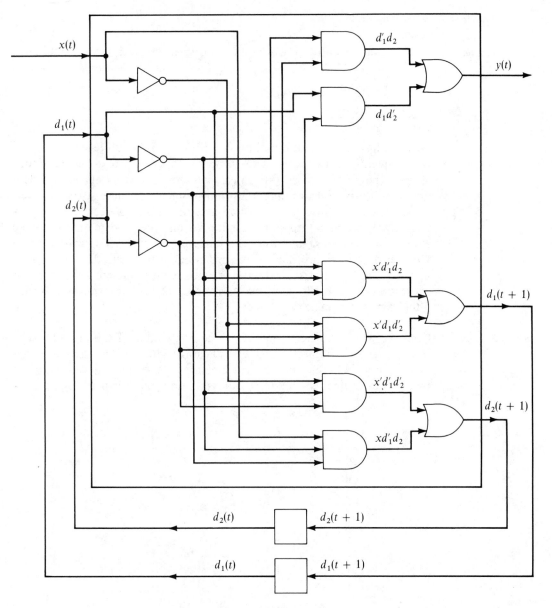

Figure 8.24

Time	t_0	t_1	t_2	t_3	t_4	t_5
Input	0	1	1	0	1	—
State	s_0	s_1	s_1	s_1	s_2	s_0
Output	0	1	1	1	1	0

Figure 8.25

8.17 Practice   Construct a sequential network for the parity-check machine of Example 7.8.

(Answer, page A-21)

MINIMIZATION

Now we have a procedure that allows us to build any finite-state machine by a sequential network using the proper arrangement of elements of four simple types: AND gates, OR gates, inverters, and delay elements. There are several aspects to the minimization problem for these sequential networks. If the machine itself is not already minimized, we can use the procedure of Section 8.1 to minimize it. We may be able to minimize the Boolean expressions, and thus the corresponding logic networks, as in Chapter 4. If there are don't care conditions to the truth functions, these may be assigned values of either 0 or 1 to aid in minimizing the Boolean expression. And finally, a different encoding of machine states as binary n-tuples may minimize the network.

8.18 Example   Consider again the finite-state machine analyzed in Example 8.15. The truth functions of Figure 8.23 have don't care conditions. If we use Karnaugh maps (Section 4.2) to examine these functions, we see that we can minimize by assigning values to the don't care conditions as shown in Figure 8.26.

$x(t)$	$d_1(t)$	$d_2(t)$	$y(t)$	$d_1(t+1)$	$d_2(t+1)$
0	0	0	0	0	1
1	0	0	0	0	0
0	0	1	1	1	0
1	0	1	1	0	1
0	1	0	1	1	0
1	1	0	1	0	0
0	1	1	1	1	0
1	1	1	1	0	1

Figure 8.26

The Boolean expressions for the logic network become

$$y(t) = d_1 + d_2$$
$$d_1(t+1) = x'd_1 + x'd_2 = x'(d_1 + d_2)$$
$$d_2(t+1) = xd_2 + x'd_1'd_2'$$

The simplified network appears in Figure 8.27.   ☐

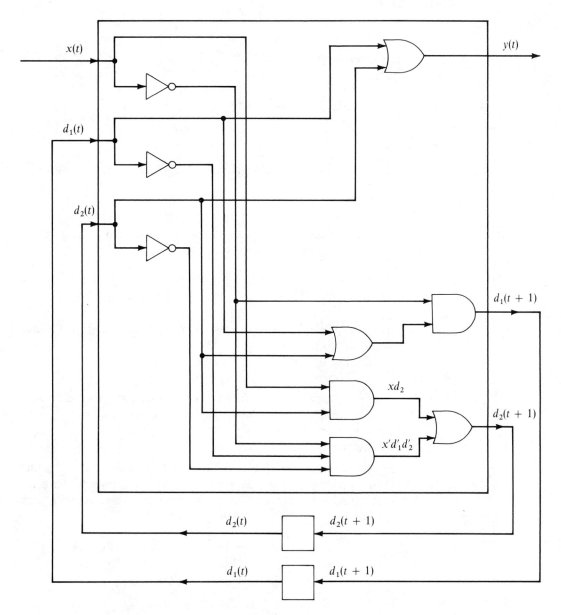

Figure 8.27

√ CHECKLIST

Definitions

delay element (*p. 296*)
combinational network (*p. 297*)
sequential network (*p. 297*)
ceiling function (*p. 299*)

Technique

Construct sequential networks for finite-state machines.

Main Idea

Only four simple types of logic elements are needed to construct a network for any finite-state machine.

EXERCISES, SECTION 8.2

1. Construct a sequential network for each of the finite-state machines below.
   ⋆ (a) The machine of Example 7.9.
   (b) The machine of Practice 7.4.
   (c) The serial binary adder of Figure 7.7.

2. Construct a sequential network for each of the finite-state machines below. Make use of don't care conditions to simplify the networks.
   (a) The machine of Practice 7.5.
   (b) The machine of Exercise 1(a), Section 7.1.

3. Using the finite-state machine of Example 8.15 and Example 8.18, find a different encoding of states as binary n-tuples that, together with the don't care conditions, can be used to produce a simpler network than that of Figure 8.27. Draw the network.

Section 8.3   *PARALLEL AND SERIAL DECOMPOSITIONS*

In Section 1 of this chapter, we learned how to find an equivalent, minimized machine for a given finite-state machine M, which allows us to possibly reduce the number of states required, but still leaves us with a single, perhaps very complex, finite-state machine. In Section 2 we built sequential networks for finite-state machines using four basic logic elements. The sequential network replaces a single machine with a network of very simple elements, but there may be lots of these elements and their interconnections may be complex.

Between these two extremes, we now want to discuss how to replace a single finite-state machine with interconnections of a few simpler finite-state machines. The ability to do such a replacement might allow us to stock up on some standard small machines and to use them in various combinations to build new machines. (Think of writing a computer program calling for various standard subroutines.) This area of investigation is called **structure theory** or **decomposition theory** of machines, and it will occupy us for the rest of this chapter.

Suppose that M is our original machine. A decomposition for M will be a network of smaller machines. The network viewed as a whole, however, is itself a machine M'. We want M' to simulate M. In the first two types of decomposition we will consider, something stronger will happen: a submachine of M' will be isomorphic to M.

Our first two decomposition types borrow ideas from basic electronics. Given two black boxes, we can think of connecting them in parallel (Figure 8.28a) or series (Figure 8.28b). If the black boxes are finite-state machines, we want to view the resulting network as a finite-state machine.

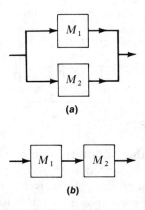

(a)

(b)

Figure 8.28

PARALLEL DECOMPOSITIONS

Let M_1 and M_2 be two finite-state machines connected in parallel as in Figure 8.28a. We want to represent the total network as a finite-state machine M_p. Clearly, the present state of M_p should depend equally upon the present states of M_1 and M_2. This goal suggests using ordered pairs of states from the two machines for the state set of M_p; in fact, we define the entire machine M_p in terms of ordered pairs of elements from M_1 and M_2 with the next-state function and output function operating componentwise.

8.19 Definition   Given two finite state machines, $M_1 = [S_1, I_1, O_1, f_{s_1}, f_{o_1}]$ and $M_2 = [S_2, I_2, O_2, f_{s_2}, f_{o_2}]$, the **parallel connection** of M_1 and M_2 is the machine $M_p = [S_1 \times S_2, I_1 \times I_2, O_1 \times O_2, f_s, f_o]$ where $f_s((s_1, s_2), (i_1, i_2)) = (f_{s_1}(s_1, i_1), f_{s_2}(s_2, i_2))$, $f_o(s_1, s_2) = (f_{o_1}(s_1), f_{o_2}(s_2))$. □

8.20 Practice   The state tables for machines M_1 and M_2 are given in Figure 8.29. Complete the state table (Figure 8.30) for the parallel connection M_p of M_1 and M_2.

(*Answer, page A-22*)

M_1	Input 0	1	Output
A	A	A	1
B	B	A	0

M_2	Input 0	1	Output
a	b	a	1
b	a	b	1

Figure 8.29

M_p	Input (0, 0)	(0, 1)	(1, 0)	(1, 1)	Output
(A, a)	(A, b)	(A, a)	(A, b)	(A, a)	$(1, 1)$
(A, b)	(A, a)	—	(A, a)	—	$(1, 1)$
(B, a)	(B, b)	(B, a)	—	(A, a)	$(0, 1)$
(B, b)	—	—	(A, a)	(A, b)	—

Figure 8.30

We want to begin with an arbitrary machine M and find two machines, M_1 and M_2, whose parallel connection M_p will simulate M—indeed, so M_p will have a submachine isomorphic to M. M_p is called a **parallel decomposition** of M. When will a parallel decomposition of M exist and if it does, how can we find M_1 and M_2?

If a parallel decomposition of M exists, then the original machine M will probably be more complex than either M_1 or M_2; yet in some sense it must carry on the activities of both M_1 and M_2, suggesting that M_1 and M_2 will be closely related to quotient machines of M. It turns out that we can temporarily ignore outputs and then later choose them so that they work. Thus, we define a partition of a machine in which all states in a block, under a given input symbol, proceed to states in the same partition block as having the **substitution property** or of being an **S.P. partition**. How to define the next-state function for a machine whose states are the blocks of an S.P. partition is clear. For π_1 and π_2, two S.P. partitions of M, we can define (except for outputs) machines M_1 and M_2 using the blocks of π_1 and π_2, respectively, as states. Any state of M belongs to a block of π_1 and a block of π_2. The next-state function for M_1 will tell us, to within a partition block of π_1, the next state of M. The next-state function for M_2 will tell us, to within a partition block of π_2, the next state of M. Suppose now that π_1 and π_2 have the additional property that the intersection of any block from π_1 with any block from π_2 contains at most one element of M. Then the next state of M can be discovered. Partitions with such a property are called **orthogonal**.

8.21 Example M is a machine given by the state table of Figure 8.31.

M	Input 0	1	Output 0
0	3	2	0
1	5	2	0
2	4	1	0
3	1	4	1
4	0	3	0
5	2	3	0

Figure 8.31

The partitions

$$\pi_1 = \{\overline{0, 1, 2}; \overline{3, 4, 5}\}$$

and

$$\pi_2 = \{\overline{0, 5}; \overline{1, 4}; \overline{2, 3}\}$$

are orthogonal S.P. partitions. Let's use π_1 and π_2 to construct a parallel decomposition of M. (Note that M is already a reduced machine.) The machines M_1 corresponding to π_1 and M_2 corresponding to π_2 have next-state functions

M_1	0	1
$A = \{0, 1, 2\}$	B	A
$B = \{3, 4, 5\}$	A	B

M_2	0	1
$a = \{0, 5\}$	c	c
$b = \{1, 4\}$	a	c
$c = \{2, 3\}$	b	b

Because M receives input of only 0 and 1, in constructing the parallel connection M_p of M_1 and M_2 to simulate M, we need only consider inputs of $(0, 0)$ and $(1, 1)$, which we will write as 0 and 1, respectively. Then the next-state function for M_p is

M_p	0	1
(A, a)	(B, c)	(A, c)
(A, b)	(B, a)	(A, c)
(A, c)	(B, b)	(A, b)
(B, a)	(A, c)	(B, c)
(B, b)	(A, a)	(B, c)
(B, c)	(A, b)	(B, b)

Now we want to define an isomorphism g from a submachine of M_p to M. Because π_1 and π_2 are orthogonal, each state of M_p corresponds to a single state of M, suggesting that g be defined by

$(A, a) \rightarrow 0$
$(A, b) \rightarrow 1$
$(A, c) \rightarrow 2$
$(B, a) \rightarrow 5$
$(B, b) \rightarrow 4$
$(B, c) \rightarrow 3$

By the construction of M_p, it should be clear that g serves as an isomorphism from M_p onto M at least as far as preserving the next-state function. We will do a sample calculation, in which f_s denotes the next-state function of M and f_p denotes the next-state function of M_p. Then $g(f_p((B, a), 0)) = g(A, c) = 2$ and $f_s(g(B, a), 0) = f_s(5, 0) = 2$. The commutative diagram appears in Figure 8.32.

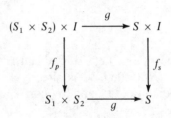

Figure 8.32

Now it is time to take care of the outputs. The output for M_p will be ordered pairs of the output symbols for M_1 and M_2. For M_p to simulate M, we must extend our definition of homomorphism to allow a mapping g' from the output symbols of M_p to those of M. If each state of M_1 and each state of M_2 has a distinct output symbol, then each ordered pair in the output of M_p is associated with at most one state of M, and therefore one output symbol of M. This defines our mapping. In some cases, however, we can limit the outputs of M_1 and M_2 to $\{0, 1\}$ and choose the output symbol for each state in such a way that we can define a mapping. For this example, we can define the function g' by

$(0, 0) \rightarrow 0$
$(0, 1) \rightarrow 0$
$(1, 0) \rightarrow 0$
$(1, 1) \rightarrow 1$

and assign the outputs for M_1 and M_2 so that $(1, 1)$ in M_p occurs exactly at (B, c), the state of M_p corresponding to 3, the only state of M with output 1. (Other output functions g' and other output assignments can be used.) The

M_1	Input 0	1	Output
A	B	A	0
B	A	B	1

M_2	Input 0	1	Output
a	c	c	0
b	a	c	0
c	b	b	1

M_p	Input 0	1	Output
(A, a)	(B, c)	(A, c)	$(0, 0)$
(A, b)	(B, a)	(A, c)	$(0, 0)$
(A, c)	(B, b)	(A, b)	$(0, 1)$
(B, a)	(A, c)	(B, c)	$(1, 0)$
(B, b)	(A, a)	(B, c)	$(1, 0)$
(B, c)	(A, b)	(B, b)	$(1, 1)$

Figure 8.33

complete tables for M_1, M_2, and M_p appear in Figure 8.33. M_p, the parallel connection of M_1 and M_2, is isomorphic to M by means of the function g (and g'). The state table for M_p is a relabeling of the original table for M. □

8.22 Practice   For Example 8.21, compute

(a) $g(f_p((A, c), 1))$

and

(b) $f_s(g(A, c), 1)$ *(Answer, page A-22)*

The following theorem on parallel decomposition of machines merely formalizes the discussion preceding Example 8.21 and the construction of Example 8.21 itself; we will omit the proof.

8.23 Theorem   A finite-state machine M has a parallel decomposition if and only if there are two orthogonal S.P. partitions on M.   □

Notice that the partition π_1 consisting of a single block—all the states of M—is trivially S.P., and so is the partition π_2 into single states of M. Furthermore, these are orthogonal partitions, but they result in a parallel decomposition where M_2 is essentially M and M_1 does nothing. For a nontrivial decomposition, we must find nontrivial, orthogonal S.P. partitions.

8.24 Example   The state table for a machine M is given in Figure 8.34. The state tables for component machines in a parallel decomposition of M appear in Figure 8.35.

	Input		
M	0	1	Output
0	0	1	0
1	2	0	0
2	4	3	0
3	0	4	1
4	2	3	1

Figure 8.34

	Input		
M_1	0	1	Output
$A = \{1, 3\}$	B	B	0
$B = \{0, 2, 4\}$	B	A	1

	Input		
M_2	0	1	Output
$a = \{1, 4\}$	c	b	0
$b = \{0, 3\}$	b	a	1
$c = \{2\}$	a	b	1

Figure 8.35

The isomorphism g (and the output function g') for these machines is given by

g	g'
$(A, a) \to 1$	$(0, 0) \to 0$
$(A, b) \to 3$	$(0, 1) \to 1$
$(B, a) \to 4$	$(1, 0) \to 1$
$(B, b) \to 0$	$(1, 1) \to 0$
$(B, c) \to 2$	

(Again other output assignments and other output functions g' can be used.) Here, g is an isomorphism from a submachine of the decomposition onto M since state (A, c) is never used. The total number of states in the decomposition of M is larger than in M itself, but the component machines are smaller. □

8.25 Practice   For a parallel decomposition of the machine M of Figure 8.36,

M	Input 0	1	Output
0	0	3	0
1	0	2	1
2	3	1	0
3	3	0	1

Figure 8.36

(a) Write the state tables of component machines.
(b) Give the isomorphism g (and the output function g').

(Answer, page A-22)

SERIAL DECOMPOSITIONS

Now let's consider two machines connected in series as in Figure 8.28b. Actually, we'll modify the wiring for a series connection. Figure 8.28b suggests that an input signal will enter M_1 only. It will take one clock pulse before the output of M_1 (via its next state) reflects the impact of this input signal, and two clock pulses before the output of M_2 reflects it. For more machines connected serially, this delay is propagated. Then if we want to interconnect serial and parallel networks, we must be concerned with timing. To avoid this problem, we'll direct the input signal to both M_1

and M_2, and compute next-state functions simultaneously. We will also read both outputs simultaneously. Our wiring diagram is thus modified to that of Figure 8.37a.

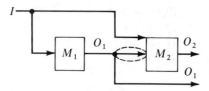

Figure 8.37 (a)

All this talk about simultaneous operations may sound as if we have gone back to the parallel case; the crucial difference is the use of M_1's output as input to M_2 (see the segment enclosed by dots). In fact, we will allow M_2 to have the maximum information from M_1 by requiring that the output of M_1 be the present state of M_1. For simplicity, we will put the same requirement on M_2, giving us Figure 8.37b.

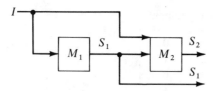

Figure 8.37 (b)

8.26 Definition   Let $M_1 = [S_1, I_1, O_1, f_{s_1}, f_{o_1}]$ and $M_2 = [S_2, I_2, O_2, f_{s_2}, f_{o_2}]$ be two finite-state machines with

$$O_1 = S_1, f_{o_1}(s_1) = s_1$$
$$O_2 = S_2, f_{o_2}(s_2) = s_2$$

and

$$I_2 = S_1 \times I_1$$

The **serial connection** of M_1 and M_2 is the machine $M_s = [S_1 \times S_2, I_1, O, f_s, f_o]$ where

$$O = S_1 \times S_2, \qquad f_o(s_1, s_2) = (f_{o_1}(s_1), f_{o_2}(s_2)) = (s_1, s_2)$$
$$f_s((s_1, s_2), i) = (f_{s_1}(s_1, i), f_{s_2}(s_2, (s_1, i))) \qquad\qquad \square$$

8.27 Practice   The state tables for machines M_1 and M_2 are given in Figure 8.38. Complete the state table (Figure 8.39) for the serial connection of M_1 and M_2.

(*Answer, page A-23*)

M_1	Input 0	1	Output
A	A	A	A
B	B	A	B

M_2	$(A, 0)$	$(A, 1)$	$(B, 0)$	$(B, 1)$	Output
a	b	a	a	b	a
b	a	b	a	b	b

Figure 8.38

M_s	Input 0	1	Output
(A, a)	(A, b)	(A, a)	(A, a)
(A, b)	—	(A, b)	(A, b)
(B, a)	—	—	(B, a)
(B, b)	(B, a)	(A, b)	(B, b)

Figure 8.39

Given a machine M, we want to find two machines whose serial connection M_s will have a submachine isomorphic to M. M_s is a **serial decomposition** of M. If M has a serial decomposition, then each state s of M is associated with a unique (s_1, s_2) state of M_s, and under an input symbol i, if $(s, i) \to r$ in M, where r is associated with (r_1, r_2) in M_s, then we must have $(s_1, i) = r_1$. Except for outputs, this association identifies a machine homomorphism from M onto M_1, making M_1 (except for outputs) a quotient machine of M. Thus, M_1 is the machine of an S.P. partition π of M. Conversely, if M has an S.P. partition π, and we make the "head machine" in a serial decomposition of M the machine for π, then we can build a "tail machine" as follows.

Find a partition π_2 of M such that π_1 and π_2 are orthogonal. (This is easy; make one block of π_2 consist of the first element in each block of π_1, the second block of π_2 consist of the second element in each block of π_1, etc. We are not requiring that π_2 be S.P.). Any state of M belongs to a block of π_1 and a block of π_2. Let M_2 have as states the blocks of π_2. Because π_1 and π_2 are orthogonal and M_2 (from its present state) knows the block of π_2 and (from M_1's output) knows the block of π_1, M_2 knows the exact, present state s of M. Consulting M for the next state of s under the present input tells us the next state function for M_2.

8.28 Example M is a machine given by the state table of Figure 8.40. $\pi_1 = \{\overline{0,3}; \overline{1,6}; \overline{2,4,5}\}$ is an S.P. partition of M. Thus, we define M_1 by Figure 8.41. Now we let $\pi_2 = \{\overline{0,1,2}; \overline{3,6,4}; \overline{5}\}$. The state table for M_2 will have the format of Figure 8.42.

M	Input 0	1	Output
0	1	6	1
1	3	3	1
2	3	5	1
3	6	6	1
4	0	2	0
5	3	4	0
6	0	0	1

Figure 8.40

M_1	Input 0	1	Output
$A = \{0, 3\}$	B	B	A
$B = \{1, 6\}$	A	A	B
$C = \{2, 4, 5\}$	A	C	C

Figure 8.41

M_2	Input $(A, 0)$	$(A, 1)$	$(B, 0)$	$(B, 1)$	$(C, 0)$	$(C, 1)$	Output
$a = \{0, 1, 2\}$							a
$b = \{3, 6, 4\}$							b
$c = \{5\}$							c

Figure 8.42

M_2	Input						Output
	$(A,0)$	$(A,1)$	$(B,0)$	$(B,1)$	$(C,0)$	$(C,1)$	
a	a	b	b	b	b	c	a
b	b	b	a	a	a	a	b
c	–	–	–	–	b	b	c

Figure 8.43

To complete the state table, consider the next state for a under input $(A, 0)$. The a–A pair uniquely selects state 0 of M, which under input 0 goes to state $1 \in \{0, 1, 2\}$. Thus, the next state for M_2 should be a. Other entries are computed in the same way; the completed table for M_2 appears in Figure 8.43. The dashes indicate don't care conditions; they occur for combinations of states of M_1 and M_2 that do not represent states of M. These conditions could be specified in any way that might economize construction of M_2. Without actually writing the machine M_s, we can see that the isomorphism function g (and output function g') given by

$$g:(A, a) \to 0 \qquad g':(A, a) \to 1$$
$$(A, b) \to 3 \qquad (A, b) \to 1$$
$$(B, a) \to 1 \qquad (B, a) \to 1$$
$$(B, b) \to 6 \qquad (B, b) \to 1$$
$$(C, a) \to 2 \qquad (C, a) \to 1$$
$$(C, b) \to 4 \qquad (C, b) \to 0$$
$$(C, c) \to 5 \qquad (C, c) \to 0$$

makes a submachine of M_s isomorphic to M.   □

8.29 Practice   In M of Example 8.28, state 4 under input 1 goes to state 2. What is the equivalent computation in M_s?   (*Answer, page A-23*)

The discussion preceding Example 8.28 was essentially a proof of the following theorem.

8.30 Theorem   A finite-state machine M has a serial decomposition if and only if there is an S.P. partition on M.   □

Again, a nontrivial decomposition would require that both M_1 and M_2 have fewer states than M itself, in which case the S.P. partition is nontrivial.

8.31 Practice   Write the state tables of component machines and give the isomorphism g

and the output function g' for a serial decomposition of the machine M of Figure 8.44.

(*Answer, page A-23*)

M	Input 0	1	Output
0	3	0	1
1	1	0	1
2	3	2	0
3	3	0	0

Figure 8.44

A parallel decomposition of a machine seems philosophically a less complex arrangement than a serial decomposition of the same machine. Practically, computers are being built that operate in a parallel mode. If all computations in a matrix multiplication, for example, are done in parallel and the answers fitted into the proper array, this is faster than the serial approach of computing one entry, then the next, and so on.

✓ CHECKLIST

Definitions

parallel connection (*p. 306*)
parallel decomposition (*p. 306*)
S.P. partition (*p. 307*)
orthogonal partitions (*p. 307*)
serial connection (*p. 312*)
serial decomposition (*p. 313*)

Techniques

Find a nontrivial, parallel decomposition of a given machine M if one exists.

Find a nontrivial, serial decomposition of a given machine M if one exists.

Main Ideas

A finite-state machine M has a nontrivial, parallel decomposition if and only if there are two nontrivial, orthogonal, S.P. partitions on M.

A finite-state machine M has a nontrivial, serial decomposition if and only if there is a nontrivial, S.P. partition on M.

EXERCISES, SECTION 8.3

★ 1. (a) Given the machine M of Figure 8.45, write the state tables for component machines (with output alphabets $\{0, 1\}$) in a parallel decomposition of M; also give the isomorphism g and the output function g'.

M	Input 0	1	Output
0	4	3	0
1	2	1	1
2	1	3	1
3	2	0	1
4	0	1	0

Figure 8.45

   (b) The following computations are done in M; do the corresponding computations in the parallel decomposition of M.

$$f_s(4, 0) = 0$$
$$f_s(0, 1) = 3$$
$$f_s(3, 0) = 2$$

2. (a) Given the machine M of Figure 8.46, write the state tables for component machines (with output alphabets $\{0, 1\}$) in a parallel decomposition of M; also give the isomorphism g and the output function g'.

M	Input 0	1	Output
0	3	3	1
1	6	4	0
2	3	2	0
3	2	0	0
4	1	1	1
5	0	5	0
6	1	6	0

Figure 8.46

(b) The following computations are done in M; do the corresponding computations in the parallel decomposition of M.

$$f_s(5, 1) = 5$$
$$f_s(3, 0) = 2$$
$$f_s(6, 0) = 1$$

3. (a) Given the machine M of Figure 8.47, write the state tables for component machines in a parallel decomposition of M; also give the isomorphism g and the output function g'. Can you use $\{0, 1\}$ as the output alphabet for each component machine?

M	Input 0	1	Output
0	1	2	0
1	1	5	1
2	5	3	1
3	4	0	0
4	4	1	0
5	5	4	0

Figure 8.47

(b) The following computations are done in M; do the corresponding computations in the parallel decomposition of M.

$$f_s(3, 0) = 4$$
$$f_s(1, 1) = 5$$
$$f_s(0, 1) = 2$$

⋆ 4. (a) Given the machine M of Figure 8.48, write the state tables for com-

M	Input 0	1	Output
0	1	2	0
1	0	4	0
2	5	3	1
3	4	2	0
4	3	1	1
5	3	1	0

Figure 8.48

ponent machines in a serial decomposition of M; also give the isomorphism g and the output function g'.

(b) The following computations are done in M; do the corresponding computations in the serial decomposition of M.

$$f_s(2, 0) = 5$$

$$f_s(4, 1) = 1$$

$$f_s(3, 1) = 2$$

5. (a) Given the machine M of Figure 8.49, write the state tables for component machines in a serial decomposition of M; also give the isomorphism g and the output function g'.

M	Input 0	1	Output
0	4	2	1
1	2	3	1
2	0	4	0
3	1	2	1
4	0	3	0

Figure 8.49

(b) The following computations are done in M; do the corresponding computations in the serial decomposition of M.

$$f_s(0, 0) = 4$$

$$f_s(2, 0) = 0$$

$$f_s(4, 1) = 3$$

Section 8.4   **CASCADE DECOMPOSITIONS**

The simulations of a machine M by either parallel or serial decompositions (Section 8.3) have two features. First, there is an isomorphism from a submachine of the decomposition onto M. Second, the decompositions are *loop-free* in that no machine output becomes the input of an earlier machine. (The individual component machines may have loop structure; indeed, if we recall how we built machines from logic elements in Section 8.2, loops were essential.) We want to relax the isomorphism property, but preserve the loop-free construction. A more general, loop-free construction, of which the parallel and serial forms are special cases, appears in Figure 8.50. It is called a **cascade connection** of machines M_1 and M_2.

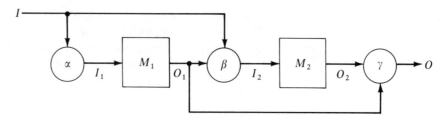

Figure 8.50

Let $M_1 = [S_1, I_1, O_1, f_{s_1}, f_{o_1}]$ and $M_2 = [S_2, I_2, O_2, f_{s_2}, f_{o_2}]$. The network as a whole, M_c, has input I, output O, and states $S_1 \times S_2$. In Figure 8.50, α, β, and γ are transformations on, respectively, I (to convert input to I_1 for use by M_1), $I \times O_1$ (to convert information to I_2 for use by M_2), and $O_1 \times O_2$ (to convert outputs to O). The output function f_o for M_c is

$$f_o(s_1, s_2) = \gamma(f_{o_1}(s_1), f_{o_2}(s_2))$$

Thus, M_c lets M_1 and M_2 each compute its output for its present state and runs the results through γ. The next-state function f_s for M_c is

$$f_s((s_1, s_2), i) = (f_{s_1}(s_1, \alpha(i)), f_{s_2}(s_2, \beta(i, f_{o_1}(s_1))))$$

Thus, M_c lets M_1 and M_2 each compute its next state based on its present state and appropriately transformed input. The cascade construction can be extended to more than two machines.

We now want to know when an arbitrary machine M can be simulated by a cascade of machines, and how these machines can be found. Before we explore this question, we need a few more algebraic ideas.

ALGEBRAIC STRUCTURES AND FINITE-STATE MACHINES

Presently, we will tie together algebraic structures, namely semigroups and monoids, and finite-state machines. But first, a quick list (without proofs) of other algebraic results we will need.

8.32 Definition   A semigroup B **divides** a semigroup A if B is a homomorphic image of a subsemigroup of A.   □

8.33 Theorem   If a group G divides a finite semigroup A, then G is a homomorphic image of a subgroup of A.   □

8.34 Definition   A group $[G, \cdot]$ is **simple** if it has no proper normal subgroups.   □

8.35 Example   (a) Any group of prime order is simple since, by Lagrange's Theorem (5.118), it has no proper subgroups of any kind.

(b) $[\mathbb{Z}_6, +_6]$ is not a simple group because $\{0, 2, 4\}$ and $\{0, 3\}$ are both proper normal subgroups.   □

8.36 Definition   A proper normal subgroup of a group $[G, \cdot]$ is **maximal** if it is not a proper subset of any other proper normal subgroup of G.   □

8.37 Example   Both $\{0, 2, 4\}$ and $\{0, 3\}$ are maximal, normal subgroups in $[\mathbb{Z}_6, +_6]$. Also, $\{0\}$ is a maximal, normal subgroup of $\{0, 2, 4\}$ and of $\{0, 3\}$.   □

Let's assume that a group $[G, \cdot]$ with identity i contains a maximal, normal subgroup G_1, that G_1 in turn contains a maximal, normal subgroup G_2, G_2 contains a maximal, normal subgroup G_3, and so on. Also assume that we can only perform this process a finite number of times. Thus,

$$\{i\} = G_k \subset G_{k-1} \cdots \subset G_3 \subset G_2 \subset G_1 \subset G_0 = G$$

with each G_i a maximal, normal subgroup of G_{i-1}, $1 \leq i \leq k$. This arrangement is called a **composition series** of G. Associated with it is a set of quotient groups G_{i-1}/G_i, $1 \leq i \leq k$, called the **factor set** of the composition.

8.38 Example $$\{0\} \subset \{0, 2, 4\} \subset \mathbb{Z}_6$$

and

$$\{0\} \subset \{0, 3\} \subset \mathbb{Z}_6$$

are two composition series for $[\mathbb{Z}_6, +_6]$. The factor set for the first series contains $\mathbb{Z}_6/\{0, 2, 4\}$, which has two elements and is isomorphic to $[\mathbb{Z}_2, +_2]$, and $\{0, 2, 4\}/\{0\}$, which has three elements and is isomorphic to $[\mathbb{Z}_3, +_3]$. The factor set for the second series contains $\mathbb{Z}_6/\{0, 3\} \simeq \mathbb{Z}_3$, and $\{0, 3\}/\{0\} \simeq \mathbb{Z}_2$.   □

In Example 8.38, each factor set for $\mathbb{Z}_6$ contains the same number of elements. Furthermore, to within isomorphism, each factor set contains the same elements. That this is typical is given by the Jordan–Hölder Theorem.

8.39 Theorem   Jordan–Hölder
Any two factor sets of a group G have the same number of elements, and the members of any factor set are isomorphic to the members of any other factor set.   □

Every finite group G must have a composition series. The Jordan–Hölder Theorem allows us to identify a set of essentially unique groups associated with G; these groups are called **composition factors** of G.

8.40 Practice   (a) A composition series for $[\mathbb{Z}_{12}, +_{12}]$ is

$$\{0\} \subset \{0, 6\} \subset \{0, 3, 6, 9\} \subset \mathbb{Z}_{12}$$

Find the members of the factor set.

(b) Two other composition series for $\mathbb{Z}_{12}$ are given below. Show that the members of the factor sets are isomorphic to the members of the factor set of part (a).

$$\{0\} \subset \{0, 4, 8\} \subset \{0, 2, 4, 6, 8, 10\} \subset \mathbb{Z}_{12}$$

$$\{0\} \subset \{0, 6\} \subset \{0, 2, 4, 6, 8, 10\} \subset \mathbb{Z}_{12} \qquad \textit{(Answer, page A-23)}$$

Semigroup of a Machine

Given a finite-state machine M, each input symbol, through applying the next-state function, has the effect of a transformation on the set S of states of M. Composition of these transformations represents the effects of strings of input symbols. (Here, we will compose from left to right to go along with reading input strings from left to right.) Thus, the set of all transformations on S accomplished by input strings is closed under composition (and associative) and is therefore a semigroup, called the **semigroup of** M, denoted by S_M. If we consider the identity transformation to be the result of applying the empty string of input symbols, then every such semigroup is actually a monoid, but for some reason, it is still called the semigroup of the machine. The mapping taking a string of input symbols to its corresponding transformation of states is a homomorphism from I^*, the free monoid generated by I (Example 5.21), onto this monoid.

8.41 Example

The next-state function for a machine M is given in Figure 8.51. The transformations on $S = \{0, 1, 2\}$ resulting from all input strings of length 0 to 3 symbols are given in Figure 8.52. Input strings of four or more characters do not produce any new transformations. S_M has eight elements; if we label the eight columns of Figure 8.52 by i, 0, 1, 2, and so on, left to right, then we have the semigroup table for S_M as given in Figure 8.53. Recall that the binary operation is composition. For example, $3 \circ 4$ in S_M is the transformation produced by an input string 0110, but since 011 has the same effect as 11, we can write 0110 as 110, which accomplishes transformation 6. Thus, $3 \circ 4 = 6$.

	Input	
M	0	1
0	2	1
1	0	1
2	2	0

Figure 8.51

		i	0	1	2	3	4	5	6
					100			111	
			010	101	000			011	110
S	λ	0	1	00	01	10	11	001	
0	0	2	1	2	0	0	1	0
1	1	0	1	2	1	0	1	0
2	2	2	0	2	0	2	1	0

Figure 8.52

∘	i	0	1	2	3	4	5	6
i	i	0	1	2	3	4	5	6
0	0	2	3	2	6	0	5	6
1	1	4	5	2	1	6	5	6
2	2	2	6	2	6	0	5	6
3	3	0	5	2	3	6	5	6
4	4	2	1	2	6	4	5	6
5	5	6	5	2	5	6	5	6
6	6	2	5	2	6	6	5	6

Figure 8.53   □

8.42 Practice

(a) Find the state transformations in the semigroup of M where the next-state function for M is given by Figure 8.54.

	Input	
M	0	1
0	0	1
1	0	0

Figure 8.54

(b) Show the semigroup table for S_M.   *(Answer, page A-23)*

Machine of a Semigroup

As a converse to the semigroup of a machine, we can view any finite semi-group S as a finite-state machine $M(S)$ whose inputs and states are the elements of S and where the semigroup table for S defines the next-state function for $M(S)$.

We can now start with a machine M, form its semigroup S_M, and then consider it as a machine $M(S_M)$. There is a relationship between the two machines; we can find a homomorphism from $M(S_M)$ onto M. Every state q in $M(S_M)$ stands for a class $[q]$ of input strings in M. Each member of $[q]$ takes the starting state s_0 of M to the same state of M, denoted by $s_0[q]$. Now we define a mapping g from the states of $M(S_M)$ to the states of M by $g(q) = s_0[q]$. To show that g preserves the next-state function, let $q \in M(S_M)$ and let $x \in I$, the input alphabet for M (here we consider only a subset of $M(S_M)$'s input alphabet). If we operate and then map, we compute the next-state in $M(S_M)$, $q \circ x$, and then $g(q \circ x) = s_0[q \circ x]$. If we map and then operate, we take $g(q) = s_0[q]$ and then apply the next-state function in M. But the next state of $s_0[q]$ under x is $s_0[q \circ x]$. The mapping g is onto because every state in M is reachable from s_0 (s_0 is reachable from s_0 by λ). We have never bothered to define the output for $M(S_M)$, but once g is determined, output for $M(S_M)$ can be defined so that corresponding states have corresponding output. Thus, we have a homomorphism from $M(S_M)$ onto M, and $M(S_M)$ simulates M.

8.43 Example   For the machine of Example 8.41, the homomorphism g from $M(S_M)$ onto M is given by

$$g: i \to 0$$
$$0 \to 2$$
$$1 \to 1$$
$$2 \to 2$$
$$3 \to 0$$
$$4 \to 0$$
$$5 \to 1$$
$$6 \to 0$$

In $M(S_M)$, the next state of 3 under 1 is 5, and $g(5) = 1$; $g(3) = 0$ and in M the next state of 0 under 1 is 1.   □

8.44 Practice   Find a homomorphism from $M(S_M)$ onto M for the machine of Practice 8.42.

(*Answer, page A-23*)

Before we get to some results on cascade decompositions, we need to define a simple finite-state machine that will be a stock item in such networks.

8.45 Definition   A **flip-flop** is a 2-state, 3-input machine defined by the state table of Figure 8.55.   □

	Input		
F	i	0	1
0	0	0	1
1	1	0	1

Figure 8.55

THEOREMS ON CASCADE DECOMPOSITIONS

The most general result on cascade decompositions was given in 1965 by K. B. Krohn and J. L. Rhodes; we will state it, but omit the proof.

8.46 Theorem   Let M be a finite-state machine with semigroup S_M. Then M can be simulated by a cascade connection of flip-flops and machines of simple groups dividing S_M.   □

Note that for any machine M, S_M will be a finite semigroup. Groups dividing S_M are, by Theorem 8.33, homomorphic images of subgroups of S_M. And by the Fundamental Homomorphism Theorem for Groups, such groups are isomorphic to quotient groups of subgroups of S_M. Thus, as component machines, we are essentially looking for (are you ready?) simple quotient groups of subgroups of the semigroup S_M! If S_M is itself a group (inputs to M produce permutations of the states), we may already know a lot about its structure and be able to find the needed substructures without much difficulty. But, in fact, we have a simpler cascade decomposition when the semigroup of M is a group.

8.47 Theorem   Let M be a finite-state machine whose semigroup is a group G. Then M can be simulated by a cascade connection of the machines of the composition factors of G.   □

Notice that the Jordan–Hölder Theorem guarantees that the composition factors of G are unique to within isomorphism.

Now for the proof of Theorem 8.47. Let M be a machine with semigroup G, a group. Since we already know that $M(G)$ simulates M, we can concentrate on a simulation of $M(G)$. Suppose that a composition series for G has the form

$$\{i\} = G_k \subset G_{k-1} \cdots \subset G_2 \subset G_1 \subset G_0 = G$$

What we will actually prove is that, given any group G^* and a normal subgroup S of G^*, we can simulate $M(G^*)$ by a cascade of $M(S)$ and $M(G^*/S)$. This result will allow us to simulate $M(G_0)$ by a cascade of $M(G_1)$ and $M(G_0/G_1)$, and then to simulate $M(G_1)$ by a cascade of $M(G_2)$ and $M(G_1/G_2)$, and so on. Putting these decompositions together, we will simulate $M(G_0)$ by a cascade of $M(G_0/G_1)$, $M(G_1/G_2)$, $\cdots M(G_k/\{i\}) = M(G_k)$ and $M(\{i\})$. But $M(\{i\})$ is a trivial machine and not needed in the network.

Thus, we let G (instead of G^*) be any group with S a normal subgroup of G. Our claim is that the cascade decomposition shown in Figure 8.56 can simulate $M(G)$ for the right choices of α, β, and γ. I will be the input alphabet for $M(G)$, which consists of the elements of G. We will choose γ so that O consists of the states of $M(G)$, that is, the elements of G; we can easily assign outputs to these states by using the output function for $M(G)$. Then we must

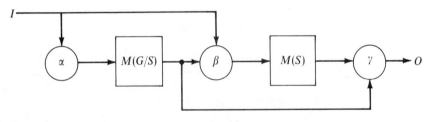

Figure 8.56

associate with each state of $M(G)$ (each $g \in G$) a state of the cascade machine so that corresponding states will produce the same output strings when given the same input strings. We assume that the output for $M(G/S)$ equals the present state, and similarly for $M(S)$.

For each coset $[g]$ in G/S, we pick a fixed representative member g'. Then for any $g \in G$, $g \in [g] = [g'] = Sg'$, and $g = sg'$ for some $s \in S$. We associate with g the state $([g], s)$ of the cascade machine. Our choice for α and γ is relatively clear; α will map $g \in G$ to $[g]$, and γ will map $([g], s)$ to sg' where g' is the fixed representative of $[g]$. Now let's figure out a definition for β that will make things work.

Suppose that $M(G)$ is presently in state g^* and receives an input of g. The next state for $M(G)$ is $g^* \cdot g$. The present state of the cascade machine must be the corresponding state for g^*, namely $([g^*], s^*)$ where $g^* = s^* g^{*\prime}$. The input to the cascade machine is g. The action of $M(G/S)$ is clear; its present state is $[g^*]$, and its input is $[g]$, so its next state is $[g^*][g] = [g^*g]$. The present state of $M(S)$ is s^*, and its input is $\beta(g, [g^*])$, a yet-to-be-determined member of S. The next state of $M(S)$ is $s^* \cdot \beta(g, [g^*])$. Thus, the next state of the cascade is $([g^* g], s^* \cdot \beta(g, [g^*]))$. γ will map this to

$$s^* \cdot \beta(g, [g^*])(g^*g)'$$

But if we want this to equal the next state of $M(G)$, g^*g, then

$$g^*g = s^* \cdot \beta(g, [g^*])(g^*g)'$$

or (since $g^* = s^*g^{*\prime}$)

$$s^*g^{*\prime}g = s^* \cdot \beta(g, [g^*])(g^*g)'$$

or

$$\beta(g, [g^*]) = (g^{*\prime}g)((g^*g)')^{-1}$$

This argument gives us our definition of β. It only remains to show that $\beta(g, [g^*])$ is indeed a member of S, and the proof of Theorem 8.47 is complete.

8.48   Practice   Show that $\beta(g, [g^*]) = (g^{*\prime}g)((g^*g)')^{-1}$ is a member of S. (Hint: first show that $g^{*\prime}g$ and $(g^*g)'$ are in the same coset of S in G.)   (*Answer, page A-23*)

8.49   Example   We know that

$$\{0\} \subset \{0, 2, 4\} \subset \mathbb{Z}_6$$

is a composition series for $[\mathbb{Z}_6, +_6]$. We can use $M(\mathbb{Z}_6/\{0, 2, 4\})$ and $M(\{0, 2, 4\})$ to simulate $M(\mathbb{Z}_6)$. The elements of $Z_6/\{0, 2, 4\}$ are $[0] = \{0, 2, 4\}$ and $[1] = \{1, 3, 5\}$, and we will take 0 and 1 as the fixed coset representatives. If, for example, $M(\mathbb{Z}_6)$ is in state 5 receiving an input of 3, its next state is $5 +_6 3 = 2$. The present state of the cascade machine is $([5], 4)$ because

$5 = 4 +_6 1$. The next state of the cascade machine is

$$([5 +_6 3], 4 +_6 \beta(3, [5]))$$

or

$$([2], 4 +_6 (1 +_6 3) +_6 (-(5 +_6 3)')) = ([2], 4 +_6 (4) +_6 (-0))$$
$$= ([2], 2)$$

and

$$\gamma([2], 2) = 2$$

because

$$2 = 2 +_6 0 \qquad\qquad \square$$

✓ CHECKLIST

Definitions

cascade connection (*p. 319*)
division of a semigroup (*p. 320*)
simple group (*p. 320*)
maximal subgroup (*p. 321*)
composition series
  of a group (*p. 321*)

factor set (*p. 321*)
composition factors
  of a group (*p. 321*)
semigroup of a machine (*p. 322*)
machine of a semigroup (*p. 323*)
flip-flop (*p. 324*)

Techniques

Given a finite-state machine, construct the table for its semigroup (monoid).

Find a homomorphism from $M(S_M)$ onto M for a given machine M.

For a group G with normal subgroup S, simulate a computation in $M(G)$ by using a cascade decomposition of $M(G)$.

Main Ideas

For any finite group G, there is a set of essentially unique groups obtained by forming the quotient groups of successive subgroups in a composition series of G.

For a given machine M, there is a homomorphism from $M(S_M)$ onto M, and, thus, $M(S_M)$ simulates M.

If M is a finite-state machine whose semigroup is a group G, then M can be simulated by a cascade connection of machines of composition factors of G.

EXERCISES, SECTION 8.4

1. If
$$\{i\} = G_k \subset G_{k-1} \subset \cdots \subset G_3 \subset G_2 \subset G_1 \subset G_0 = G$$
   is a composition series of a finite group G and if $|G_{i-1}/G_i| = m_i$, $1 \le i \le k$, show that $|G| = m_1 \cdot m_2 \cdots m_k$.

2. Find two composition series for $[\mathbb{Z}_{15}, +_{15}]$ and show that the members of the respective factor sets are isomorphic.

3. Show that the group $[\mathbb{Z}, +]$ does not have a composition series. (Hint: assume that $\{0\} = G_k \subset G_{k-1} \cdots \subset G_1 \subset \mathbb{Z}$ is a composition series for $\mathbb{Z}$. Use Theorem 5.58 to show that $\{0\}$ is not a maximal subgroup of G_{k-1}.)

4. In each case below, the next-state function for a machine M is given. Find the state transformations in the semigroup S_M and show the semigroup table for S_M. Is S_M a group?

★ (a)

M	Input 0	1
0	1	2
1	2	1
2	2	2

(b)

M	Input 0	1
0	2	1
1	1	0
2	0	1

(c)

M	Input 0	1
0	1	2
1	2	1
2	0	0

5. Find a homomorphism for $M(S_M)$ onto M for each machine of Exercise 4.

6. Let M be any finite-state machine with k states, and suppose that each input symbol performs a permutation on the states of M. Show that S_M is a group.

M	Input	
	0	1
0	0	1
1	1	0

Figure 8.57

7. (a) The next-state function for a machine M is given in Figure 8.57. Show that the semigroup S_M is isomorphic to $[\mathbb{Z}_2, +_2]$.
   (b) For the machine M of part (a), show that the semigroup of the parallel connection of M and M is isomorphic to $[\mathbb{Z}_2 \times \mathbb{Z}_2, +]$.
   (c) Let S_{M_1} and S_{M_2} be the semigroups of two machines M_1 and M_2. Show that the semigroup of the parallel connection of M_1 and M_2 is isomorphic to the semigroup $S_{M_1} \times S_{M_2}$.

8. In $M(\mathbb{Z}_6)$, if the present state is 3 and the input is 4, the next state is $3 +_6 4 = 1$. Simulate this computation in the cascade connection of $M(\mathbb{Z}_6/\{0, 2, 4\})$ and $M(\{0, 2, 4\})$ (see Example 8.49).

9. Use the composition series
$$\{0\} \subset \{0, 3\} \subset \mathbb{Z}_6$$
to simulate the following computations in $M(\mathbb{Z}_6)$ by means of the cascade connection of $M(\mathbb{Z}_6/\{0, 3\})$ and $M(\{0, 3\})$ (see Example 8.49).
   ★ (a) $f_s(2, 5) = 2 +_6 5 = 1$
   (b) $f_s(4, 1) = 4 +_6 1 = 5$

10. Use the composition series
$$\{0\} \subset \{0, 5, 10\} \subset \mathbb{Z}_{15}$$
to simulate the following computations in $M(\mathbb{Z}_{15})$ by means of the cascade connection of $M(\mathbb{Z}_{15}/\{0, 5, 10\})$ and $M(\{0, 5, 10\})$ (see Example 8.49).
   (a) $f_s(6, 7) = 13$
   (b) $f_s(8, 11) = 4$

FOR FURTHER READING

Good discussions on the material in the first two sections of this chapter can be found in [1]. Decomposition theory is covered in [2], [3], and [4].

1. Zvi Kohavi, *Switching and Finite Automata Theory*. New York: McGraw-Hill, 1970.

2. J. Hartmanis and R. E. Stearns, *Algebraic Structure Theory of Sequential Machines*. Englewood Cliffs, N.J.: Prentice-Hall, 1966.

3. Michael A. Arbib, *Theories of Abstract Automata*. Englewood Cliffs, N.J.: Prentice-Hall, 1969.

4. Leonard S. Bobrow and Michael A. Arbib, *Discrete Mathematics*. Philadelphia: Saunders, 1974.

Chapter 9

COMPUTABILITY

In this chapter we consider an adequate model, or simulator, of an algorithm or effective procedure, the Turing machine. We will also discuss a universal simulator for all such machines (the theoretical inspiration for the stored-program computer). We will find that there are problems for which no algorithmic solution can ever be found. Finally, for those problems for which solution algorithms do exist, we will discuss a way to measure the relative efficiency of these algorithms.

Section 9.1   *TURING MACHINES—SIMULATION I*

ALGORITHMS

A Case Study

Kleene's Theorem (Chapter 7) tells us that finite-state machines are only capable of recognizing regular sets. Because $S = \{0^n 1^n \mid n \geq 0\}$ is not regular, no finite-state machine can recognize it. We probably consider ourselves to be finite-state machines and imagine that our brains, being composed of a large number of cells, can only take on a finite, although immensely large, number of configurations, or states. We feel sure, however, that if someone presented us with an arbitrarily long string of 0's followed by an arbitrarily long string of 1's, we could detect whether the number of 0's and 1's was the same. Let's think of some techniques we might use.

For small strings of 0's and 1's, we could perhaps just look at the strings and decide. Thus, we can tell without great effort that $000111 \in S$ and that $00011 \notin S$. However, for the string 00000000000000001111111111111111, we must devise another procedure to attack the problem, and we would probably resort to counting. We count the number of 0's received and when we get to the first 1, we write that number down (or remember it) for future reference; then we begin counting 1's. (This process is what we did mentally for smaller strings.) But we have now made use of some extra memory because when we finish counting 1's, we have to retrieve the number representing the sum of the 0's to make a comparison. But such "information retrieval" is what the finite-state machine cannot do; its only capacity for remembering input is to have a given input symbol send it to a particular state. Suppose we attempt to build a finite-state machine to recognize S. We could count the number of 0's seen by having each new 0 move us to a new state of the machine. However, since the number of states of any given machine is a finite number, this plan fails if the number of 0's read in is larger than this finite number, so our machine clearly could not process $0^n 1^n$ for all n. In fact, if we think of solving this problem on an actual digital computer, we encounter the same difficulty. If we set a counter as we read in the 0's, we might get an overflow because our counter can only go so high. To process $0^n 1^n$ for arbitrarily large n requires that we have unlimited auxiliary memory available to store the value of our counter, which in practice cannot happen.

Another way we humans might consider attacking the problem of recognizing S is to wait until the entire string had been presented to us. We would then go to one end of the string and cross out a 0, go to the other end and cross out a 1, go back and forth to cross out another 0-1 pair, and continue this operation until we run out of 0's or 1's. The string belongs to S if and only if we run out of both at the same time. Although this approach sounds rather different from the first one, it still requires remembering each of the inputs in that we must go back and read them once the string is complete. The finite-state machine, of course, cannot reread input.

We have come up with two computational procedures to decide, given a string of 0's and 1's, whether that string belongs to S. Both procedures required some form of additional memory unavailable in a finite-state machine. Evidently, the finite-state machine is not a model of the most general form of computational procedure. Before we consider a better model, we'll try to elaborate on what we mean by a computational procedure.

Intuitive Description

We will use the terms "algorithm," "effective procedure," and "computational procedure" interchangeably, and we will not give a formal definition for any of them. Instead, we will appeal to a common, intuitive understanding of an algorithm or effective procedure as a "recipe" for carrying out a task. (One of the first steps in writing a computer program to solve a problem is to decide upon an algorithm to use, so algorithms have both practical and theoretical significance.) We will assume that any input to which an algorithm is to be applied has been encoded into numeric form, usually nonnegative integers, just as input for an actual digital computer program is encoded and then stored in binary form.

An algorithm is characterized by certain properties. It consists of a list of precise instructions in some language, say English; each instruction must be finite and the list itself must be finite. Each instruction must be one that can be mechanically carried out. If the algorithm is a solution method for a particular problem, then, when applied to appropriate input, the algorithm must stop (halt) producing the correct answer if an answer exists. If an answer does not exist, the algorithm may halt and declare that an answer does not exist, or it may go on indefinitely, searching for an answer.

THE TURING MACHINE

To simulate more general computational procedures than the finite-state machine can handle, we use a Turing machine, proposed by the British mathematician Alan M. Turing in 1936. A Turing machine is essentially a finite-state machine with the added ability to reread its input and also to erase and write over its input, and with unlimited auxiliary memory—thus overcoming deficiences already noted about finite-state machines.

A Turing machine consists of a finite-state machine and a tape divided into cells, each cell containing, at most, one symbol from an allowable finite alphabet. At any one instant, only a finite number of cells on the tape are nonblank. We use the special symbol b to denote a blank cell. The finite-state unit, through its read–write head, reads one cell of the tape at any given moment (see Figure 9.1). By the next clock pulse, depending upon the present state of the unit and the symbol read, the unit either does nothing (halts) or completes three actions. These actions are (1) print a symbol from the alphabet on the cell read (it might be the same symbol that's already there); (2) go to the next state (it might be the same state as before); and (3) move the read–write head one cell left or right.

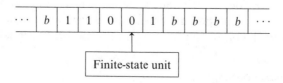

Figure 9.1

We can describe the action of any particular Turing machine by a set of quintuples of the form (s, i, i', s', d) where s and i indicate the present state and the tape symbol being read, and i' denotes the symbol printed, s' denotes the new state, and d denotes the direction of the move of the read–write head (R for right, L for left). Thus, a machine in the configuration illustrated by Figure 9.2a, if acting according to the instructions contained in the quintuple $(2, 1, 0, 1, R)$, would move to the configuration illustrated by Figure 9.2b.

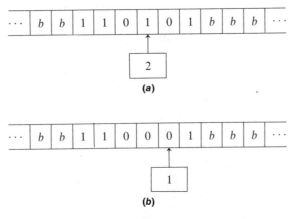

Figure 9.2

9.1 Definition

Let S be a finite set of states, and I a finite set of tape symbols (the **tape alphabet**) including a special symbol b. A **Turing machine** is a set of quintuples of the form (s, i, i', s', d) where $s, s' \in S$, $i, i' \in I$, and $d \in \{R, L\}$ and where no two quintuples begin with the same s and i symbols.   □

The restriction that no two quintuples begin with the same s and i symbols ensures that the action of the Turing machine is deterministic and completely specified by its present state and symbol read. If a Turing machine gets into a configuration for which its present state and symbol read are not the first two symbols of any quintuple, the machine halts.

Just as in the case of ordinary finite-state machines, we specify a fixed starting state, denoted by 0, in which the machine begins any computation. We also assume an initial configuration for the read–write head, namely, that it is positioned over the farthest left nonblank symbol on the tape. (If the tape is initially all blank, the read–write head can be positioned anywhere to start.)

9.2 Example

A Turing machine is defined by the set of quintuples

$$(0, 0, 1, 0, R)$$

$$(0, 1, 0, 0, R)$$

$$(0, b, 1, 1, L)$$

$$(1, 0, 0, 1, R)$$

$$(1, 1, 0, 1, R)$$

The action of this Turing machine when processing a particular initial tape is shown by the sequence of configurations of Figure 9.3.

334

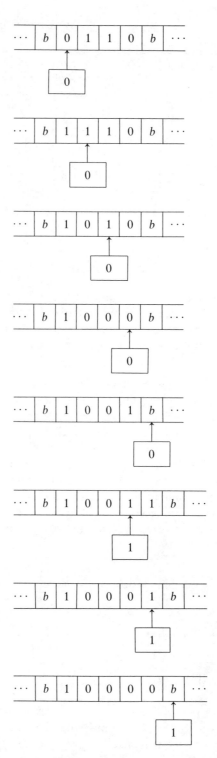

Figure 9.3

Since there are no quintuples defining action to be taken when in state 1 reading b, the machine halts with final tape

	b	1	0	0	0	0	b	
$\cdots$								$\cdots$

□

The tape serves as a memory medium for a Turing machine, and, in general, the machine can reread cells of the tape. It can also write on the tape, and thus the nonblank portion of the tape can be as long as desired although there are still only a finite number of nonblank cells at any time. Hence, the machine has available an unbounded, although finite, amount of storage. The limitations of finite-state machines observed early in this chapter do not exist for the Turing machine, so Turing machines should have considerably higher capabilities than finite-state machines. In fact, a finite-state machine is a very special case of a Turing machine, one that always prints the old symbol on the cell read, always moves to the right, and halts on the symbol b.

9.3 Practice   Given the Turing machine

$(0, 0, 0, 1, R)$

$(0, 1, 0, 0, R)$

$(0, b, b, 0, R)$

$(1, 0, 1, 0, R)$

$(1, 1, 1, 0, L)$

(a) What is the final tape, given the initial tape

$\cdots$	b	1	0	b	$\cdots$

(Since it is tedious to draw all the little squares, you don't need to do so; just write down the contents of the final tape.)

(b) Describe the behavior of the machine when started on the tape

$\cdots$	b	0	1	b	$\cdots$

(c) Describe the behavior of the machine when started on the tape

$\cdots$	b	0	0	b	$\cdots$

(Answer, page A-24)

Practice 9.3b and c illustrate two ways in which a Turing machine can fail to halt: by endlessly cycling or by moving forever along the tape.

Turing Machines as Set Recognizers

Although the Turing machine computations we have seen so far are not particularly meaningful, we will use Turing machines to do two kinds of jobs. First, we'll use the Turing machine as a recognizer, much as we considered finite-state machines as recognizers in Chapter 7. We can even give a very similar definition, provided we first define a final state for a Turing machine. A **final state** in a Turing machine is one that is not the first symbol in any quintuple. Thus, upon entering a final state, whatever the symbol read, the Turing machine halts.

9.4 Definition

A Turing machine T with tape alphabet I **recognizes (accepts)** a subset S of I^* if T, beginning in standard initial configuration on a tape containing a string α of tape symbols, halts in a final state if and only if $\alpha \in S$. $\qquad \square$

Note that Definition 9.4 leaves open two alternative behaviors for T when applied to a string α of tape symbols not in S. T may halt in a nonfinal state, or T may fail to halt at all.

We can now build a Turing machine to recognize our old friend $S = \{0^n 1^n \mid n \geq 0\}$. The machine is based on our second approach to this recognition problem, sweeping back and forth across the input crossing out pairs of 0's and 1's.

9.5 Example

We want to build a Turing machine that will recognize $S = \{0^n 1^n \mid n \geq 0\}$. We will use one additional special symbol, call it X; so the tape alphabet $I = \{0, 1, b, X\}$. State 6 is the only final state. The quintuples making up T are given below, together with a description of their function.

$(0, b, b, 6, R)$ Recognizes the empty tape, which is in S.

$(0, 0, X, 1, R)$ Erases the leftmost 0 and begins to move right.

$\left.\begin{array}{l}(1, 0, 0, 1, R) \\ (1, 1, 1, 1, R) \\ (1, X, X, 2, L) \\ (1, b, b, 2, L)\end{array}\right\}$ Moves right in state 1 until it reaches the end of the string; then moves left in state 2.

$(2, 1, X, 3, L)$ Erases the rightmost 1 and begins to move left.

$(3, 1, 1, 3, L)$ Moves left over 1's.

$(3, 0, 0, 4, L)$ Goes to state 4 if more 0's are left.

$(3, X, X, 5, R)$ Goes to state 5 if no more 0's in string.

$(4, 0, 0, 4, L)$ Moves left over 0's.

$(4, X, X, 0, R)$ Finds left end of string and begins sweep again.

$(5, X, X, 6, R)$ No more 1's in string, machine accepts.

Figure 9.4 shows key configurations in the machine's behavior on the tape

···	b	0	0	0	1	1	1	b	···

which, of course, it should accept. □

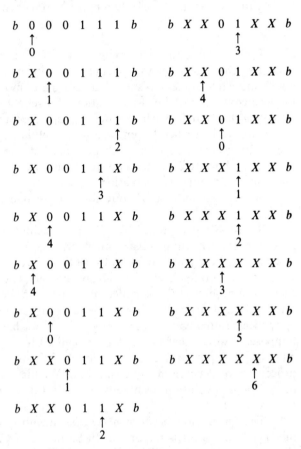

Figure 9.4

9.6 Practice   For the Turing machine of Example 9.5, describe the final configuration upon processing input tapes

(a)

···	b	0	0	1	1	1	b	···

(b)

···	b	0	0	0	1	1	b	···

(c)

···	b	0	0	0	0	1	1	b	···

(Answer, page A-24)

9.7 Practice   Design a Turing machine to recognize the set of all strings of 0's and 1's ending in 00. (This set can be described by the regular expression $(0 \vee 1)^*00$, so you should be able to use a Turing machine that changes no tape symbols and always moves to the right.)   (*Answer, page A-24*)

9.8   Practice   Modify the Turing machine of Example 9.5 to recognize $\{0^n 1^{2n} | n \geq 0\}$.

(*Answer, page A-24*)

The set of quintuples defining a given Turing machine can always be altered so that whenever the machine is about to halt in a nonfinal state, it goes instead to some new state that causes it to move forever right. Therefore, we could have defined recognition of a set S by a Turing machine T by saying that T halts when begun on a tape containing a string α if $\alpha \in S$, and T fails to halt when begun on a tape containing a string α if $\alpha \notin S$. We can, in other words, force "halt in a nonfinal state" to become "do not halt." We cannot, however, do the opposite because we may not be able to recognize every "do not halt" situation to convert it to halting in a nonfinal state; we will say more about our ability to recognize halting situations in Section 9.2.

If a set S is recognized by a Turing machine T, we have an effective procedure for generating a list of members of S. We can make lots of copies of T, say T_1, T_2, T_3, and so on, and start them up on tapes containing strings $\alpha_1, \alpha_2, \alpha_3$, and so forth. Periodically, we go back and check the progress of each T_i. Whenever a T_i has halted in a final state, we know that $\alpha_i \in S$. For those T_i's that are still computing, we can't generally conclude whether $\alpha_i \in S$ because the computation may or may not halt at some future time. Suppose, however, that the set S is recognized by a Turing machine T that halts on all input strings (as in Example 9.5). We then have an effective procedure for deciding membership in S. We pick any string α and run T on α. The computation eventually halts. If T halts in a final state, $\alpha \in S$; otherwise, $\alpha \notin S$.

This distinction between *generating* members of a set and *deciding* membership in a set is important. In Section 2 of Chapter 10, we will see that there are sets for which we can build Turing machines that recognize them, so that we have an effective procedure for generating their members, but no Turing machine allowing us to decide membership, that is, no Turing machine that halts on all input.

Turing Machines as Function Computers

The second job for which we will use Turing machines is to compute functions. Given a particular Turing machine T and a string α of tape symbols, we begin T in standard initial configuration on a tape containing α. If T eventually halts with a string β on the tape, we may consider β as the value of a function evaluated at α. Using function notation, $T(\alpha) = \beta$. The domain of the function T consists of all strings α for which T eventually halts. We can also think of T as computing **number-theoretic functions**, functions from a subset of $\mathbb{N}^k$ into $\mathbb{N}$ for any $k \geq 1$. We will think of a string of $n + 1$ 1's as the unary representation of the nonnegative integer n; we'll denote this "encoding" of n by $\bar{n}$. Then a tape containing the string $\bar{n}_1 * \bar{n}_2 * \cdots * \bar{n}_k$ can be thought of as the representation of the k-tuple $(n_1, n_2, \ldots, n_k)$ of nonnegative integers. If T begun in the standard initial configuration on such a tape eventually halts with a final tape that is the representation $\bar{m}$ of a nonnegative integer m, then T has acted as a k-variable function T^k, where $T^k(n_1, n_2, \ldots, n_k) = m$. If T begun in standard initial configuration on such a tape either fails to halt or halts with the final tape not a representation of a nonnegative integer, then the function T^k is undefined at $(n_1, n_2, \ldots, n_k)$.

There is thus an infinite sequence $T^1, T^2, \ldots, T^k, \ldots$, of number-theoretic functions computed by T associated with each Turing machine T. For each k, the function T^k is a **partial function** on $\mathbb{N}^k$, meaning that its domain may be a proper subset of $\mathbb{N}^k$. A special case of a partial function on $\mathbb{N}^k$ is a **total function** on $\mathbb{N}^k$ where the function is defined for all k-tuples of nonnegative integers.

9.9 Example Let a Turing machine T be given by the quintuples

$(0, 1, 1, 0, R)$
$(0, b, 1, 1, R)$

If T is begun in standard initial configuration on the tape

$\cdots$	b	1	1	1	b	$\cdots$

then T will halt with final configuration

$b\ 1\ 1\ 1\ 1\ b$
$\uparrow$
1

Therefore, T defines a 1-variable function T^1 that maps $\bar{2}$ to $\bar{3}$. In general, T maps $\bar{n}$ to $\overline{n+1}$, so $T^1(n) = n + 1$, a total function of one variable. □

In Example 9.9, we began with a Turing machine and observed a particular function it computed. But we can also begin with a number-theoretic function and try to find a Turing machine to compute it.

9.10 Definition A **Turing computable function** is a number-theoretic function computed by some Turing machine. □

A Turing computable function f can in fact be computed by an infinite number of Turing machines. Once a machine T is found to compute f, we can always include extraneous quintuples in T, producing other machines that also compute f.

9.11 Example We want to find a Turing machine that computes the function f defined below:

$$f(n_1, n_2) = \begin{cases} n_2 - 1 & \text{if } n_2 \neq 0 \\ \text{undefined} & \text{if } n_2 = 0 \end{cases}$$

Thus, f is a partial function of two variables. Let's consider the Turing machine given by the following quintuples. The actions performed by various sets of quintuples are described.

$\left.\begin{array}{l}(0, 1, 1, 0, R) \\ (0, *, *, 1, R)\end{array}\right\}$ Passes right over $\bar{n}_1$ to $\bar{n}_2$.

$(1, 1, 1, 2, R)$ Counts first 1 in $\bar{n}_2$.

$(2, b, b, 3, R)$ $n_2 = 0$; halts.

$\left.\begin{array}{l}(2, 1, 1, 4, R) \\ (4, 1, 1, 4, R) \\ (4, b, b, 5, L)\end{array}\right\}$ Finds the right end of $\bar{n}_2$.

$(5, 1, b, 6, L)$ Erases last 1 in $\bar{n}_2$.

$\left.\begin{array}{l}(6, 1, 1, 6, L) \\ (6, *, b, 7, L)\end{array}\right\}$ Passes left to $\bar{n}_1$, erasing *.

$(7, 1, b, 7, L)$ Erases $\bar{n}_1$.

$(7, b, b, 8, L)$ $\bar{n}_1$ erased; halts with $\overline{n_2 - 1}$ on tape.

If T is begun on the tape

$\cdots$	b	1	1	*	1	1	1	1	b	$\cdots$

then T will halt with final configuration

$b\,b\,b\,b\,b\,1\,1\,1\,b$
$\uparrow$
8

This configuration agrees with the requirement that $f(1, 3) = 2$. If T is begun on the tape

$\cdots$	b	1	1	*	1	b	$\cdots$

then T will halt with final configuration

$b\,1\,1 * 1\,b\,b$
$\uparrow$
3

Because the final tape is not $\bar{m}$ for any nonnegative integer m, the function computed by T is undefined at $(1, 0)$—just as we want. It is easy to see that this Turing machine computes f, and that f is, therefore, a Turing computable function. $\square$

9.12 Practice   Design a Turing machine to compute the function

$$f(n) = \begin{cases} n - 2 & \text{if } n \geq 2 \\ 1 & \text{if } n < 2 \end{cases}$$

(Answer, page A-24)

The domain of a Turing computable function f is a set that can be generated by the procedure of building many copies of a machine T to compute f, feeding some input into them, and periodically checking to see which have halted on a final tape representing a nonnegative integer. If T is still running on a given input, we can't in general decide whether T will eventually halt, so we cannot say whether the input is in the domain of the function. If T halts on all input, we can run T on a given input and then simply look at the final tape, decide whether it represents a nonnegative integer, and so determine if the input belongs to the domain of f. As in the case of set recognition, there are Turing computable functions (so we can *generate* the domain) not computable by Turing machines that halt on all input (so we cannot *decide* what belongs to the domain).

Other Formulations

There are a number of alternative definitions of the Turing machine. We could, for instance, allow the machine's tapes to have several tracks, so that the machine could read more than one symbol at a time (see Figure 9.5a), or we could allow it to have multiple tapes and finite-state units that operate independently (see Figure 9.5b). We could also require that the tape we use be a singly infinite tape; that is, instead of extending indefinitely in both directions, the tape extends indefinitely in only one direction and the other end is fixed (see Figure 9.6).

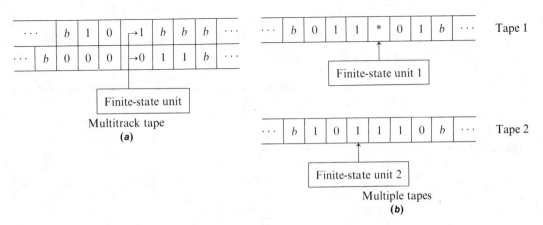

Multitrack tape
(a)

Multiple tapes
(b)

Figure 9.5

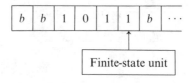

Figure 9.6

We can even require that the Turing machine be allowed only two tape symbols, the blank symbol plus one additional symbol. All of these definitions of a Turing machine are equivalent to each other and to the original definition in that for any job done by one kind of Turing machine, a machine of any other kind can be found to do the same job (perhaps requiring some encoding and decoding of tape symbols to and from a new alphabet). A Turing machine computation is a pretty versatile idea; we'll see just how versatile shortly.

THE CHURCH–TURING THESIS

Is the Turing machine a better model of an effective procedure than the finite-state machine? Although our concept of effective procedure is an intuitive one, we are quite likely to agree that any Turing computable function f is a function whose values can be found by an effective procedure or algorithm. In fact, if f is computed by the Turing machine T, then the set of quintuples of T is itself the algorithm; as a finite list of finite instructions that can be mechanically carried out, it satisfies the various conditions outlined as common to anyone's notion of an algorithm. Therefore, we are probably willing to accept the proposal illustrated by Figure 9.7. Here we show "computable by effective procedure" as a "cloudy" intuitive idea, and "Turing computable" as a mathematically precise, well-defined idea.

Given the simplicity of the Turing machine definition, it is a little startling to contemplate Figure 9.8, which asserts that any function computable by any means we might think of as an effective procedure is also Turing computable. Combining Figures 9.7 and 9.8, we get the Church–Turing Thesis (Figure 9.9), named after Turing and another famous mathematician, Alonzo Church.

9.13 Church–Turing Thesis   A number-theoretic function is computable by an effective procedure if and only if it is Turing computable.   □

Because the Church–Turing Thesis equates an intuitive idea with a mathematical idea, it can never be formally proved and must remain a "thesis," not a "theorem." What, then, is its justification?

One piece of evidence is that whenever a procedure everyone could agree was an effective procedure (according to his or her own insights into

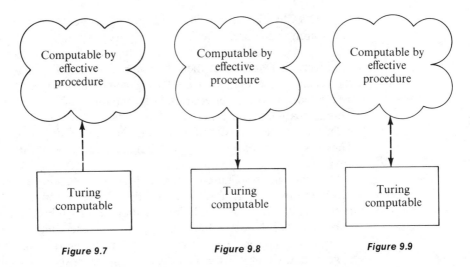

Figure 9.7   Figure 9.8   Figure 9.9

this idea) has been proposed to compute a function, someone has designed a Turing machine to compute that function. (Of course there is always the nagging thought that someday this might not happen.)

Another piece of evidence is that other mathematicians, several of them at about the same time that Turing developed the Turing machine, also proposed models of effective procedures. On the surface, each proposed model seems quite unrelated to any of the others. Because all of these models are formally defined, just as Turing computability is, it is possible to consider on a formal, mathematical basis whether any of them are equivalent. All of these models, as well as Turing computability, have been proven equivalent; that is, they all define the same class of functions, which suggests that Turing computability embodies everyone's concept of effective procedure. Figure 9.10 illustrates what has been done; here, the solid lines represent mathematical proofs and the dotted lines correspond to the Church–Turing Thesis. Dates indicate when the various models were proposed.

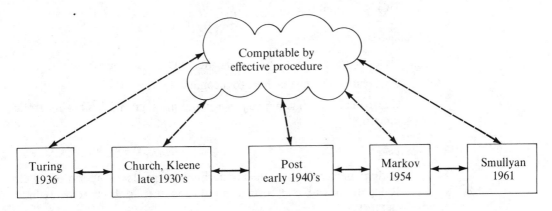

Figure 9.10

The Church–Turing Thesis is now widely accepted as a working tool in research in the area of effective computability. If, in a research paper, a procedure is described to compute a function and the procedure seems intuitively to be effective, then the Church–Turing Thesis is invoked, and the function is declared to be Turing computable (or one of the names associated with the other equivalent formulations of Turing computability). This invocation means that the author presumably could, if pressed, produce a Turing machine to compute the function, but, again, the Church–Turing Thesis is so universally accepted that no one bothers with these details anymore.

Although the Church–Turing Thesis is stated in terms of number-theoretic functions, it is subject to a broader interpretation. Any effective procedure involving manipulation of a finite set of symbols can be translated into a number-theoretic function by a suitable encoding of the symbols as nonnegative integers. Thus, if we accept the Church–Turing Thesis, we can, in general, say that if there is an effective procedure to do a symbol manipulation task, there is a Turing machine to do it. We will accept and make use of the Church–Turing Thesis in Section 9.2.

By accepting the Church–Turing Thesis, we have accepted the Turing machine as the ultimate model of an effective computational device. Its capabilities exceed those of any actual computer that, after all, does not have the unlimited tape storage of a Turing machine. It is remarkable that Turing proposed this concept in 1936, well before the advent of the modern computer.

✔ CHECKLIST

Definitions

Turing machine (*p. 333*)
Recognition of a set by a Turing machine (*p. 336*)
Number-theoretic function (*p. 338*)
Turing computable function (*p. 339*)

Techniques

Describe the action of a given Turing machine on a given initial tape.

Construct a Turing machine to recognize a given set.

Construct a Turing machine to compute a given number-theoretic function.

Main Ideas

Turing machines with their deterministic mode of operation, their ability to reread and rewrite input, and their unbounded auxiliary memory.

A finite-state machine is a special case of a Turing machine.

Turing machines can be used as set recognizers and as function computers.

The existence of an effective procedure to generate the members of a set does not mean an effective procedure to decide membership in the set exists.

The Church–Turing Thesis equates a function computable by an effective procedure with a Turing computable function. Because this thesis expresses a relationship between an intuitive idea and a formally defined one, it can never be proved but, nonetheless, has been widely accepted.

EXERCISES, SECTION 9.1

★ 1. Given the Turing machine

$(0, 0, 0, 0, L)$
$(0, 1, 0, 1, R)$
$(0, b, b, 0, L)$
$(1, 0, 0, 1, R)$
$(1, 1, 0, 1, R)$

(a) What is its behavior when started on the tape

$\cdots$	b	1	0	0	1	1	b	$\cdots$

(b) What is its behavior when started on the tape

$\cdots$	b	0	0	1	1	1	b	$\cdots$

2. Given the Turing machine

$(0, 1, 1, 0, R)$
$(0, 0, 0, 1, R)$
$(1, 1, 1, 1, R)$
$(1, b, 1, 2, L)$
$(2, 1, 1, 2, L)$
$(2, 0, 0, 2, L)$
$(2, b, 1, 0, R)$

(a) What is its behavior when started on the tape

$\cdots$	b	1	0	1	0	b	$\cdots$

(b) What is its behavior when started on the tape

$\cdots$	b	1	0	1	b	$\cdots$

3. Given the set of quintuples describing a Turing machine T and the initial tape configuration, create a computer program that will write out the sequence of successive tape configurations. Assume that T is described by ≤ 100 quintuples, that the number of cells used on the tape is ≤ 70, and that T always halts.

★ 4. Find a Turing machine that recognizes the set of all unary strings consisting of an even number of 1's (this includes the empty string).

5. Find a Turing machine to accept the set of strings of well-balanced parentheses. (Note (( )(( ))) is well balanced and (( )(( )) is not.)

6. Find a Turing machine that recognizes $\{0^{2n}1^n2^{2n} \mid n \geq 0\}$.

7. Find a Turing machine that recognizes $\{w * w^R \mid w \in \{0, 1\}^*$ and w^R is the reverse of the string $w\}$.

8. Find a Turing machine that recognizes $\{w_1 * w_2 \mid w_1, w_2 \in \{0, 1\}^*$ and $w_1 \neq w_2\}$.

9. Find a Turing machine that recognizes the set of palindromes on $\{0, 1\}^*$, that is, the set of all strings in $\{0, 1\}^*$ that read the same forwards and backwards, such as 101.

10. Find a Turing machine that converts a string of 0's and 1's representing a nonzero binary number into a string of that number of 1's. As an example, the machine should, when started on a tape containing $\cdots b\, 1\, 0\, 0\, b \cdots$, halt on a tape containing $\cdots b\, 1\, 1\, 1\, 1\, b \cdots$.

11. Find a Turing machine that, given an initial tape containing a nonempty string of 1's, marks the right end of the string with a * and puts a copy of the string to the right of the *. As an example, the machine should, when started on a tape containing $\cdots b\, 1\, 1\, 1\, b \cdots$, halt on a tape containing $\cdots b\, 1\, 1\, 1 * 1\, 1\, 1\, b \cdots$.

★ 12. What number-theoretic function of three variables is computed by the Turing machine given below?

$(0, 1, b, 0, R)$
$(0, *, b, 1, R)$
$(1, 1, 1, 2, R)$
$(2, *, *, 3, R)$
$(3, 1, 1, 2, L)$
$(2, 1, 1, 4, R)$
$(4, 1, 1, 4, R)$
$(4, *, 1, 5, R)$
$(5, 1, b, 5, R)$
$(5, b, b, 6, R)$

13. Find a Turing machine to compute the function

$$f(n) = 2n$$

14. Find a Turing machine to compute the function

$$f(n) = \begin{cases} \dfrac{n}{3} & \text{if 3 divides } n; \\ \text{undefined} & \text{otherwise} \end{cases}$$

15. Find a Turing machine to compute the function

$$f(n_1, n_2) = \max(n_1, n_2)$$

★ 16. Find a Turing machine to compute the function

$$f(n_1, n_2) = n_1 + n_2$$

17. Do Exercise 13 again, this time making use of the machines T_1 and T_2 of Exercises 11 and 16, respectively, as "subroutines". (Formally, the states of these machines would have to be renumbered as the quintuples are inserted into the "main program", but you may omit this tiresome detail and merely "call T_1," etc.)

18. Describe verbally the actions of a Turing machine that computes the function $f(n_1, n_2) = n_1 \cdot n_2$, that is, design the algorithm but do not bother to create all the necessary quintuples. You may make use of Exercises 11 and 16.

★ 19. (a) In this section, an effective procedure was described for generating the members of a set S recognized by a Turing machine T. Why does this procedure require that we have more than one copy of T available?
      (b) Describe an effective procedure for generating the members of the range set of a Turing computable function.

20. We can prove that there exist functions $f: \mathbb{N} \to \mathbb{N}$ that are not computable by any effective procedure. We will need three results from set theory:

   1. Any set equivalent to $\mathbb{N}$ is said to have **cardinality** $\aleph_0$ (cardinality is roughly a measure of the number of elements in a set).
   2. The cardinality of the set of all functions from $\mathbb{N}$ to $\mathbb{N}$ is $\aleph_0^{\aleph_0}$.
   3. $\aleph_0^{\aleph_0} > \aleph_0$.

   Show that the cardinality of the set of all algorithms is $\aleph_0$, thus proving that there are more functions than algorithms. This argument is nonconstructive; as soon as we try to describe a function that is not computable by an algorithm, say by giving an equation, we are in fact describing an algorithm for computing the function, thus making it effectively computable!

Section 9.2   *THE UNIVERSAL TURING MACHINE—SIMULATION II;*
UNSOLVABILITY

THE UNIVERSAL TURING MACHINE

The Turing machine, according to the Church–Turing Thesis, is a mathematical structure simulating algorithms or effective procedures. Once again, we see Simulation I at work, one structure serving as a model to capture the common features of a variety of examples. Simulation II considers whether one instance of the structure can simulate another instance of that structure. Given a particular Turing machine, can another one be found to do the same job? There are several levels at which we can consider this question, but the answer is yes in any case.

Suppose we have a machine T, and we view T as a function computer for the function f (the same ideas hold if we view T as a recognizer). One trivial way to produce a new machine T' that also computes f is to add extraneous quintuples to the definition of T, ones representing state–input pairs that will never be encountered while processing f. Clearly, this produces an infinite number of new machines simulating T. Another possibility is to develop some new algorithm for the computation of f. As an example, a Turing machine to compute $f(n) = n + 1$ could proceed by first copying $\bar{n}$ over and then erasing n 1's. (This procedure is not the most efficient one, but it works.) Finally, we could use a machine that computes f but operates on a different encoding of the input, say, in binary, rather than unary form.

As we've seen, it's an easy job given a Turing machine T to find another machine T' simulating T. We'll concentrate on a more sweeping idea. Can we possibly construct one machine that will simulate all others, a sort of super machine? Notice that we ourselves have been able to simulate the actions of any given machine by simply running it according to the instructions programmed into its quintuples. The process of "look at tape symbol and present state, consult the list of quintuples until you find the right one and do what it says, or if you can't find the right one, halt" is an effective procedure. If we invoke the Church–Turing Thesis, then because there exists an effective procedure that runs any Turing machine on any input, there must be a single Turing machine that will simulate the actions of any Turing machine on any input. Such a machine is called a **universal Turing machine**, denoted by U. We'll give a brief discussion of how U can be designed.

For U to simulate the actions of machine T on input string α, U must know both the description of T, that is, what its quintuples are, and what the string α is. Thus, the ordered pair (T, α) serves as input to U, but we must decide how to encode (T, α) onto U's tape. One thought is to simply write the quintuples of T, followed by α, on the tape. So to simulate T, given by the quintuples $(0, 0, 0, 1, R)$ and $(0, 1, 1, 1, L)$, acting on the string $\cdots b \, 0 \, 1 \, b \cdots$, we might write

$$\cdots b \, 0 \, 0 \, 0 \, 1 \, R * 0 \, 1 \, 1 \, 1 \, L * * 0 \, 1 \, b \cdots$$

on U's tape. However, U is to be a single Turing machine, with a finite tape alphabet, which can simulate *any* Turing machine T acting on *any* input string α. If we allow every state of every T to be its own representation on U's tape, we have no bound on the number of tape symbols U may need. The same thing happens if we allow every symbol in every possible input string α to go directly onto U's tape.

Instead, we must encode T's quintuples and α's symbols into, say, a unary representation, where, again, n is represented by $n + 1$ 1's. If we encode L as 1 and R as 11 and use * to separate symbols, ** to separate quintuples, and *** to separate T's quintuples from α, then the string

$$\cdots b \, 1 * 1 * 1 * 1 \, 1 * 1 \, 1 * * 1 * 1 \, 1 * 1 \, 1 * 1 \, 1 * 1 * * * 1 * 1 \, 1 \, b \cdots$$

would represent our (T, α). U must also know the current position of T's read–write head. We will mark this with a P just before the tape symbol read. And finally, U needs some working space to keep track of T's present state and symbol read. Thus, the string of Figure 9.11 could be on U's tape at the beginning of its simulation of T acting on α.

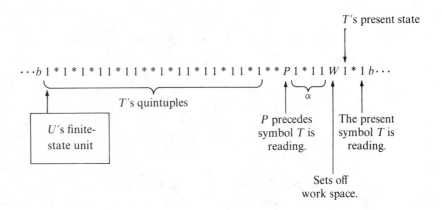

Figure 9.11

U's activities are carried on in cycles, each cycle representing a single step in T's computation on α. A single cycle involves the following actions. U searches through T's quintuples comparing the first two symbols of each quintuple with the two symbols in its work space. (Here we are thinking of unencoded symbols; a single symbol in a quintuple can be more than one symbol when encoded onto U's tapes.) If it finds no match after searching all of T's quintuples, then T must halt, so U halts also. If a match is found in a given quintuple, U will change the symbol following P to the third symbol in that quintuple, this being the new symbol printed on T's tape. It will then change the symbol following W to the fourth symbol in that quintuple, this being T's new state. It will move P some number of cells left or right, depending upon the last symbol in the given quintuple. Finally, it copies the symbol that now follows P into the second block behind W to show what new tape symbol T is reading. After one cycle of our example case, U's tape would contain the string of Figure 9.12.

$$\cdots b\,1*1*1*1\,1*1\,1**1*1\,1*1\,1*1\,1*1***1\,P\,1\,1\,W\,1\,1*1\,1\,b\cdots$$

Figure 9.12

U must be prepared at any time to create more room in the middle of its tape by moving some symbols farther to the right. This step will be necessary if, for example, T's new state has a longer unary representation than the old one and we need more room right after W (this happened above), or if T ever moves off the original nonblank portion of its tape and we need more room between *** and W.

We have described the general actions we want U to carry out. To write the actual quintuples for U so that it would perform these actions would be tedious but not very difficult. For example, we can see that the first phase of a cycle, searching for the correct quintuple, would involve much sweeping back and forth across U's tape, setting markers to keep our place, and then replacing the markers with the original symbols.

The universal Turing machine provides us with a single machine capable of simulating any other Turing machine when presented with that machine's program (its set of quintuples) and data. Rather then "building" an individual Turing machine for each job we want to do, we could simply "build" the universal Turing machine and then program it for various jobs. Here is a case where a theoretical idea developed in the late 1930s by Turing served as a blueprint for reality a decade later. Early computers were individually wired to do distinct tasks. To perform a new job, a computer had to have new circuitry and become essentially a new machine. John Von-Neumann in 1947 turned the idea of the universal Turing machine into the real-life, stored-program computer, a single computer that could accept programs as part of its input, thereby simulating the actions of each of the individually wired machines. Today's computers are, of course, stored-program computers.

DECISION PROBLEMS

We have spent quite a bit of time discussing what Turing machines can do. By the Church–Turing Thesis, they can do a great deal indeed, although not very efficiently. It is even more important, however, to consider what

Turing machines *cannot* do. Because a Turing machine's abilities exceed those of an actual computer in terms of what tasks can be done, if we find something no Turing machine can do, then a real computer can't do it either. In fact, by invoking the Church–Turing Thesis, no algorithm exists to do it. The type of task we generally have in mind in such a situation is determining the truth or falsity of statements from some large class of statements. The question of whether an algorithm exists to perform this type of task is called a decision problem.

9.14 Definition   A **decision problem** asks whether an algorithm exists to decide the truth or falsity of statements from some large class of statements.   □

The solution to a decision problem answers the question of whether an algorithm exists. A **positive solution** consists of proving that such an algorithm exists, and it is generally given by actually producing an algorithm that works. A **negative solution** consists of proving that no such algorithm exists. Note that this statement is much stronger than simply saying that a lot of people have tried but no one has come up with an algorithm; this result might simply mean that the algorithm is hard. It must be shown that it is impossible for anyone ever to come up with an algorithm. In the case of a negative solution to a decision problem, the problem is said to be **unsolvable** or **undecidable**, which is confusing terminology because the decision problem itself—the question of whether an algorithm exists to do a task—has been solved; what must forever be "unsolvable" is the task itself.

Examples

We will look at examples of some decision problems that have been answered.

9.15 Example   Consider the following decision problem.

Does there exist an algorithm to decide, given integers a, b, and c, whether $a^2 = b^2 + c^2$? Clearly, this question is a solvable decision problem. The algorithm consists of multiplying b by itself, multiplying c by itself, adding the two results, and comparing it with the result of multiplying a by itself.   □

Obviously, Example 9.15 is a rather trivial decision problem. Historically, much of mathematics has concerned itself at least indirectly with finding positive solutions to decision problems, that is, producing algorithms. Negative solutions to decision problems arose only in the twentieth century.

9.16 Example   One of the earliest decision problems to be formulated was Hilbert's Tenth Problem, tenth in a list of problems David Hilbert posed to the International Congress of Mathematicians in 1900. The problem is: Does an algorithm exist to decide for any polynomial equation $P(x_1, x_2, \ldots, x_n) = 0$ with integral coefficients whether it has integral solutions? For polynomial equations of the form $ax + by + c = 0$, where a, b, and c are integers, it is known that integer solutions exist if and only if the greatest common divisor of a and b also divides c. Thus, for particular subclasses of polynomial equations, there might be algorithms to decide whether integer solutions exist, but the decision problem as stated applies to the whole class of polynomial equations with integer coefficients. The general belief when this problem was posed, and for some time after, was that surely an algorithm existed and that no one had found one merely implied that it must be a difficult algorithm. In the mid-1930's, results such as those in the next example began to cast doubt on this point of view. It was not until 1970, however, that the problem was finally shown to be unsolvable. □

9.17 Example   Mathematical logic deals with **formal theories**, wherein certain strings of symbols are identified as **axioms**, and **rules of inference** are given whereby a new string can be obtained from old strings. Any string that is the last one in a finite list of strings consisting of either axioms or strings obtainable by the rules of inference from earlier strings in the list is said to be a **theorem**. The decision problem for a formal theory is: Does an algorithm exist to decide, given a string in the formal theory, whether that string is a theorem of the theory?

The work of Church and Kurt Gödel showed that any formal theory axiomatizing properties of arithmetic (making commutativity of addition an axiom, for example) which is not completely trivial (not everything is a theorem) is undecidable. Their work can be considered good news in a mathematical sense because it means that ingenuity in answering number theory questions will never be replaced by application of a mechanical procedure. □

9.18 Example   A group can be defined by giving an alphabet of symbols (strings or words from the alphabet are the group members) together with a set of relations specifying certain words as equivalent. Thus, an abelian group could have the relation $ab \sim ba$. Using this relation, the word $\alpha = abab$ can be transformed to the word $\beta = aabb$. The **word problem** for groups asks: Does an algorithm exist to decide, given the alphabet and relations for a group and two words from the group, whether one word can be transformed to the other? The word problem was shown to be unsolvable in 1955. □

9.19 Example   A particular Turing machine T begun on a tape containing a string α will either eventually halt or else never halt. The **halting problem** for Turing machines is a decision problem. It asks: Does an algorithm exist to decide, given a Turing machine T and string α, whether T begun on a tape containing α will eventually halt? Turing proved the unsolvability of the halting problem in the late 1930s. □

The Halting Problem

We will prove the unsolvability of the halting problem after two observations. First, it might occur to us that "run T on α" or, equivalently, "run the universal machine on an encoding of (T, α)" would constitute an algorithm to see whether T halted on α. If within 25 steps of T's computation T has halted, then we know T halts on α. But if within 25,000 steps T has not halted, what can we conclude? T may still eventually halt; how long should we wait? This so-called algorithm will not always give us the answer to our question.

A second observation is that the halting problem asks for one algorithm to be applied to a large class of statements. The halting problem asks: Does an algorithm exist to decide, given a (T, α) pair, whether T halts when begun on a tape containing α? The algorithm comes first, and that single algorithm has to give the correct answer for all (T, α) pairs. Consider the following statement that sounds very similar: Given a (T, α) pair, does an algorithm exist to decide whether T halts when begun on a tape containing α? Here, the (T, α) pair comes first and an algorithm can be chosen based on the particular (T, α); for a different (T, α), there can be a different algorithm. This problem is solvable. Suppose someone gives us a (T, α). Two effective procedures are (1) Say "yes"; (2) Say "no." Since T acting on α either does or does not halt, one of these two algorithms correctly answers the question. This solution may seem trivial or even sneaky, but consider again the problem statement. Given a (T, α), does an algorithm exist to decide, and so forth. There *does exist* an algorithm; it is either to say yes or to say no, we are not required to choose which one is correct!

This turnabout of words changes the unsolvable halting problem into a trivially solvable problem. It also points out the character of a decision problem, asking whether a single algorithm exists to solve a large class of problems. An unsolvability result has both a good side and a bad side. That no algorithm exists to solve a large class of problems guarantees jobs for creative thinkers who cannot be replaced by Turing machines. But that the class of problems considered is so large might make the result too general to be of interest; in practical terms, we don't expect a single algorithm to be able to do everything.

We will state the halting problem again and then prove its unsolvability.

9.20 Definition   **Halting Problem**
Does an algorithm exist to decide, given a Turing machine T and string α, whether T begun on a tape containing α will eventually halt? $\square$

9.21 Theorem   The Halting Problem is unsolvable.

Proof:
Assume that the halting problem is solvable and that a single algorithm

exists that can act on any (T, α) pair as input and will eventually give us a decision as to whether T acting on α halts. We have already seen in building the universal Turing machine how to encode T as a string s_T of symbols. By the Church–Turing Thesis, we are assuming the existence of a single Turing machine X that acts on a tape containing (s_T, α) for any T and α and eventually halts, at the same time telling us whether T on α halts or not. To be definite, suppose that X begun on (s_T, α) halts with a 1 left on the tape if and only if T begun on α halts, and X begun on (s_T, α) halts with a 0 left on the tape if and only if T begun on α fails to halt; these are the only two possibilities.

Now we add to X's quintuples to create a new machine Y. Y modifies X's behavior so that whenever X halts with a 1 on its tape, Y goes to a state that moves it endlessly to the right so that it never halts. If X halts with a 0 on its tape, so does Y. Finally, we modify Y to get a new machine Z that acts on input β, copies β, and then turns the computation over to Y so that Y acts on (β, β).

Now by the way Z is constructed, if Z acting on s_Z halts, it is because Y acting on (s_Z, s_Z) halts and that is because X acting on (s_Z, s_Z) halts with a 0 on the tape; but if this happens, it implies that Z begun on s_Z fails to halt! Therefore,

$$Z \text{ on } s_Z \text{ halts} \Rightarrow Z \text{ on } s_Z \text{ fails to halt} \tag{1}$$

This implication is very strange; let's see what happens if Z on s_Z does not halt. By the way Z is constructed, if Z acting on s_Z does not halt, neither does Y acting on (s_Z, s_Z) and this happens exactly when X acting on (s_Z, s_Z) halts with a 1 on the tape; but this result implies that Z begun on s_Z halts! Therefore,

$$Z \text{ on } s_Z \text{ fails to halt} \Rightarrow Z \text{ on } s_Z \text{ halts} \tag{2}$$

Together, implications (1) and (2) provide an airtight contradiction, so our assumption that the halting problem is solvable is incorrect.   □

The proof of the unsolvability of the halting problem depends on two ideas. One is that of encoding a Turing machine into a string description, and the other is that of having a machine look at and act upon its own description. Notice also that neither (1) nor (2) alone in the proof is sufficient to prove the result. Both are needed to contradict the original assumption of the solvability of the halting problem. If you have never encountered a proof of this nature before, it can be rather boggling, so perhaps you should read it again!

Reducibility

As we have mentioned, proving a decision problem unsolvable may be somewhat difficult since it involves showing the impossibility of existence of an algorithm to do something. Once a given problem has been found to be unsolvable (as we have now shown the halting problem to be), it may be used to prove that related problems are also unsolvable. Suppose that A and B are two decision problems. If by assuming that B is solvable we can deduce that A must be solvable also, we say that A is **reducible** to B. If we already know that A is unsolvable, then so is B.

9.22 Example   Consider the blank-tape halting problem: Does an algorithm exist to decide, given any Turing machine T, whether T begun on a blank tape will eventually halt?

Because the class of questions we are asking is more restricted than before, we might think that an algorithm could exist to do this job. However, we can show that the halting problem is reducible to the blank-tape halting problem. Suppose we had an algorithm P that was the solution to the blank-tape halting problem. We want to show the existence of an algorithm to solve the halting problem. For any machine T and string α, we can tack some quintuples on the front of T to create a machine T_α that first prints α on its tape (regardless of what was already there) and after that, turns the computation over to T acting on α. Thus, T_α halts when started on a blank tape if and only if T halts when started on a tape containing α. An algorithm to solve the halting problem is: given a (T, α) pair, create T_α (this procedure is effective) and apply algorithm P. P will then tell us whether T acting on α halts. Therefore, the halting problem is reducible to the blank-tape halting problem, and the blank-tape halting problem is unsolvable.   □

Example 9.22 showed that the halting problem is reducible to the blank-tape halting problem. It is easy to see that the blank-tape halting problem is reducible to the halting problem because an algorithm to decide whether any (T, α) pair halts can clearly make this decision when α is the empty string. If two decidability problems are each reducible to the other, then they are said to be of the same **degree of unsolvability**. There are different degrees of unsolvability, suggesting levels of difficulty in finding answers to the corresponding decision problems.

An unsolvability result of more practical interest concerns protection in a computer operating system; unauthorized access by one computer user to files created by another user, for example, is to be prevented. Here, the decision problem asks: Does an algorithm exist to decide, given a protection system in a certain status and a particular accessing mechanism, whether the system is safe against that accessing mechanism? There is a model of a protection system for which this decision problem has been proved unsolvable by showing that the halting problem reduces to it (Michael A. Harrison, Walter L. Ruzzo, Jeffrey D. Ullman, "Protection in Operating Systems," *Communications of the ACM* 19, no. 8 (August 1976):461–470).

9.23 Practice   Consider the following decision problem. Does an algorithm exist to decide, given a Turing machine T and string α, whether T begun on a tape containing α will eventually halt with the tape blank? Prove that this problem is unsolvable by showing that the halting problem is reducible to it.

(*Answer, page A-24*)

✔ CHECKLIST

Definitions

universal Turing machine (*p. 348*)
decision problem (*p. 351*)
unsolvable decision problem (*p. 351*)
halting problem (*p. 353*)

Techniques

Prove that a decision problem is unsolvable by showing that a decision problem already known to be unsolvable is reducible to it.

Main Ideas

The universal Turing machine simulates the actions of any Turing machine on any tape.

A decision problem asks whether an algorithm exists to decide the truth or falsity of statements from a large class of statements; if no algorithm exists, the decision problem is unsolvable.

Unsolvable decision problems have arisen in a number of contexts in this century.

The proof that the halting problem is unsolvable.

Reducibility of one decision problem to another.

EXERCISES, SECTION 9.2

★ 1. Let s be a fixed symbol. The **printing problem** asks: Does an algorithm exist to decide, given any Turing machine T and string α, whether T begun on a tape containing α will ever print s on the tape? Prove that the printing problem is unsolvable. (Hint: show that the halting problem is reducible to the printing problem.)

2. The **uniform halting problem** asks: Does an algorithm exist to decide, given any Turing machine T, whether T halts for every input tape? Prove that the uniform halting problem is unsolvable. (Hint: show that the blank-tape halting problem is reducible to the uniform halting problem.)

3. Suppose we have a Turing machine M. The **equivalence problem** asks: Does an algorithm exist to decide, given any Turing machine T, whether T computes the same number-theoretic function of one variable as M; in other words, whether $T(\bar{n}) = M(\bar{n})$ for all $n \in \mathbb{N}$? Prove that the equivalence problem is unsolvable. (Hint: show that the blank-tape halting problem is reducible to the equivalence problem.)

Section 9.3   *COMPUTATIONAL COMPLEXITY*

In this chapter, through the Church–Turing Thesis, we have formalized the notion of algorithm. We have also learned that there are problems for which a solution algorithm does not exist. However, in this section we will concentrate on problems solvable by algorithm, and on how good the algorithms are.

TIME COMPLEXITY

Suppose we have an algorithm A solving problem P. Here we may be thinking of an algorithm as a Turing machine program, or as an actual computer program. In either case, we are interested in how fast our algorithm works. By being tricky, is it possible to devise an algorithm A' to solve P that is faster than A? What do we mean by "fast" anyway? We will think of algorithm A' as "faster" (more efficient) than algorithm A if the number of basic operations A' has to do is smaller than the number of basic operations A has to do. Of course, A' and A must involve comparable basic operations, and we must be comparing A' and A in the same environment; we cannot compare the number of steps in a Turing machine computation with the number of steps in a computation done in some higher-level programming language. We will use Turing machine computations as our environment; by the Church–Turing Thesis, we can express any algorithm as a Turing machine computation. There are other models of computation that could be used, such as a RAM. A RAM (random access machine) works somewhat like a Turing machine, but its allowable operations resemble more closely those in actual programming languages; thus, there are arithmetic operations, branching instructions, and so forth.

We have alluded to efficiency of an algorithm earlier in this book (Section 2.3, in connection with Hamiltonian circuits, and Section 7.3, in connection with regular expressions). The efficiency of an algorithm is also known as its *complexity* although this reference is not to the amount of convolution in its logic but merely to some sort of measure as to the amount of work the algorithm must do. An algorithm that is straightforward in its logic may still require a very large number of steps to carry out on a Turing machine, for example.

Remember that we have used Turing machines for two kinds of jobs, function computation and set recognition. Suppose that the number-theoretic function f is Turing computable and is computed by a Turing machine T. We will consider only cases where T halts on all input since we want to count the number of steps in T's computation of f and we don't want T to go on indefinitely. As T does a computation of f, we encode the input in some way on T's tape, start T in standard initial configuration, and count the number of steps (clock pulses) in the computation until T halts. We would expect that the number of steps required for T to process f at some value would be a function of the length of the input.

9.24 Example   Consider the Turing machine T of Example 9.9 that computes the total function $f(n) = n + 1$. For any number m, m is encoded as a string of $m + 1$ 1's. T passes over each existing 1 and adds a new 1 at the end. Thus, T requires $(m + 1) + 1$ steps. For inputs other than a string of 1's, T halts immediately. Another way to say all this is that on any input of length n, T requires at most $n + 1$ steps, so the function $t(n) = n + 1$ describes the maximum number of steps in any computation by T as a function of the length of the input to the computation. □

9.25 Definition   Let T be a Turing machine that halts on all input. If the maximum number of steps for a computation by T on any input of length n is $t(n)$, then T is of **time complexity** $t(n)$. □

Thus, the Turing machine of Example 9.9 is of time complexity $t(n) = n + 1$.

9.26 Practice   Find an expression for the time complexity of the Turing machine of Example 9.11.   (*Answer, page A-24*)

Time complexity also applies to a Turing machine used as a set recognizer; again we assume that the machine halts on all inputs.

9.27 Example   Consider the Turing machine of Example 9.5 that recognizes $S = \{0^n 1^n \mid n \geq 0\}$. The maximum number of steps in a computation occurs when the input belongs to S. Suppose an input of length n is a member of S. The computation first moves the read–write head beyond the right end of the input, which requires n steps. The read–write head then sweeps back and forth across that part of the input not replaced by X's. This process requires successively $n, n - 1, n - 2, \ldots, 1$ steps. Recognition requires one final step. The total number of steps is thus

$$n + (n + n - 1 + \cdots + 1) + 1$$

which equals

$$n + \frac{n(n + 1)}{2} + 1 \qquad \text{(see Practice 1.26)}$$

$$= (n + 1)\left(1 + \frac{n}{2}\right)$$

$$= \frac{n^2 + 3n + 2}{2}$$

This Turing machine is of time complexity

$$t(n) = \frac{n^2 + 3n + 2}{2} \qquad\qquad\qquad \square$$

9.28 Practice   Verify that the Turing machine of Example 9.5 requires $t(4)$ steps to recognize the input $\ldots b\, 0\, 0\, 1\, 1\, b \ldots$, but uses less than $t(4)$ steps to reject $\ldots b\, 0\, 0\, 0\, 1\, b \ldots$, where $t(n) = \dfrac{n^2 + 3n + 2}{2}$.   *(Answer, page A-24)*

Another measure of efficiency, which we won't consider here, is the **space complexity** of a Turing machine, a measure of the amount of tape the machine uses as a function of input length.

Intractable Problems

Although in our examples we have developed precise expressions for the time complexity of Turing machines (or algorithms), what really interests us is the general order of magnitude of the complexity of a Turing machine.

9.29 Definition   Let f and g be two functions from $\mathbb{N} \to \mathbb{N}$. Then f and g are of the same **order of magnitude** if (1) there exists positive constants c_1 and n_1 such that $f(n) \leq c_1 g(n)$ for all $n \geq n_1$; and (2) there exist positive constants c_2 and n_2 such that $g(n) \leq c_2 f(n)$ for all $n \geq n_2$. $\square$

9.30 Example   Let $f(n) = 3n^2$ and $g(n) = 200n^2 + 140n + 7$. Then f and g are of the same order of magnitude because $3n^2 \leq 200n^2 + 140n + 7$ for all n, and $200n^2 + 140n + 7 \leq 100(3n^2)$ for all $n \geq 2$. Both f and g are of the same order of magnitude as n^2. A polynomial function is the order of its term of highest degree. The functions $f(n) = n$ and $g(n) = n^2$ are not of the same order because we cannot force $n^2 \leq cn$ (this implies $n \leq c$) for all n greater than or equal to some n_0—no matter how large a value we choose for c. $\square$

The algorithm given in Example 9.5 is of order n^2, and that given in Example 9.9 is of order n. Of course, these algorithms do two different jobs, so it is not of much interest to compare them. But suppose we have two algorithms to do the same job and their time complexities are of different orders of magnitude, say A is of order n and A' of order n^2 (see Exercise 1, this section). Even if each step in a computation takes only .0001 seconds, this difference will affect total computation time as n grows larger. Figure 9.13 compares total computation time for A and A' under various values of input length. Now suppose a third algorithm A'' exists whose time complexity is not even given by a polynomial function but by an exponential function, say 2^n. Figure 9.14 extends Figure 9.13 to include the exponential case.

Order · · · · · · · · Size of input

		10	50	100
A	n	.001 sec	.005 sec	.01 sec
A'	n^2	.01 sec	.25 sec	1 sec

Figure 9.13

Order · · · · · · · · Size of input

		10	50	100
A	n	.001 sec	.005 sec	.01 sec
A'	n^2	.01 sec	.25 sec	1 sec
A''	2^n	.1024 sec	3570 years	4×10^{16} centuries

Figure 9.14

Note that the exponential case grows at a fantastic rate! Even if we assume that each computation step takes much less time than .0001 seconds, the relative growth rates between polynomial and exponential functions still follow this same pattern.

Because of this immense growth rate, algorithms not of polynomial order are generally not useful for large values of n. In fact, problems for which no polynomial time algorithms exist are called **intractable**. There may be extenuating circumstances. When an algorithm has time complexity $t(n) = 2^n$, for instance, at least one input of length n requires 2^n steps, but the average case may run much faster. And for small values of n, an exponential function can have smaller values than a polynomial function of large degree. In general, however, a choice between possible algorithms for a given problem, or attempts to improve a given algorithm, should concentrate on the order of magnitude of the time complexity functions involved.

The Set P

9.31 Definition   **P** is the collection of all sets recognizable by Turing machines of polynomial time complexity (*P* stands for *polynomial time*).   □

Consideration of set recognition in Definition 9.31 is not as restrictive as it may seem. First, since a Turing machine for which a time complexity can be determined must halt on all input, it *decides* membership in a set. Many problems can be posed as set decision problems through a suitable encoding of the objects involved in the problem.

For example, we may describe a given graph by some encoding of each of its nodes, and then an encoding of the ordered pairs of nodes connected by arcs. This encoding would result in a string of symbols effectively describing the graph. We then pose the Hamiltonian circuit problem as a set decision problem by considering the set of all strings of allowable symbols that are descriptions of graphs having Hamiltonian circuits. If we can build a Turing machine to decide membership in this set, we then have an algorithm to solve the decision problem that asks whether an arbitrary graph has a Hamiltonian circuit. The particular encoding scheme we use determines the length of the input string for a given instance of a problem, and thus may affect the time complexity of an algorithm to solve the problem. However, if there are two encodings for a given problem such that input under each encoding can be transformed in polynomial time to corresponding input under the other encoding, then if one encoding results in a set belonging to *P*, so does the other.

The situation is equally pleasant with respect to alternative models of algorithms (Turing machine, RAMs, etc.). A problem solvable by an algorithm with polynomial time complexity on one model is solvable by an algorithm with polynomial time complexity on another model. Thus, we can speak of a *problem* belonging to *P*, meaning that a polynomial time-bounded algorithm exists for its solution, without having to specify the computational device that carries out the algorithm or the details of the encoding of the problem for that device. (Do not infer from what we've said here that the Hamiltonian circuit problem belongs to *P*; we have merely said that the problem can be formulated so that it is possible to *ask* whether it belongs to *P*, but no one yet knows the answer.)

NONDETERMINISM

The decision problem for Hamiltonian circuits, unlike problems such as the halting problem, is not unsolvable. An algorithm exists to test whether an arbitrary graph has a Hamiltonian circuit, namely, the trial-and-error approach of testing all possible paths. As we trace out paths that are candidates for success, we may have a choice of possible next moves every time we come to a node. We can simulate this type of behavior by using a **nondeterministic Turing machine (NDTM)**. A nondeterministic Turing machine is defined just like an ordinary Turing machine except that for each state–input pair there is a set of applicable quintuples and so, possibly, a choice for the Turing machine's behavior at that point. Each choice (each quintuple) specifies the symbol to be printed, the next state, and the direction of motion of the read–head. We can think of the NDTM as pursuing all of its possible sequences of action in parallel. An NDTM T **recognizes**, or **accepts**, a string α of tape symbols if T, begun in standard configuration on α, has some sequence of moves leading to a halt in a final state. T recognizes the set of all recognized strings.

9.32   Definition   Let T be an NDTM. For every recognized input string α of length n, there is at least one sequence of moves leading to a final state; for each accepted string, consider only the shortest sequence of moves leading to acceptance. If the maximum number of steps used in any such sequence accepting a string of length n is $t(n)$, then T is of **time complexity** $t(n)$.   □

9.33   Definition   **NP** is the collection of all sets recognizable by NDTMs of polynomial time complexity. (*NP* stands for *n*ondeterministic *p*olynomial time.)   □

Any ordinary (deterministic) Turing machine is a trivial NDTM, so it is clear that $P \subseteq NP$. Whether P is a proper subset of NP is a question that occupies us for the rest of this section.

We should note that time complexity for an NDTM gives an upper bound on time required to *accept* any (acceptable) string α of length n, and time complexity for an ordinary Turing machine gives an upper bound on the time required to *process* any string α of length n. In other words, in an ordinary (deterministic) Turing machine, any input can be fed in and the answer "yes—accepted" or "no—rejected" will be obtained within $t(n)$ units of time; in an NDTM, any input can be fed in and if it is an accepted input, the answer "yes—accepted" will be obtained within $t(n)$ units of time, but if it is a nonaccepted string, there is no bound on the amount of processing time.

Is Nondeterminism Faster?

As in the corresponding situation for finite-state machines (Theorem 7.37), any set recognizable by an NDTM T can also be recognized by a deterministic Turing machine T^*. We can think of T^* acting on a given input α as simulating one after another the possible sequences of moves T could make on α until α is accepted or all possible sequences have been tried and α is rejected. Therefore, although nondeterminism gains us no new capabilities, we would expect it to gain us some lower time complexity.

Thus, if the time complexity for T is $t(n)$, we would expect the time complexity $t^*(n)$ for T^* to be higher for two reasons. T^* cannot execute sequences of moves in parallel as T can; it must do them in a serial fashion. Also, T^* gives us more information about an input α of length n; although T may give an answer within $t(n)$ units of time only if α is accepted, T^* always gives us an answer (yes or no) about any input α within $t^*(n)$ units of time. There is one detail we glossed over a moment ago in discussing T^*'s simulation of T on α. T may have sequences of moves that do not halt; if T^* begins simulating one of these sequences, how does it know when to give up and try another sequence? If T has time complexity $t(n)$, then T^* need not pursue any sequence of moves for longer than $t(n)$ units of time. If α is accepted by T, there is some sequence that will do the job within this time. We may imagine T's possible actions on a given α as something like the tree shown in Figure 9.15; as T^* simulates T, it need not look below $t(n)$ levels, and it can trace out each branch of the tree that far. Because there is a bound b on the maximum number of possible moves T can make at any point, there are at most b branches of the tree at any node. Thus, the tree can have at most $b^{t(n)}$ separate paths each of length at most $t(n)$, so we would expect some exponential expression such as $t(n)b^{t(n)}$ to be the time complexity for T^*. We should never need more time than this, but probably some input of length n for some n would require this much time.

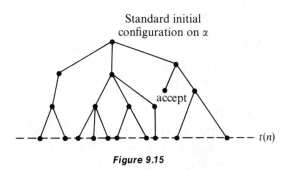

Standard initial
configuration on α

accept

$t(n)$

Figure 9.15

The above argument seems to convince us that, in most cases, if a set is accepted by an NDTM of time complexity $t(n)$, it will probably require a deterministic machine of time complexity that looks like $t(n)b^{t(n)}$, a function of a higher order of magnitude. Such a result has not been proven, however. No one has found any set S recognizable by an NDTM with time complexity $t(n)$ for which no deterministic machine of complexity $t(n)$ exists to recognize S. Although there are certainly sets for which such a deterministic machine has not been found, it has not been established that one cannot exist. In particular, whether P is a proper subset of NP is an open question.

There are many quite famous problems, such as the Hamiltonian circuit problem, that have been shown to be in *NP*, that is, they are representable as *NP* sets, but for which no polynomial-bounded, deterministic solution algorithm has been found. This fact lends weight to the speculation that *P* is indeed a proper subset of *NP*. This view is the prevailing one in computer science circles today, strengthened by work begun in 1971 on a class of problems known as **NP-complete problems**. Roughly, if a problem is *NP*-complete, it is *NP* and at least as hard to solve as any other *NP* problem in that if it could be shown to belong to *P*, then every *NP* problem would belong to *P* and *P* would equal *NP*.

Since 1971 many problems from many different fields (graph theory, number theory, algebra, data storage, network theory, etc.) have been shown to be *NP*-complete. The Hamiltonian circuit problem is *NP*-complete as is the problem of deciding whether two arbitrary regular expressions represent the same set. The *NP*-complete problems are quite diverse, and the search for efficient (polynomially bounded) solution procedures has been extensive. Remember that if an efficient procedure could be found to solve any one of them, such a procedure would exist for all the others as well. In view of the so-far unsuccessful search for an efficient solution procedure for even one such problem, it seems likely that $P \neq NP$. On the practical side, perhaps one should not look too long for a quick and easy algorithm to solve any *NP* problem one may encounter.

↙ CHECKLIST

Definitions

time complexity for a Turing machine (*p. 358*)
order of magnitude (*p. 359*)
intractable problem (*p. 360*)
P (*p. 361*)
nondeterministic Turing machine (NDTM) (*p. 362*)
recognition of a set by an NDTM (*p. 362*)
time complexity for an NDTM (*p. 362*)
NP (*p. 362*)
NP-complete problem (*p. 364*)

Main Ideas

Time complexity for a Turing machine is a measure of its efficiency in processing inputs.

Problems for which no polynomial time solution algorithms exist are considered intractable even though special cases may exist for which they can be efficiently solved.

Time complexity for an NDTM is a measure of its efficiency in recognizing problem solutions.

Nondeterminism offers parallel processing that suggests a gain in efficiency over a deterministic solution, but such a gain has not been proved.

$P \subseteq NP$, but it is unknown whether $P \subset NP$.

The NP-complete problems, of which there are many, are as difficult to solve efficiently as any NP problem.

EXERCISES, SECTION 9.3

1. The algorithm given in Example 9.9 that computes $f(n) = n + 1$ is of order n. Another Turing machine to compute $f(n)$ could proceed by first copying $\bar{n}$ over and then erasing n 1's. Show that the algorithm of this Turing machine is of order n^2.

★ 2. Explain intuitively why nondeterminism might be expected to improve the efficiency of a solution algorithm for each of the following problems. (Each of these problems is NP-complete.)

  (a) The **satisfiability problem**: given a set of variables, and a Boolean expression over these variables (see Definition 4.1), does there exist a truth assignment to the variables making the expression true? (This problem was essentially the first ever shown to be NP-complete).

  (b) The **clique problem**: given a graph G and a positive integer k, does G contain a complete subgraph with k vertices?

  (c) The **set packing problem**: given a collection C of finite sets and a positive integer k, does C contain at least k mutually disjoint sets?

For Further Reading

Entry [1] gives an interesting and easy-to-read presentation of Turing machines and unsolvability results. Turing machines are also discussed thoroughly in [2] and [3]. A briefer coverage of Turing machines appears at the end of [4]. The little book [5] gives a good treatment of Turing machines and the halting problem. A well-written and practical introduction to issues in computational complexity is [6]; [7] covers similar topics from a more theoretical viewpoint. P and NP are discussed in [3], and [8] is devoted entirely to this topic.

1. Marvin L. Minsky, *Computation: Finite and Infinite Machines*. Englewood Cliffs, N.J.: Prentice-Hall, 1967.

2. Fred Hennie, *Introduction to Computability.* Reading, Mass.: Addison-Wesley, 1977.

3. John E. Hopcroft and Jeffrey D. Ullman, *Introduction to Automata Theory, Languages, and Computation.* Reading, Mass.: Addison-Wesley, 1979.

4. Alfred B. Manaster, *Completeness, Compactness and Undecidability.* Englewood Cliffs, N.J.: Prentice-Hall, 1975.

5. J. Loeckx, *Computability and Decidability.* Lecture Notes in Economics and Mathematical Systems, vol. 68. New York: Springer-Verlag, 1972.

6. Sara Baase, *Computer Algorithms: Introduction to Design and Analysis.* Reading, Mass.: Addison-Wesley, 1978.

7. Alfred V. Aho, John E. Hopcroft, and Jeffrey D. Ullman, *The Design and Analysis of Computer Algorithms.* Reading, Mass.: Addison-Wesley, 1974.

8. Michael R. Garey and David S. Johnson, *Computers and Intractability.* San Francisco: W. H. Freeman and Co., 1979.

Chapter 10

FORMAL LANGUAGES

This chapter presents several classes of formal grammars, that is, rules for generating "legal sentences" in a "language." It also discusses recognizing devices for languages. If the syntax for a higher-level programming language is described by a formal grammar, which frequently happens, then the compiler serves as a recognizing device for that language.

Section 10.1 **CLASSES OF LANGUAGES**

GRAMMARS AND LANGUAGES

Suppose we come upon the English language sentence "The walrus talks loudly." Although we might be surprised at the meaning, or **semantics**, of the sentence, we accept its form, or **syntax** as valid, meaning that the various parts of speech (noun, verb, etc.) are strung together in a reasonable way. In contrast, we reject "Loudly walrus the talks" as an illegal combination of parts of speech, or as syntactically incorrect. Our feeling for correct syntax in English is just that, a feeling, based on years of experience. We must also worry about correct syntax in programming languages, but there, unlike natural languages (English, French, etc.), legal combinations of symbols are specified in detail. Let's give a formal definition of what is meant by a language; the definition will be general enough to include both natural and programming languages.

10.1 Definition An **alphabet,** or **vocabulary,** V, is a finite, nonempty set of symbols. A **word** over V is a finite-length string of symbols from V. V^* is the set of all words over V. A **language** over V is any subset of V^*. □

The set V^* is infinite; in fact, it is equivalent to $\mathbb{N}$ because we can list in some lexicographical order all strings from V of length one, then all strings of length two, and so on. A language L over V can therefore be infinite. How can we describe a given L, in the sense of specifying exactly those words belonging to L? If L is finite, we can just list its members, but if L is infinite, can we find a finite description of L? Not always; there are many more languages than possible finite descriptions. We will only consider languages that can be finitely described, but even here we can think of two possibilities. We may be able to describe an algorithm to *decide* membership in L; that is, given any word in V^*, we can apply our algorithm and receive a yes or no answer as to whether the word belongs to L. Or we may only be able to describe a procedure allowing us to *generate* members of L, that is, crank out one at a time a list of all the members of L.

10.2 Definition A language L over V is **recursive** if there exists an effective procedure to decide membership in L. L is **recursively enumerable** (r.e.) if there exists an effective procedure to generate a list of members of L. ☐

Because a language is just a subset of V^*, we will also speak of *sets* as being recursive, or recursively enumerable.

Any recursive language is also recursively enumerable because a procedure for generating the language is to list those members of V^* belonging to L. However, an r.e. language may not be recursive as we will see in the next section. At any rate, we will be interested in r.e. languages, and we will describe such a language by defining its generative process, or giving a **grammar** for the language.

10.3 Definition A **phrase-structure grammar (Type 0 grammar)** G is a 4-tuple, $G = (V, V_T, S, P)$, where

> V is a vocabulary.
>
> V_T is a nonempty subset of V called the set of **terminals**.
>
> S is an element of $V - V_T$ called the **start symbol**.
>
> P is a finite set of **productions** of the form $\alpha \to \beta$ where α is a word over V containing at least one nonterminal symbol and β is a word over V. ☐

10.4 Example Here is a very simple grammar. $G = (V, V_T, S, P)$ where $V = \{0, 1, S\}$, $V_T = \{0, 1\}$, $P = \{S \to 0S, S \to 1\}$. ☐

The productions of a grammar allow us to transform some words over V into others; the productions could be called rewriting rules.

10.5 Definition   Let G be a grammar, $G = (V, V_T, S, P)$, and let w_1 and w_2 be words over V. Then w_1 **directly generates (directly derives)** w_2, written $w_1 \Rightarrow w_2$, if $\alpha \rightarrow \beta$ is a production of G, w_1 contains an instance of α, and w_2 is obtained from w_1 by replacing that instance of α with β. If $w_1, w_2, \ldots, w_n$ are words over V and $w_1 \Rightarrow w_2$, $w_2 \Rightarrow w_3, \ldots, w_{n-1} \Rightarrow w_n$, then w_1 **generates (derives)** w_n, written $w_1 \overset{*}{\Rightarrow} w_n$. (By convention, $w_1 \overset{*}{\Rightarrow} w_1$.)   □

10.6 Example   In the grammar of Example 10.4, $00S \Rightarrow 000S$ because the production $S \rightarrow 0S$ has been used to replace the S in $00S$ with $0S$. Also $00S \overset{*}{\Rightarrow} 00000S$.   □

10.7 Practice   Show that in the grammar of Example 10.4, $0S \overset{*}{\Rightarrow} 00001$.   *(Answer, page A-25)*

10.8 Definition   Given a grammar G, the **language L generated by G**, sometimes denoted $L(G)$, is the set

$$L = \{w \in V_T^* \mid S \overset{*}{\Rightarrow} w\}$$

In other words, L is the set of all strings of terminals generated from the start symbol.   □

Notice that once a string w of terminals has been obtained, no productions can be applied to w, and w cannot generate any other words. The language L of a given grammar is an r.e. set because the following effective procedure generates a list of the members of L. Begin with the start symbol S and systematically apply some sequence of productions until a string w_1 of terminals has been obtained; $w_1 \in L$. Go back to S and repeat this procedure using a different sequence of productions to generate another word $w_2 \in L$, and so forth. However, this procedure doesn't quite work because we might get started on an infinite sequence of direct derivations that never leads to a string of terminals and so never contributes a word to our list. Instead, we need to run a number of derivations from S simultaneously, checking on each one after each step, and for any that terminate, adding the final word to the list of members of L. That way we cannot get stuck waiting indefinitely while unable to do anything else.

10.9 Practice   Describe the language generated by the grammar G of Example 10.4.

(Answer, page A-25)

Languages derived from grammars such as we have defined are called **formal languages**. If the grammar is defined first, the language follows as an outcome of the definition. It may be that the language, as a well-defined set of strings, is given first and we seek a grammar that generates it.

10.10 Example   Let L be the set of all nonempty strings consisting of an even number of 1's. Then L is generated by the grammar $G = (V, V_T, S, P)$ where $V = \{1, S\}$, $V_T = \{1\}$, and $P = \{S \rightarrow SS, S \rightarrow 11\}$. A language can be generated by more than one grammar. L is also generated by the grammar $G' = (V', V'_T, S', P')$ where $V' = \{1, S\}$, $V'_T = \{1\}$, and $P' = \{S \to 1S1, S \to 11\}$.   □

10.11 Practice   (a) Find a grammar that generates the language $L = \{0^n 10^n \mid n \geq 0\}$.
(b) Find a grammar that generates the language $L = \{0^n 10^n \mid n \geq 1\}$.

(*Answer, page A-25*)

Trying to describe concisely the language generated by a given grammar and defining a grammar to generate a given language can both be quite difficult. We'll look at another example where the grammar is a bit more complicated than any we've seen so far.

10.12 Example   Let $L = \{a^n b^n c^n \mid n \geq 1\}$. A grammar generating L is $G = (V, V_T, S, P)$ where $V = \{a, b, c, S, B, C\}$, $V_T = \{a, b, c\}$, and P consists of the following productions:

1. $S \to aSBC$
2. $S \to aBC$
3. $CB \to BC$
4. $aB \to ab$
5. $bB \to bb$
6. $bC \to bc$
7. $cC \to cc$

It is fairly easy to see how to generate any particular member of L using these productions. Thus, a derivation of the string $a^2 b^2 c^2$ is

$$S \Rightarrow aSBC$$
$$\Rightarrow aaBCBC$$
$$\Rightarrow aaBBCC$$
$$\Rightarrow aabBCC$$
$$\Rightarrow aabbCC$$
$$\Rightarrow aabbcC$$
$$\Rightarrow aabbcc$$

In general, $L \subseteq L(G)$ where the outline of a derivation for any $a^n b^n c^n$ is given below; the numbers refer to the productions used.

$$S \overset{*}{\underset{(1)}{\Rightarrow}} a^{n-1} S (BC)^{n-1}$$
$$\underset{(2)}{\Rightarrow} a^n (BC)^n$$
$$\overset{*}{\underset{(3)}{\Rightarrow}} a^n B^n C^n$$
$$\underset{(4)}{\Rightarrow} a^n b B^{n-1} C^n$$
$$\overset{*}{\underset{(5)}{\Rightarrow}} a^n b^n C^n$$
$$\overset{*}{\underset{(6)}{\Rightarrow}} a^n b^n c C^{n-1}$$
$$\overset{*}{\underset{(7)}{\Rightarrow}} a^n b^n c^n$$

We must also show that $L(G) \subseteq L$, which involves arguing that some productions must be used before others and that the general derivation shown above is the only sort that will lead to a string of terminals. □

In trying to invent a grammar to generate the L of Example 10.12, an initial attempt might be to use productions of the form $B \rightarrow b$ and $C \rightarrow c$ instead of productions (4)–(7). Then we would indeed have $L \subseteq L(G)$, but $L(G)$ would also include words such as $a^n(bc)^n$. In devising a grammar to generate a given language, we may have to be a bit tricky.

Formal languages are our last Simulation I example. They were developed in the 1950s in an attempt to model natural languages, such as English, with an eye toward automatic translation. However, since the language already exists and is quite complex, defining a formal grammar to generate a natural language is very difficult. Attempts to do this for English have been only partially successful.

10.13 Example   We can describe a formal grammar that will generate a very restricted class of English sentences. The terminals in the grammar are the words "the, a, legal river, walrus, talks, flows, loudly, swiftly," and the nonterminals are the words "sentence, noun-phrase, verb-phrase, article, noun, verb, adverb." The start symbol is "sentence" and the productions are sentence → noun-phrase verb-phrase, noun-phrase → article noun, verb-phrase → verb adverb, article → the, article → a, noun → river, noun → walrus, verb → talks, verb → flows, adverb → loudly, adverb → swiftly.

Here is a derivation of "the walrus talks loudly" in this grammar.

$$
\begin{aligned}
\text{sentence} &\rightarrow \text{noun-phrase verb-phrase} \\
&\rightarrow \text{article noun verb-phrase} \\
&\rightarrow \text{the noun verb-phrase} \\
&\rightarrow \text{the walrus verb-phrase} \\
&\rightarrow \text{the walrus verb adverb} \\
&\rightarrow \text{the walrus talks adverb} \\
&\rightarrow \text{the walrus talks loudly}
\end{aligned}
$$

A few other sentences making various degrees of sense, such as "a walrus flows loudly," are also part of the language defined by this grammar. The difficulty of specifying a grammar for English as a whole is more apparent when we consider that a phrase such as "time flies" can be an instance of either a noun followed by a verb or of a verb followed by a noun! This situation is "ambiguous" (see Exercise 12, this section). □

Programming languages are less complex than natural languages, and some programming languages, such as ALGOL, have been defined as formal languages.

10.14 Example   A section of formal grammar to generate identifiers in a programming language could be presented as follows.

$$\text{identifier} \rightarrow \text{letter}$$
$$\text{identifier} \rightarrow \text{identifier letter}$$
$$\text{identifier} \rightarrow \text{identifier digit}$$
$$\text{letter} \rightarrow a$$
$$\text{letter} \rightarrow b$$
$$\vdots$$
$$\text{letter} \rightarrow z$$
$$\text{digit} \rightarrow 0$$
$$\vdots$$
$$\text{digit} \rightarrow 9$$

Here, the set of terminals is $\{a, \ldots, z, 0, \ldots, 9\}$ and "identifier" is the start symbol. A shorthand that avoids listing all these productions is called **Backus–Naur form** or BNF. The productions listed above can be given in BNF by three lines.

$$\langle\text{identifier}\rangle ::= \langle\text{letter}\rangle \,|\, \langle\text{identifier}\rangle\langle\text{letter}\rangle \,|\, \langle\text{identifier}\rangle\langle\text{digit}\rangle$$

$$\langle\text{letter}\rangle ::= a\,|\,b\,|\,c\,|\cdots|\,z$$

$$\langle\text{digit}\rangle ::= 0\,|\,1\,|\cdots|\,9$$

In BNF, nonterminals are identified by $\langle\ \rangle$, the production arrow becomes $::=$, and $|$ stands for "or," identifying various productions having the same left-hand symbol. $\qquad\qquad\square$

CLASSES OF GRAMMARS

We will identify several types of grammars. First, we'll look at one more example.

10.15 Example Let L be the empty string λ together with the set of all strings consisting of an odd number n of 0's, $n \geq 3$. The grammar $G = (V, V_T, S, P)$ generates L where $V = \{0, A, B, E, F, W, X, Y, Z, S\}$, $V_T = \{0\}$, and the productions are

$S \rightarrow FA$	$FX \rightarrow F0W$
$S \rightarrow FBA$	$YA \rightarrow Z0A$
$FB \rightarrow F0EB0$	$W0 \rightarrow 0W$
$EB \rightarrow 0$	$0Z \rightarrow Z0$
$EB \rightarrow XBY$	$WBZ \rightarrow EB$
$0X \rightarrow X0$	$F \rightarrow \lambda$
$Y0 \rightarrow 0Y$	$A \rightarrow \lambda$

The derivation $S \Rightarrow FA \overset{*}{\Rightarrow} \lambda\lambda = \lambda$ produces λ. The derivation

$$
\begin{aligned}
S &\Rightarrow FBA \\
&\Rightarrow F0EB0A \\
&\Rightarrow F0XBY0A \\
&\overset{*}{\Rightarrow} FX0B0YA \\
&\overset{*}{\Rightarrow} F0W0B0Z0A \\
&\overset{*}{\Rightarrow} F00WBZ00A \\
&\Rightarrow F00EB00A \\
&\Rightarrow F00000A \\
&\overset{*}{\Rightarrow} 00000
\end{aligned}
$$

produces five 0's. Notice how X and Y, and also W and Z, march back and forth across the strings of 0's, adding one more 0 on each side. This activity is highly reminiscent of a Turing machine read–write head sweeping back and forth across its tape and enlarging the printed portion of the tape. □

The above grammar allows erasing productions ($F \to \lambda$, $A \to \lambda$). To generate any language containing λ, we have to be able to erase somewhere. In the following grammar types, we will limit erasing, if it occurs at all, to a single production of the form $S \to \lambda$ where S is the start symbol, and we will not allow S to appear on the right-hand side of any other productions. This restriction allows us to crank out λ from S as a special case and then get on with other derivations, in none of which is any erasing allowed. Let's call this the **erasing convention**. The following definition defines three special types of grammars by putting further restrictions on the productions allowed.

10.16 Definition   A grammar G is **context-sensitive** if it obeys the erasing convention and for every production $\alpha \to \beta$ (except $S \to \lambda$), the word β is at least as long as the word α. A grammar G is **context-free** if it obeys the erasing convention and for every production $\alpha \to \beta$, α is a single nonterminal. A grammar G is **regular** if it obeys the erasing convention and for every production $\alpha \to \beta$ (except $S \to \lambda$), α must be a single nonterminal and β must be of the form t or tW where t is a terminal symbol and W is a nonterminal symbol.   □

In a context-free grammar, a single, nonterminal symbol on the left of a production can be replaced wherever it appears by the right side of the production. In a context-sensitive grammar, a given nonterminal symbol might be replaceable only if it is part of a particular string, or context, hence, the names context-free and context-sensitive. It is clear that any regular grammar is also context-free, and any context-free grammar is also context-sensitive. The grammar of Example 10.4 is regular, both the grammars of Example 10.10 are context-free but not regular, the grammar of Example 10.12 is context-sensitive but not context-free. The grammars of Example

10.13 and of Example 10.14 are context-free but not regular. And the grammar of Example 10.15 is a Type 0 grammar, but it is not context-sensitive.

10.17 *Definition* A language is **type 0 (context-sensitive, context-free, regular)** if it can be generated by a type 0 (context-sensitive, context-free, regular) grammar.   □

Because of the relationships among the four grammar types, we can classify languages according to Figure 10.1. Thus, any regular language is also context-free because any regular grammar is also a context-free grammar, and so on. However, although it turns out to be true, we do not know from what we have done that these sets are properly contained in one another. For example, the language L described in Example 10.15 was generated there by a grammar that was type 0 but not context-sensitive, but that does not imply that L itself falls into that category. Different grammars can generate the same language.

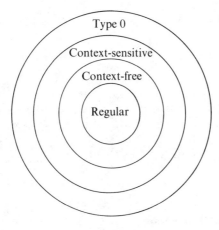

Figure 10.1

10.18 *Definition* Two grammars are **equivalent** if they generate the same language.   □

10.19 *Example* The language L of Example 10.15 can be described by the regular expression $\lambda \vee (000)(00)^*$, so L is a regular set. Example 10.15 gave a grammar G to generate L. We will give three more grammars equivalent to G.

$G_1 = (V, V_T, S, P)$ where $V = \{0, A, B, S\}$, $V_T = \{0\}$, and the productions are

$$S \rightarrow \lambda$$
$$S \rightarrow ABA$$
$$AB \rightarrow 00$$
$$0A \rightarrow 000A$$
$$A \rightarrow 0$$

G_1 is context-sensitive but not context-free.

$$G_2 = (V, V_T, S, P) \text{ where } V = \{0, A, S\}, V_T = \{0\},$$

and the productions are

$S \to \lambda$

$S \to 00A$

$A \to 00A$

$A \to 0$

G_2 is context-free but not regular.

$G_3 = (V, V_T, S, P)$ where $V = \{0, A, B, C, S\}$, $V_T = \{0\}$, and the productions are

$S \to \lambda$

$S \to 0A$

$A \to 0B$

$B \to 0$

$B \to 0C$

$C \to 0B$

G_3 is regular.

Thus, L is a regular language. That a regular set turned out to be a regular language is not coincidental as we will see in the next section. □

10.20 Practice   Give the derivation of 00000 in G_1, G_2, and G_3.   (*Answer, page A-25*)

CONTEXT-FREE GRAMMARS

We will elaborate on the proper containment of the sets of Figure 10.1 in the next section. Meanwhile, we'll concentrate on context-free grammars for three reasons. Context-free grammars seem to be the easiest for us to work with since they involve replacing only one symbol at a time. Furthermore, many programming languages are defined such that sections of syntax, if not the whole language, can be described by context-free grammars. And finally, a derivation in a context-free grammar has a lovely graphical representation called a **parse tree**.

10.21 Example   The grammar of Example 10.14 is context-free. The word *d2q* can be derived as follows: identifier $\Rightarrow$ identifier letter $\Rightarrow$ identifier digit letter $\Rightarrow$ letter digit letter $\Rightarrow$ *d* digit letter $\Rightarrow$ *d* 2 letter $\Rightarrow$ *d2q*. We can represent this derivation as a tree with the start symbol for the root. When a production is applied to a node, that node is replaced at the next lower level of the tree by the symbols in the right-hand side of the production used. A tree for the above derivation appears in Figure 10.2.   □

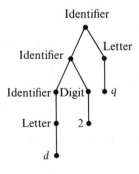

Figure 10.2

10.22 Practice   Draw a parse tree for the word *m*34*s* in the grammar of Example 10.14.

(Answer, page A-25)

Suppose that a programming language has been described by a context-free grammar *G*. The programmer uses the rules of *G* to generate legitimate strings of symbols, that is, words in the language. Here we may think of a "word" as corresponding to a program instruction. Thus, a word consists of various subwords, for example, identifiers, operators, and key words for the language (such as IF and DO in FORTRAN, or REPEAT and UNTIL in PASCAL). The program instructions are fed into the compiler for the language so that the program can be translated into machine language code for the computer. The compiler must decide whether each program instruction is legitimate in the language. This question is really a two-level one: are the subwords themselves legitimate strings, and is the program instruction a legitimate way of grouping the subwords together?

Usually, the set of legitimate subwords of a language can be described by a regular expression, and then a finite-state machine can be used to detect the subwords; this phase of the compilation process is handled by the lexical analyzer or scanner portion of the compiler. If all goes well, the scanner then passes the program instruction, in the form of a string of legitimate subwords, to the syntax analyzer. The syntax analyzer determines whether the string is correct by trying to parse it (construct its parse tree). Various parsing techniques, which we won't go into, have been devised. One obvious approach is to construct a tree by beginning with the start symbol, applying productions, and ending with the string to be tested. This procedure is called **top-down parsing**. The other alternative is to begin with the string, see what productions were used to create it, apply productions "backwards," and end with the start symbol. This process is called **bottom-up parsing**. The trick in either approach is how to decide exactly which productions should be used.

10.23 Example   Consider the context-free grammar G given by $G = (V, V_T, S, P)$ where $V = \{a, b, c, A, B, C, S\}$, $V_T = \{a, b, c\}$, and the productions are

$$S \rightarrow B \qquad S \rightarrow A \qquad B \rightarrow ab$$
$$B \rightarrow C \qquad A \rightarrow abc \qquad C \rightarrow c$$

Suppose we want to test the string abc. A derivation for abc is $S \Rightarrow A \Rightarrow abc$. If we try a top-down parse, we might begin with

$$S$$
$$|$$
$$B$$

and then we have to detect that this will not work. If we try a bottom-up parse, we might begin with

$$BC$$
$$|$$
$$abc$$

and then we have to detect that this will not work. Parsing techniques try to automate this process.   □

Overall, we see once again a distinction between *generating* members of a set, which the programmer does, and *deciding* membership in a set, which the compiler does. Since we ask the compiler to decide membership in a set, the set must be a recursive language. It turns out that context-free languages are indeed recursive, another point in their favor.

✓ *CHECKLIST*

Definitions

vocabulary (*p. 368*)
word (*p. 368*)
language (*p. 368*)
recursive language (*p. 369*)
recursively enumerable
   language (*p. 369*)
phrase-structure grammar
   (type 0 grammar) (*p. 369*)
terminal (*p. 369*)
production (*p. 369*)
start symbol (*p. 369*)

language generated
   by a grammar G (*p. 370*)
Backus–Naur form (*p. 373*)
context-sensitive grammar (*p. 374*)
context-free grammar (*p. 374*)
regular grammar (*p. 374*)
equivalent grammars (*p. 375*)
parse tree (*p. 376*)
top-down parsing (*p. 377*)
bottom-up parsing (*p. 377*)

Techniques

Describe $L(G)$ for a given grammar G.

Define a grammar to generate a given language L.

Construct parse trees in a context-free grammar.

Main Ideas

A grammar G is a generating mechanism for its language $L(G)$; $L(G)$ is thus an r.e. language.

Formal languages were developed as an attempt to describe correct syntax for natural languages; although this attempt has largely failed because of the complexity of natural languages, it has been quite successful in the case of higher-level programming languages.

Special classes of grammars are defined by putting restrictions on the allowable productions.

Derivations in context-free grammars can be illustrated by parse trees.

A compiler for a context-free programming language checks correct syntax by parsing.

EXERCISES, SECTION 10.1

★ 1. Describe $L(G)$ for each of the following grammars G.
   (a) $G = (V, V_T, S, P)$ where $V = \{a, A, B, C, S\}$, $V_T = \{a\}$, and P consists of

$$S \to A$$
$$A \to BC$$
$$B \to A$$
$$A \to a$$
$$aC \to \lambda$$

   (b) $G = (V, V_T, S, P)$ where $V = \{0, 1, A, B, S\}$, $V_T = \{0, 1\}$, and P consists of

$$S \to 0A$$
$$S \to 1A$$
$$A \to 1BB$$
$$B \to 01$$
$$B \to 11$$

(c) $G = (V, V_T, S, P)$ where $V = \{0, 1, A, B, S\}$, $V_T = \{0, 1\}$, and P consists of

$$S \rightarrow 0$$
$$S \rightarrow 0A$$
$$A \rightarrow 1B$$
$$B \rightarrow 0A$$
$$B \rightarrow 0$$

(d) $G = V, V_T, S, P)$ where $V = \{0, 1, A, S\}$, $V_T = \{0, 1\}$, and P consists of

$$S \rightarrow 0S$$
$$S \rightarrow 11A$$
$$A \rightarrow 1A$$
$$A \rightarrow 1$$

2. (a) Which of the grammars of Exercise 1 are regular? Which are context-free?

   (b) Find regular grammars to generate each of the languages of Exercise 1.

3. Find a grammar that generates the set of all strings of well-balanced parentheses.

★ 4. Find a context-free grammar that generates the language
$$L = \{0^n 1^n \mid n \geq 0\}.$$

5. Find a grammar that generates the language $L = \{0^{2^i} \mid i \geq 0\}$.

6. Find a context-free grammar that generates the language L where L consists of the set of all nonempty strings of 0's and 1's with an equal number of 0's and 1's.

7. Find a context-free grammar that generates the language L where L consists of the set of all nonempty strings of 0's and 1's with twice as many 0's as 1's.

★ 8. Find a context-free grammar that generates the language $L = \{w w^R \mid w \in \{0, 1\}^*$, and w^R is the reverse of the string $w\}$.

9. Find a grammar that generates the language $L = \{ww \mid w \in \{0, 1\}^*\}$.

10. Find a grammar that generates the language $L = \{a^{n^2} \mid n \geq 1\}$. (By Exercise 10 of Section 10.2, L is not a context-free language, so your grammar cannot be too simple.)

11. Draw parse trees for the following words:
    (a) The word 111111 in the grammar G of Example 10.10.
    (b) The word 111111 in the grammar G' of Example 10.10.
    (c) The word 011101 in the grammar of Exercise 1(b).
    (d) The word 00111111 in the grammar of Exercise 1(d).

★ 12. Consider the context-free grammar $G = (V, V_T, S, P)$ where $V = \{0, 1, A, S\}$, $V_T = \{0, 1\}$, and P consists of

$$S \to A1A$$
$$A \to 0$$
$$A \to A1A$$

Draw two distinct parse trees for the word 01010 in G. A grammar in which a word has more than one parse tree is **ambiguous**.

13. Show that for any context-free grammar G there exists a context-free grammar G' in which for every production $\alpha \to \beta$, β is a string at least as long as α, $L(G') \subseteq L(G)$, and $L(G) - L(G')$ is a finite set.

Section 10.2   LANGUAGE RECOGNIZERS

RECOGNIZERS FOR REGULAR LANGUAGES

In Chapter 7 we defined regular sets and showed that a set is regular if and only if it can be recognized by some finite-state machine. In this chapter we have defined regular languages. Next we will see that a set is regular if and only if it is a regular language. It will follow that

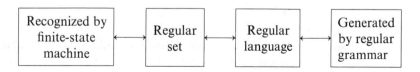

Therefore, regular grammars serve as generating devices for exactly those sets for which finite-state machines serve as recognition devices.

First, we show, given a finite-state machine M, how to construct a regular grammar G such that $L(G)$ is the set recognized by M. By throwing in a new start state, we could create a new machine $\bar{M}$ that recognizes the same set as M but whose start state is never revisited. Thus, we may as well assume that M has this property. If M recognizes λ, then the start state S is a final state. The grammar G uses the states and input symbols of M as its vocabulary, with the set of input symbols as the set of terminals. The start state S of M is the start symbol of G, and the productions of G are defined so that derivations in G simulate computations by M. Thus, $A \to a$ belongs to P if and only if M in state A with input symbol a goes to a final state, and $A \to aB$ is in P if and only if M in state A with input symbol a goes to state B. $S \to \lambda$ is in G if and only if S is a final state in M. G is then regular, and it is not hard to see that $L(G)$ is the set recognized by M.

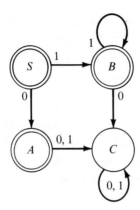

Figure 10.3

10.24 Example   Let M be given by the state graph of Figure 10.3. Following the procedure above, let $G = (V, V_T, S, P)$ where $V = \{S, A, B, C, 0, 1\}$, $V_T = \{0, 1\}$, and P consists of

$S \rightarrow \lambda$	$S \rightarrow 1$
$S \rightarrow 0$	$S \rightarrow 1B$
$S \rightarrow 0A$	$B \rightarrow 1$
$A \rightarrow 1C$	$B \rightarrow 1B$
$A \rightarrow 0C$	$B \rightarrow 0C$
$C \rightarrow 1C$	
$C \rightarrow 0C$	

G is regular. M recognizes the regular set $0 \vee 1^*$, and this is also $L(G)$. For example, a derivation of 111 in G is $S \Rightarrow 1B \Rightarrow 11B \Rightarrow 111$.   □

10.25 Practice   Let M be given by the state graph of Figure 10.4. Find a regular grammar generating the set recognized by M. Describe the set by a regular expression.

(*Answer, page A-25*)

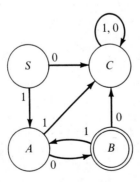

Figure 10.4

Conversely, suppose we are given a regular grammar $G = (V, V_T, S, P)$ and we want to construct a finite-state machine M to recognize $L(G)$. We will first construct a nondeterministic machine $\bar{M}$, which by Theorem 7.37 is all that's necessary. We let the states of $\bar{M}$ be the nonterminals of G together with a new final state F. If the production $S \to \lambda$ is in P, then we also let state S be final. V_T is the input alphabet of $\bar{M}$ and S is the start state. If $\bar{M}$ is in state F with input symbol a, then the set of next states of $\bar{M}$ is empty. If $\bar{M}$ is in any state $A \neq F$ with input symbol a, then the set of next states of $\bar{M}$ includes F if $A \to a$ is a production in P, and it includes all states B such that $A \to aB$ is a production in P. Computations in $\bar{M}$ simulate derivations in G, and again it can easily be shown that $\bar{M}$ recognizes the set $L(G)$.

10.26 Example   Let $G = (V, V_T, S, P)$ where $V = \{S, A, B, 0, 1\}$, $V_T = \{0, 1\}$, and P consists of

$$S \to \lambda$$
$$S \to 0A$$
$$A \to 1$$
$$A \to 0A$$
$$A \to 1B$$
$$B \to 0B$$

The state graph of a nondeterministic machine $\bar{M}$ recognizing $L(G)$ is given in Figure 10.5. $L(G) = \lambda \vee 00^*1$.   □

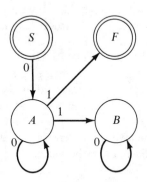

Figure 10.5

10.27 Practice   Let $G = (V, V_T, S, P)$ where $V = \{S, A, B, 0, 1\}$, $V_T = \{0, 1\}$, and P consists of

$$S \to 1A$$
$$A \to 1A$$
$$A \to 0B$$
$$B \to 0B$$
$$B \to 0$$

(a) Find the state graph of a (nondeterministic) finite-state machine recognizing $L(G)$.

(b) Describe $L(G)$ by a regular expression.   (*Answer, page A-25*)

We have now demonstrated the following theorem.

10.28   Theorem   Any set recognized by a finite-state machine is generated by a regular grammar, and conversely. Hence, regular sets coincide with regular languages.

□

We know that $S = \{0^n 1^n \mid n \geq 0\}$ is not a regular set, so by Theorem 10.28 it is not a regular language. By Exercise 4 of Section 10.1, S is a context-free language, which proves that the set of regular languages is a proper subset of the set of context-free languages.

RECOGNIZERS FOR TYPE 0 *LANGUAGES*

From Theorem 10.28 we know that the most restricted class of languages coincides with the class of sets recognized by a computational device of limited capacity. Let's consider the other end of the spectrum. As far as languages are concerned, the most general language is a type 0 language. As far as computational devices are concerned, the most general is the Turing machine. How do sets recognized by Turing machines compare with type 0 languages? As it happens, these also coincide. This result is easy to see in one direction if we employ the Church–Turing thesis. A type 0 language L is one generated by a phrase-structure grammar $G = (V, V_T, S, P)$. As discussed in the previous section, L is an r.e. set, and there is an effective procedure for generating the members of L. By the Church–Turing thesis, there must be a Turing machine T_1 generating L in the sense that T_1 begins on a blank tape and every so often signals that the word currently on its tape is a member of L. We can then use a second Turing machine T_2 that, when begun with a string $\alpha \in V$ on its tape, constantly compares α with the contents of T_1's tape. Whenever α matches a word on T_1's tape that T_1 says is a member of L, T_2 halts; otherwise, T_2 does not halt. Thus, T_2 recognizes the set L.

To go in the other direction, we want to show that given any Turing machine T, there is a phrase-structure grammar G such that $L(G)$ is exactly the set recognized by T. We will not give the proof although it is not very difficult and involves creating productions of G that manipulate symbols in a string to simulate the action of T on a tape containing that string. The result we've been discussing is stated below.

10.29   Theorem   Any set recognized by a Turing machine is generated by a type 0 grammar, and conversely.

□

RECURSIVE AND R.E. SETS

As we have just noted, an r.e. set S is recognized by some Turing machine. We also know from Chapter 9 that if a set S is recognized by a Turing machine, then S is r.e. The r.e. sets therefore coincide with sets recognized by Turing machines. However, we have hinted that there is a distinction between a Turing machine generating a set as opposed to deciding membership in a set or, equivalently, between a set being r.e. as opposed to being recursive. Now we'll prove this distinction.

An R.E. Set That's Not Recursive

First, we let S be a subset of V^* for some vocabulary V. Then we note that S is a recursive set if and only if both S and the complement of S, $S' = V^* - S$, are r.e. If S is recursive, then S is r.e. But we also have an effective procedure to generate S'. We generate the members of V^* and, given $x \in V^*$, apply the process to decide whether $x \in S$. If the answer is no, then $x \in S'$. Therefore, S' is r.e. Now suppose both S and S' are r.e., and we are given a member $x \in V^*$. We generate members of S and of S' and compare x against these lists as they are generated. Eventually, x must appear in the list for S or in the list for S', and, at that point, we know whether $x \in S$. This process gives us an effective procedure to decide membership in S, so S is recursive.

One final idea we need is that of an "effective enumeration" of Turing machines. In discussing the universal Turing machine, we saw how to effectively describe any Turing machine T as a finite string of symbols from the set $\{*, 1\}$. If we order these symbols so that $* < 1$, then we can lexicographically order all strings over $\{*, 1\}$. We can eliminate those strings that are not of the proper form to represent the quintuples of any Turing machine. The result is an enumeration $T_0, T_1, T_2, \ldots$ of all Turing machines. Furthermore, it is an effective enumeration, meaning that given a nonnegative integer i, we can construct the quintuples for Turing machine T_i, and, given any Turing machine T, we can find an index i in the enumeration such that $T = T_i$. Now we have everything we need to prove our next theorem.

10.30   Theorem   There is an r.e. set that is not recursive.

Proof:
Let $V = \{1\}$. For each $x \in V^*$, if x is a string of i 1's, we let T_x denote the Turing machine T_i in the above enumeration. Let the set S be defined by

$$S = \{x \mid x \in V^* \text{ and } T_x \text{ halts when begun on a tape containing } x\}$$

Then S is a subset of V^*. S is r.e.; an effective procedure to generate S consists of generating V^* and, for each $x \in V^*$, constructing the Turing machine T_x. T_x is then begun on a tape containing x. If T_x ever halts, then $x \in S$. Once again we resort to our usual trick to avoid waiting forever behind some one nonhalting computation, that is, we run several computations at once, going back to check on the progress of old ones and adding new ones.

Now we consider S'.

$$S' = \{x \,|\, x \in V^* \text{ and } T_x(x) \text{ does not halt.}\}$$

If we assume that S' is r.e., then S' is recognized by some Turing machine, say T_y for some $y \in V^*$. By the definition of recognition,

$$T_y(y) \text{ halts} \Leftrightarrow y \in S'$$

But by the definition of S',

$$y \in S' \Leftrightarrow T_y(y) \text{ does not halt}$$

Since together these two conclusions are contradictory, S' is not an r.e. set. (Notice that this argument is closely related to the one used to prove the unsolvability of the halting problem.) Therefore, S is r.e. but not recursive.

□

Context-Sensitive Languages are Recursive

Any type 0 language L is an r.e. set, but if L is not a recursive set, it will not be suitable as a programming language because the compiler cannot decide whether an arbitrary string of symbols belongs to the language or not. Any context-free language is recursive. In fact, any context-sensitive language is recursive.

10.31 Theorem Let L be a context-sensitive language; then L is recursive.

Proof:
Let L be generated by the context-sensitive grammar $G = (V, V_T, S, P)$. We want an effective procedure to decide, given any word $x \in V^*$, whether $x \in L$. If $x = \lambda$, we check whether the production $S \to \lambda$ is in P. If it is, $x \in L$; otherwise, $x \notin L$. Now suppose $x \neq \lambda$, and the length of x, $|x|$, is some positive integer k. Now we define a sequence of sets W_i as follows:

$$W_0 = \{S\}$$

For each positive integer i,

$$W_i = W_{i-1} \cup \{w \,|\, w \in V^*, |w| \leq k, \text{ and there is some word } v \text{ in } W_{i-1} \text{ such that } v \Rightarrow w\}$$

Therefore, W_i consists of all words in L of length $\leq k$ derivable from S in no more than i steps. It is clear that $x \in L$ if and only if $x \in W_i$ for some i.

For each i, $W_{i-1} \subseteq W_i$. If this containment is always proper, the W sets continue to have more and more elements. But since V is a finite set, there is a limit to the number of words in V^* (hence in L) of length $\leq k$. Therefore, there is a j such that $W_{j-1} = W_j$. Once this happens, then clearly $W_{j-1} = W_j = W_{j+1} = \cdots$. To determine if $x \in L$, we compute $W_0 = \{S\}$, W_1,

$W_2, \ldots$ until two sets W_{j-1} and W_j are equal (which always happens). Then $x \in L$ if and only if x has appeared in one of the finite sets in this finite collection.

The flowchart in Figure 10.6 illustrates the procedure.   □

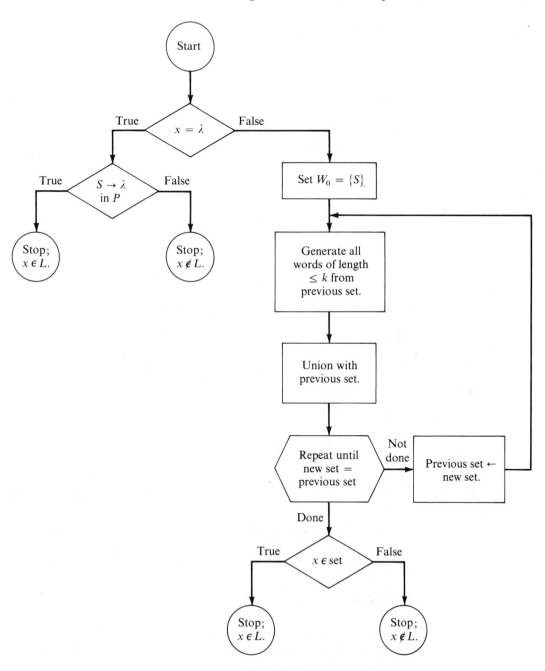

Figure 10.6

10.32 Example Consider the context-sensitive grammar of Example 10.12 where the productions are

$$S \rightarrow aSBC$$
$$S \rightarrow aBC$$
$$CB \rightarrow BC$$
$$aB \rightarrow ab$$
$$bB \rightarrow bb$$
$$bC \rightarrow bc$$
$$cC \rightarrow cc$$

Let's use the proof of Theorem 10.31 to test whether $x = abbc$ is a member of $L(G)$; $|x| = 4$. (Of course, since we already know that $L(G) = \{a^n b^n c^n\}$, we already know the answer.) We construct the W sets.

$$W_0 = \{S\}$$
$$W_1 = \{S, aSBC, aBC\}$$
$$W_2 = \{S, aSBC, aBC, abC\}$$
$$W_3 = \{S, aSBC, aBC, abC, abc\}$$
$$W_4 = \{W_3\}$$

Since x has not appeared, $x \notin L$. □

10.33 Practice Consider the context-sensitive grammar G_1 of Example 10.19 with productions

$$S \rightarrow \lambda$$
$$S \rightarrow ABA$$
$$AB \rightarrow 00$$
$$0A \rightarrow 000A$$
$$A \rightarrow 0$$

Use the procedure of Theorem 10.31 to determine whether $0000 \in L$.

(Answer, page A-26)

Although every context-sensitive language is recursive, there are recursive languages that are not context-sensitive. The proof is rather similar to that of Theorem 10.30 and is left as an exercise (see Exercise 9, this section).

THE GRAND FINALE

We conclude by adding more detail to the language diagram of Figure 10.1. The only proper containment shown in Figure 10.7 that we have not men-

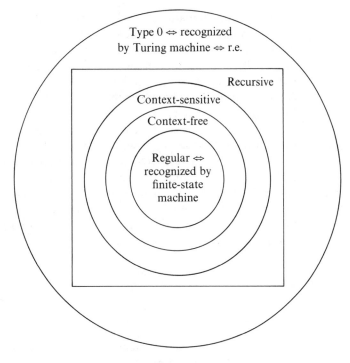

Type 0 ⇔ recognized
by Turing machine ⇔ r.e.

Recursive

Context-sensitive

Context-free

Regular ⇔
recognized by
finite-state
machine

Figure 10.7

tioned is that the context-free languages are properly contained in the context-sensitive languages. However, the context-sensitive language $L = \{a^n b^n c^n \mid n \geq 1\}$ of Example 10.12 is not context-free (see Exercise 10 of this section).

There are also computational devices midway in capabilities between finite-state machines and Turing machines that recognize exactly the context-free languages and the context-sensitive languages, respectively. The type of device that recognizes the context-free languages is called a **pushdown automaton**, or **pda**. A pda consists of a finite-state unit that reads input from a tape and controls activity in a stack. The stack operates like a set of plates stacked on a spring in a cafeteria. Symbols from some alphabet can be added to the top of the stack by a "push" instruction, which pushes previously stored items farther down in the stack. Only the topmost item on the stack is accessible at any moment, and it is removed from the stack by a "pop" instruction.

The finite-state unit, as a function of the input symbol read, the present state, and the top symbol on the stack, has a finite number of possible next moves. The moves are of the following types: go to a new state, pop the top symbol off the stack, and read the next input symbol; go to a new state and, after popping the top symbol off the stack, push a finite number of symbols onto the stack, and read the next input symbol; ignore the input symbol being read, manipulate the stack as above, but do not read the next input symbol. Because there is a choice of moves, the pda is a nondeterministic device, and, unlike the situation for finite-state machines, nondeterminism has an effect on the capabilities of this type of machine. A pda recognizes the set of all inputs for which there exists a sequence of moves causing it to empty its stack. It can be shown that any set recognized by a pda is a context-free language, and conversely.

The type of device that recognizes the context-sensitive languages is called a **linear bounded automaton**, or **lba**. An lba is a Turing machine whose read–write head is restricted to that portion of the tape containing the original input, and that at each step has a choice of possible next moves. An lba is therefore a nondeterministic device, and it recognizes the set of all inputs for which there exists a sequence of moves causing it to halt in a final state. It can be shown that any set recognized by an lba is a context-sensitive language, and conversely.

✔ *CHECKLIST*

Techniques

Given a finite-state machine M, find a regular grammar generating the set recognized by M.

Given a regular grammar G, construct a (nondeterministic) finite-state machine to recognize $L(G)$.

Given a context-sensitive language $L(G)$ over a vocabulary V, decide whether $x \in V^*$ is a member of $L(G)$.

Main Ideas

A language is regular if and only if it is recognized by a finite-state machine.

A language is type 0 if and only if it is recognized by a Turing machine and if and only if it is an r.e. set.

There is an r.e. set that is not recursive.

Any context-sensitive language is recursive, but not conversely.

EXERCISES, SECTION 10.2

★ 1. Let M be given by the state graph of Figure 10.8. Find a regular grammar generating the set recognized by M. Describe the set by a regular expression.

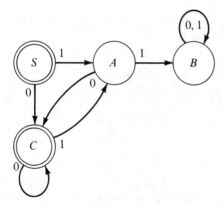

Figure 10.8

2. Let M be given by the state graph of Figure 10.9. Find a regular grammar generating the set recognized by M. Describe the set by a regular expression.

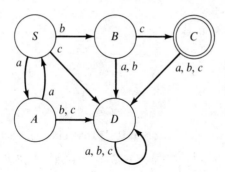

Figure 10.9

★ 3. Let $G = (V, V_T, S, P)$ where $V = \{S, A, B, C, 0, 1\}$, $V_T = \{0, 1\}$, and P consists of

$$
\begin{array}{ll}
S \to \lambda & S \to 1C \\
S \to 0A & C \to 1 \\
A \to 0B & C \to 0C \\
A \to 0 & \\
B \to 0A &
\end{array}
$$

Find the state graph of a (nondeterministic) finite-state machine recognizing $L(G)$. Describe $L(G)$ by a regular expression.

4. Let $G = (V, V_T, S, P)$ where $V = \{S, A, B, C, D, E, a, b, c\}$, $V_T = \{a, b, c\}$, and P consists of

$$
\begin{array}{ll}
S \to \lambda & A \to bC \\
S \to a & C \to cD \\
S \to aA & C \to bE \\
S \to aB & D \to cD \\
B \to aB & D \to bE \\
B \to a & E \to a
\end{array}
$$

Find the state graph of a (nondeterministic) finite-state machine recognizing $L(G)$. Describe $L(G)$ by a regular expression.

★ 5. It is stated in the proof of Theorem 10.31 that because V is finite there is a limit to the number of words in V^* of length $\leq k$. Suppose V has n elements. What is the number of words in V^* of length $\leq k$?

6. Where does the proof of Theorem 10.31 break down if L is not context-sensitive?

7. Let $G = (V, V_T, S, P)$ where $V = \{S, A, B, 0, 1\}$, $V_T = \{0, 1\}$, and P consists of

$$
\begin{array}{l}
S \to 0 \\
S \to 1A \\
1A \to 1B \\
B \to 0B1 \\
B \to 1
\end{array}
$$

Use the proof of Theorem 10.31 to decide whether 1010 belongs to $L(G)$.

8. Let $G = (V, V_T, S, P)$ where $V = \{S, A, +, a, b, c\}$, $V_T = \{+, a, b, c\}$, and P consists of

$$
\begin{array}{l}
S \to S+S \\
+S \to Sa \\
S \to a \\
S \to b \\
S \to bA \\
bA \to bc
\end{array}
$$

Use the proof of Theorem 10.31 to decide whether the following words belong to $L(G)$

★ (a) cb

(b) bbc

9. This exercise is to prove that the set of context-sensitive languages is properly contained in the set of recursive languages.

Let x_i denote the ith string in a lexicographical ordering of all strings over $\{0, 1\}$. Also we can encode all phrase-structure grammars with terminals from $\{0, 1\}$ as strings in $\{0, 1\}^*$. For example, we can let the terminal 0 be represented as 01, the terminal 1 as 01^2, the production arrow by 01^3, a comma by 01^4, and variables by 01^5, 01^6, and so on. A grammar is represented by the string consisting of the code for productions separated by commas. We can eliminate those strings in $\{0, 1\}^*$ that are not of the proper form to represent context-sensitive grammars. The result is an effective enumeration $G_1, G_2, \ldots$ of context-sensitive grammars. Now define a set $S = \{x_i | x_i \notin L(G_i)\}$. Show that S is a recursive language in $\{0, 1\}^*$ that is not context-sensitive.

10. The following is the "pumping lemma" for context-free languages. Let L be any context-free language. Then there exists some constant k such that for any word w in L with $|w| \geq k$, w can be written as the string $w_1 w_2 w_3 w_4 w_5$ with $|w_2 w_3 w_4| \leq k$ and $|w_2 w_4| \geq 1$. Furthermore, the word $w_1 w_2^i w_3 w_4^i w_5 \in L$ for each $i \geq 0$.
    (a) Use the pumping lemma to show that $L = \{a^n b^n c^n | n \geq 1\}$ is not context-free.
    (b) Use the pumping lemma to show that $L = \{a^{n^2} | n \geq 1\}$ is not context-free.

FOR FURTHER READING

Entries [1], [2], [3], and [4] are all comprehensive works on formal languages and their corresponding machine recognizers. An excellent intuitive treatment of recursive and recursively enumerable sets is given in the first part of [5]. Entry [6] discusses syntax analysis and parsing for programming languages using context-free grammars.

1. John E. Hopcroft and Jeffrey D. Ullman, *Introduction to Automata Theory, Languages, and Computation.* Reading, Mass.: Addison-Wesley, 1979.

2. Michael A. Harrison, *Introduction to Formal Language Theory.* Reading, Mass.: Addison-Wesley, 1978.

3. Richard Y. Kain, *Automata Theory: Machines and Languages.* New York: McGraw-Hill, 1972.

4. Peter J. Denning, Jack B. Dennis, and Joseph E. Qualitz, *Machines, Languages, and Computation.* Englewood Cliffs, N.J.: Prentice-Hall, 1978.

5. Neil D. Jones, *Computability Theory.* New York: Academic Press, 1973.

6. Alfred V. Aho and Jeffrey D. Ullman, *Principles of Compiler Design.* Reading, Mass.: Addison-Wesley, 1977.

ANSWERS TO SELECTED EXERCISES

CHAPTER 1

Section 1.1

1. (a) Antecedent: sufficient water
      Consequent: healthy plant growth
   (b) Antecedent: further technological advances
      Consequent: increased availability of microcomputers
   (c) Antecedent: errors will be introduced
      Consequent: there is a modification of the program
   (d) Antecedent: fuel savings
      Consequent: good insulation or all windows are storm windows

2. (a) ii
   (b) i and iii
   (c) iii

3. (a) A: prices go up; B: housing will be plentiful; C: housing will be expensive

   $$[A \Rightarrow B \wedge C] \wedge (C' \Rightarrow B)$$

   (b) A: going to bed; B: going swimming; C: changing clothes

   $$[(A \vee B) \Rightarrow C] \wedge (C \Rightarrow B)'$$

   (c) A: it will rain; B: it will snow

   $$(A \vee B) \wedge (A \wedge B)'$$

   (d) A: Janet wins; B: Janet loses; C: Janet will be tired

   $$(A \vee B) \Rightarrow C$$

   (e) A: Janet wins; B: Janet loses; C: Janet will be tired

   $$A \vee (B \Rightarrow C)$$

4. (a)

A	B	$A \Rightarrow B$	A'	$A' \vee B$	$(A \Rightarrow B) \Leftrightarrow A' \vee B$
T	T	T	F	T	T
T	F	F	F	F	T
F	T	T	T	T	T
F	F	T	T	T	T

Tautology

(b)

A	B	C	$A \wedge B$	$(A \wedge B) \vee C$	$B \vee C$	$A \wedge (B \vee C)$	$(A \wedge B) \vee C \Rightarrow A \wedge (B \vee C)$
T	T	T	T	T	T	T	T
T	T	F	T	T	T	T	T
T	F	T	F	T	T	T	T
T	F	F	F	F	F	F	T
F	T	T	F	T	T	F	F
F	T	F	F	F	T	F	T
F	F	T	F	T	T	F	F
F	F	F	F	F	F	F	T

5. 1.0, 2.4, 7.2, 5.3

6. For example: $(A$ OR $B)$ AND NOT $(A$ AND $B)$ AND NOT C

7. $2^{2^4} = 2^{16}$

Section 1.2

1. (a) Converse: Healthy plant growth implies sufficient water.
   Contrapositive: If there is not healthy plant growth, then there is not sufficient water.
   (b) Converse: Increased availability of microcomputers implies further technological advances.
   Contrapositive: If there is not increased availability of microcomputers, then there are no further technological advances.
   (c) Converse: If there is a modification of the program then errors will be introduced.
   Contrapositive: No modification of the program implies that errors will not be introduced.
   (d) Converse: Good insulation or all windows are storm windows implies fuel savings.
   Contrapositive: Poor insulation and some windows not storm windows implies no fuel savings.

3. $(2n + 1)(2m + 1)$ (m and n are integers.)
   $= 4nm + 2m + 2n + 1$
   $= 2(2nm + m + n) + 1$ ($2nm + m + n$ is an integer.)

6. $\left. \begin{array}{c} x < y \Rightarrow x^2 < xy \\ \text{and} \\ xy < y^2 \end{array} \right\}$ (Multiplication of both sides of an inequality by positive numbers x and y.)
   $\Rightarrow x^2 < y^2$ (property of $<$)

   $x^2 < y^2 \Rightarrow y^2 - x^2 > 0$ (definition of $<$)
   $\Rightarrow (y + x)(y - x) > 0$ (factoring)
   $\left. \begin{array}{c} \Rightarrow (y + x) < 0 \text{ and } (y - x) < 0 \\ \text{or} \\ (y + x) > 0 \text{ and } (y - x) > 0 \end{array} \right\}$ (A positive number is the product of two negatives or two positives.)
   $\Rightarrow (y + x) > 0 \text{ and } (y - x) > 0$ (Cannot have $(y + x) < 0$ because x
   $\Rightarrow y - x > 0$ and y are positive.)
   $\Rightarrow y > x$ (definition of $>$)

8. Proof is by induction.

$P(1)$: $2 = 1(1 + 1)$ true

Assume $P(k)$: $2 + 4 + 6 + \cdots + 2k = k(k + 1)$

Show $P(k + 1)$: $2 + 4 + 6 + \cdots + 2(k + 1) = (k + 1)((k + 1) + 1)$

$2 + 4 + 6 + \cdots + 2(k + 1)$ left side of $P(k + 1)$
$= 2 + 4 + 6 + \cdots + 2k + 2(k + 1)$
$= k(k + 1) + 2(k + 1)$ using $P(k)$
$= (k + 1)(k + 2)$ factoring
$= (k + 1)((k + 1) + 1)$ right side of $P(k + 1)$

9. Proof is by induction.

$P(1)$: $2^{2 \cdot 1} - 1$ is divisible by 3
    or $4 - 1$ is divisible by 3; true

Assume $P(k)$: $2^{2k} - 1$ is divisible by 3, which means
    $2^{2k} - 1 = 3m$ for some integer m, or
    $2^{2k} = 3m + 1$

Show $P(k + 1)$: $2^{2(k + 1)} - 1$ is divisible by 3.
    $2^{2(k + 1)} - 1 = 2^{2k + 2} - 1 = 2^2 \cdot 2^{2k} - 1 = 2^2(3m + 1) - 1$
    $= 12m + 4 - 1 = 12m + 3 = 3(4m + 1)$ where
    $4m + 1$ is an integer

Section 1.3

1. If $A = \{x \mid x = 2^n$ for n a positive integer$\}$, then $16 \in A$. But if $A = \{x \mid x = 2 + n(n - 1)$ for n a positive integer$\}$, then $16 \notin A$.

2. (a) F; $\{1\} \in S$ but $\{1\} \notin R$ (b) T   (c) F; $\{1\} \in S$, not $1 \in S$
   (d) F; 1 is not a set; the correct statement is $\{1\} \subseteq U$ (e) T   (f) F; $1 \notin S$ (g) T
   (h) T   (i) T   (j) F; $3 \notin U$ and $\pi \notin U$ (k) T   (l) T

3. (a) T   (b) F   (c) F   (d) T   (e) T   (f) F   (g) F   (h) T   (i) F   (j) F   (k) T
   (l) T

5. (a) $\mathcal{P}(S) = \{\phi, \{\phi\}, \{\{\phi\}\}, \{\{\phi, \{\phi\}\}\}, \{\phi, \{\phi\}\}, \{\phi, \{\phi, \{\phi\}\}\}, \{\{\phi\}, \{\phi, \{\phi\}\}\},$
       $\{\phi, \{\phi\}, \{\phi, \{\phi\}\}\}\}$
   (b) $\mathcal{P}(\mathcal{P}(S)) = \{\phi, \{\phi\}, \{\{a\}\}, \{\{b\}\}, \{\{a, b\}\}, \{\phi, \{a\}\}, \{\phi, \{b\}\}, \{\phi, \{a, b\}\},$
       $\{\{a\}, \{b\}\}, \{\{a\}, \{a, b\}\}, \{\{b\}, \{a, b\}\}, \{\phi, \{a\}, \{b\}\}, \{\phi, \{a\}, \{a, b\}\},$
       $\{\phi, \{b\}, \{a, b\}\}, \{\{a\}, \{b\}, \{a, b\}\}, \{\phi, \{a\}, \{b\}, \{a, b\}\}\}$

7. (a) Binary operation
   (b) No; $0 \circ 0 \notin N$
   (c) Binary operation
   (d) No; ln x undefined for $x \leq 0$
   (e) Unary operation
   (f) No; closure fails
   (g) No; uniqueness fails (assuming there are two people in Arkansas of the same height)
   (h) Binary operation
   (i) No; operation undefined for $x = 0$
   (j) Binary operation

10. (a) $\{0, 1, 2, 3, 6, 7, 8, 9\}$
    (b) $\{2, 3\}$
    (c) $\{0, 1, 3, 4, 7, 9\}$
    (d) $\{2, 6, 8\}$
    (e) $\{2, 3\}$
    (f) $\{(1, 2), (1, 3), (1, 4), (4, 2), (4, 3), (4, 4), (5, 2), (5, 3), (5, 4), (9, 2), (9, 3), (9, 4)\}$

11. (a) B' (b) $B \cap C$ (c) $A \cap B$ (d) $B' \cap C$ (e) $B' \cap C'$ (or $(B \cup C)'$ or $B' - C$)

12. (a) $B \subseteq A$ (b) $A \subseteq B$ (c) $A = \phi$ (d) $B \subseteq A$ (e) $A = B$

16. (a) $(A \cup B) \cap (A \cup B') = A \cup (B \cap B')$ 3a
$$\qquad\qquad\qquad\qquad\qquad = A \cup \phi \qquad\quad \text{5b}$$
$$\qquad\qquad\qquad\qquad\qquad = A \qquad\qquad\quad \text{4a}$$

Section 1.4

1. (a) 42 (b) 6720 (c) 120 (d) 36

3. $2 \cdot 4 \cdot 4 = 32$

5. $45 \cdot 13 = 585$

6. $26 \cdot 26 \cdot 26 \cdot 1 \cdot 1 = 17,576$

8. $26 + 26 \cdot 10 = 286$

11. $P(52, 4) = 6,497,400; 4 \cdot 3 \cdot 2 \cdot 1 = 24$

14. $C(17, 5) \cdot C(23, 7) = (6188)(245, 157)$

17. $C(12, 3) = 220; C(12, 3) - C(8, 3) = 164$

Section 2.1

1. (a) $(1, -1), (-3, 3)$ (b) $19, 41$ (c) $(3, 4, 5), (0, 5, 5), (8, 6, 10)$
   (d) $(-3, -5), (-4, 1/2), (1/2, 1/3)$

2. (a) $(2, 6), (3, 17), (0, 0)$ (b) $(2, 12)$ (c) none (d) $(1, 1), (4, 8)$

3. (a) Reflexive, transitive (b) Reflexive, symmetric, transitive (c) Symmetric
   (d) Transitive (e) Reflexive, symmetric, transitive
   (f) Reflexive, symmetric, transitive (g) Transitive

6. (a) Yes, yes (b) Yes, yes (c) No, yes (d) No, yes

8. (a)   (b)

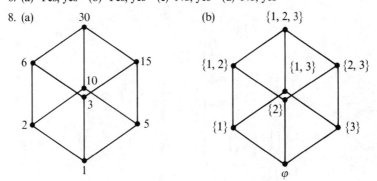

The two graphs are identical in structure.

12. (a) When; no; all but the last

(b)

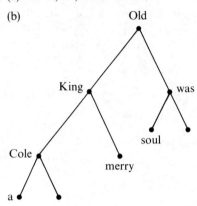

Maximal elements: a, merry, soul

Section 2.2

1. (a) Function
   (b) Not a function; undefined at $x = 0$
   (c) Function; onto
   (d) Bijection; $f^{-1}: \{p, q, r\} \rightarrow \{1, 2, 3\}$
       where $f^{-1} = \{(q, 1), (r, 2), (p, 3)\}$
   (e) Function; one-to-one
   (f) Bijection; $h^{-1}: R^2 \rightarrow R^2$
       where $h^{-1}(x, y) = (y - 1, x - 1)$

4. (a) $2^3 = 8; 6$
   (b) n^m
   (c) $n(n - 1)(n - 2) \cdots (n - (m - 1)) = \dfrac{n!}{(n - m)!}$

6. $\{m, n, o, p\}$; $\{n, o, p\}$; $\{o\}$

7. (a) 25 (b) 6 (c) 9

8. (a) $\begin{pmatrix} 1 & 2 & 3 \\ 1 & 2 & 3 \end{pmatrix}$ $\begin{pmatrix} 1 & 2 & 3 \\ 1 & 3 & 2 \end{pmatrix}$ $\begin{pmatrix} 1 & 2 & 3 \\ 3 & 2 & 1 \end{pmatrix}$

   $\begin{pmatrix} 1 & 2 & 3 \\ 2 & 1 & 3 \end{pmatrix}$ $\begin{pmatrix} 1 & 2 & 3 \\ 2 & 1 & 3 \end{pmatrix}$ $\begin{pmatrix} 1 & 2 & 3 \\ 2 & 3 & 1 \end{pmatrix}$

   (b) $n!$

10. (a) $(1, 2, 5, 3, 4)$
    (b) $(1, 7, 8) \circ (2, 4, 6)$
    (c) $(1, 5, 2, 4) \circ (3, 6)$
    (d) $(2, 3) \circ (4, 8) \circ (5, 7)$

11. (a) $\begin{pmatrix} 1 & 2 & 3 & 4 \\ 2 & 1 & 4 & 3 \end{pmatrix}$ (b) For example, $\begin{pmatrix} 1 & 2 & 3 & 4 \\ 3 & 1 & 2 & 4 \end{pmatrix}$

Section 2.3

2. (a)

This is a tree.

(b)

(c) For example,

3. (a)

(b)

(c)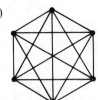

(d) $\dfrac{n(n-1)}{2}$

(e) The number of arcs is $C(n,2) = \dfrac{n(n-1)}{2}$. (Other proof methods include induction on the number of nodes.)

5. 4, 5, 6; length 2; for example (naming the nodes),

1–2–1–2–2–1–4–5–6

10.

Item		Pointer
1		7
2		8
3		9
4		0
5		0
6		0
7	2	0
8	3	0
9	4	10
10	5	11
11	6	0

Total number of entries in array is $2 \times 11 = 22$.

Total number of entries in an adjacency matrix is $6 \times 6 = 36$.

12. (a) Euler path   (b) Euler path, Hamiltonian circuit   (c) Hamiltonian circuit
    (d) Neither

17. For example,

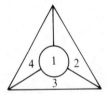

Dual: $N = 4, A = 6, R = 3$, so $N - A + R = 4 - 6 + 3 = 1$

Section 3.1

1.

+	0	1	a	a'
0	0	1	a	a'
1	1	1	1	1
a	a	1	a	1
a'	a'	1	1	a'

$\cdot$	0	1	a	a'
0	0	0	0	0
1	0	1	a	a'
a	0	a	a	0
a'	0	a'	0	a'

4. (a) $x' + x = x + x'$ (1a)
              $= 1$ (5a)

and

$x' \cdot x = x \cdot x'$ (1b)
       $= 0$ (5b)

Therefore $x = (x')'$ by uniqueness of the complement.

(b) $x + (x \cdot y) = x \cdot 1 + x \cdot y$ (4b)

$\qquad\qquad = x(1 + y)$ (3b)

$\qquad\qquad = x(y + 1)$ (1a)

$\qquad\qquad = x \cdot 1$ (Practice 3.7)

$\qquad\qquad = x$ (4b)

$x \cdot (x + y) = x$ follows by duality

(d) $x \cdot [y + (x \cdot z)] = x \cdot y + x \cdot (x \cdot z)$ (3b)

$\qquad\qquad\qquad = x \cdot y + (x \cdot x) \cdot z$ (2b)

$\qquad\qquad\qquad = x \cdot y + x \cdot z$ (dual of idempotent property)

$x + [y \cdot (x + z)] = (x + y) \cdot (x + z)$ follows by duality

9. (a) $(x - y) - z = x - (y - z)$; no because, for example,
$(2 - 3) - 4 \neq 2 - (3 - 4)$

(b) $x - y = y - x$; no because, for example, $3 - 2 \neq 2 - 3$

10. (a) Associative (b) Commutative (c) Neither
(d) Commutative, associative (e) Commutative

Section 3.2

1. (a)

(b)

(c)

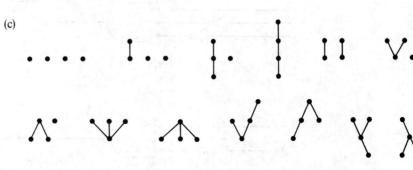

16

2. For example,

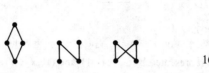

$f(1) = 1'$
$f(2) = 2'$
$f(3) = 3'$

Then $1 \prec 2$ and $f(1) \prec' f(2)$, but $f(3) \prec' f(2)$ and not $3 \prec 2$.

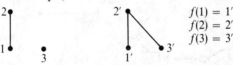

5. (a) For any $y \in b$, $y = f(x)$ for some $x \in B$. Then $y \& f(0) = f(x) \& f(0) = f(x + 0) = f(x) = y$, and $f(0) = \emptyset$ because the zero element in any Boolean algebra is unique (see Exercise 5, Section 3.1)

(b) $f(1) = f(0') = [f(0)]'' = \emptyset'' = \mathbf{I}$

6. $f\left(\begin{bmatrix} a_{11} & a_{12} \\ a_{21} & a_{22} \end{bmatrix} \cdot \begin{bmatrix} b_{11} & b_{12} \\ b_{21} & b_{22} \end{bmatrix}\right) = f\left(\begin{bmatrix} a_{11}b_{11} + a_{12}b_{21} & a_{11}b_{12} + a_{12}b_{22} \\ a_{21}b_{11} + a_{22}b_{21} & a_{21}b_{12} + a_{22}b_{22} \end{bmatrix}\right)$

$= \begin{bmatrix} a_{11}b_{11} + a_{12}b_{21} & 0 \\ 0 & a_{21}b_{12} + a_{22}b_{22} \end{bmatrix}$

but

$f\left(\begin{bmatrix} a_{11} & a_{12} \\ a_{21} & a_{22} \end{bmatrix}\right) \cdot f\left(\begin{bmatrix} b_{11} & b_{12} \\ b_{21} & b_{22} \end{bmatrix}\right) = \begin{bmatrix} a_{11} & 0 \\ 0 & a_{22} \end{bmatrix} \cdot \begin{bmatrix} b_{11} & 0 \\ 0 & b_{22} \end{bmatrix} = \begin{bmatrix} a_{11}b_{11} & 0 \\ 0 & a_{22}b_{22} \end{bmatrix}$

Section 4.1

3. (a) $x_1 x_2 + x_1' x_2$

(b) $x_1 x_2 x_3' + x_1 x_2' x_3 + x_1' x_2 x_3 + x_1' x_2' x_3'$

4. (a) $x_1 x_2 x_3' + x_1 x_2' x_3'$

(b)

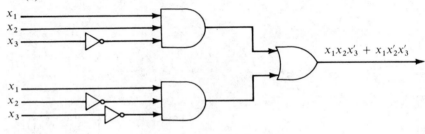

(c) $x_1 x_2 x_3' + x_1 x_2' x_3' = x_1 x_3' x_2 + x_1 x_3' x_2'$
$= x_1 x_3' (x_2 + x_2')$
$= x_1 x_3' 1$
$= x_1 x_3'$

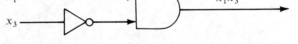

7. (a) $(x_1' + x_2)(x_1 + x_2)$

(b) $(x_1' + x_2' + x_3')(x_1' + x_2 + x_3)(x_1 + x_2' + x_3)(x_1 + x_2 + x_3')$

9. Network is represented by $(x_1'(x_2' x_3)')'$ and $(x_1'(x_2' x_3)')' = x_1 + x_2' x_3$

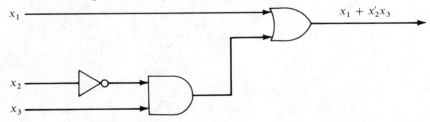

11. x_1 = neutral
x_2 = park
x_3 = seat belt

x_1	x_2	x_3	$f(x_1, x_2, x_3)$
1	1	1	–
1	1	0	–
1	0	1	1
1	0	0	0
0	1	1	1
0	1	0	0
0	0	1	0
0	0	0	0

$(x_1 + x_2)x_3$

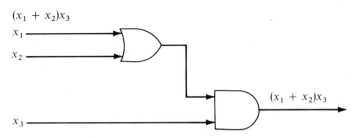

Section 4.2

1. (a)

	x_1x_2	x_1x_2'	$x_1'x_2'$	$x_1'x_2$
x_3	1	1	1	1
x_3'	1			1

$$x_3 + x_2$$

(b)

	x_1x_2	x_1x_2'	$x_1'x_2'$	$x_1'x_2$
x_3x_4		1		
x_3x_4'		1	1	1
$x_3'x_4'$	1	1	1	
$x_3'x_4$		1		

$$x_1x_3'x_4' + x_1'x_3x_4' + x_2'x_4' + x_1x_2'$$

(c)

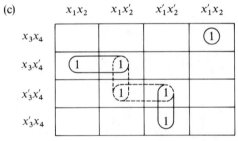

$$x_1'x_2x_3x_4 + x_1x_3x_4' + x_1'x_2'x_3 + x_1x_2'x_4'$$

or

$$x_1'x_2x_3x_4 + x_1x_3x_4' + x_1'x_2'x_3 + x_2'x_3'x_4'$$

2. (a)

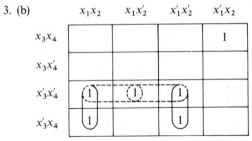

$$x_1x_2 + x_2x_3$$

3. (b)

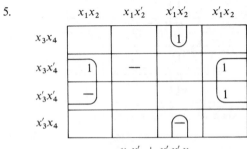

$$x_1'x_2x_3x_4 + x_1'x_2'x_3 + x_1x_2x_3' + x_1x_3'x_4'$$

or

$$x_1'x_2x_3x_4 + x_1'x_2'x_3 + x_1x_2x_3' + x_2'x_3'x_4'$$

5.

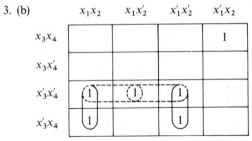

$$x_2x_4' + x_1'x_2'x_4$$

6.

x_1	x_2	x_3	
1	1	1	1,2,3
1	0	1	1,4
0	1	1	2,5,6
1	1	0	3,7
0	0	1	4,5
0	1	0	6,7

x_1	x_2	x_3	
1	–	1	1
–	1	1	1,2
1	1	–	2
–	0	1	1
0	–	1	1
0	1	–	2
–	1	0	2

x_1	x_2	x_3	
–	–	1	
–	1	–	

	111	101	011	110	001	010
—1	✓	✓	✓		✓	
–1–	✓		✓	✓		✓

—1 and –1– are essential. The minimal form is $x_3 + x_2$.

9. (a)

x_1	x_2	x_3	x_4	
0	1	1	1	1
1	0	1	0	
0	0	1	1	1,2
0	1	0	0	3
0	0	0	1	2,4
0	0	0	0	3,4

x_1	x_2	x_3	x_4	
0	–	1	1	
0	0	–	1	
0	–	0	0	
0	0	0	–	

	0111	1010	0011	0100	0001	0000
1010		✓				
0–11	✓		✓			
00–1			✓		✓	
0–00				✓		✓
000–					✓	✓

1010, 0–11, 0–00 are essential. Either 00–1 or 000– can be used as the fourth term. The minimal sum-of-products form is

$$x_1 x_2' x_3 x_4' + x_1' x_3 x_4 + x_1' x_3' x_4' + x_1' x_2' x_4 \quad \text{or}$$

$$x_1 x_2' x_3 x_4' + x_1' x_3 x_4 + x_1' x_3' x_4' + x_1' x_2' x_3'$$

Section 5.1

1. (a) Not commutative; $a \cdot b \neq b \cdot a$
   Not associative; $a \cdot (b \cdot d) \neq (a \cdot b) \cdot d$

(b)

·	p	q	r	s
p	p	q	r	s
q	q	r	s	p
r	r	s	p	q
s	s	p	q	r

Commutative

3. (a) Semigroup
   (b) Not a semigroup; not associative
   (c) Not a semigroup; S not closed under
   (d) Monoid; $i = 1 + 0\sqrt{2}$
   (e) Group; $i = 1 + 0\sqrt{2}$
   (f) Group; $i = 1$
   (g) Monoid; $i = 1$

6.

∘	R_1	R_2	R_3	F_1	F_2	F_3
R_1	R_2	R_3	R_1	F_3	F_1	F_2
R_2	R_3	R_1	R_2	F_2	F_3	F_1
R_3	R_1	R_2	R_3	F_1	F_2	F_3
F_1	F_2	F_3	F_1	R_3	R_1	R_2
F_2	F_3	F_1	F_2	R_2	R_3	R_1
F_3	F_1	F_2	F_3	R_1	R_2	R_3

Identity element is R_3; inverse for F_1 is F_1; inverse for R_2 is R_1.

8. (a) $i_L = i_L \cdot i_R = i_R$ so $i_L = i_R$ and this element is an identity in $[S, \cdot]$.

   (b) For example,

·	a	b
a	a	b
b	a	b

   (c) For example,

·	a	b
a	a	a
b	b	b

   (d) For example, $[R^+, +]$.

10. (a) $0_L = 0_L 0_R = 0_R$
    (b) For (a), zero is 0
        For (d), zero is $0 + 0\sqrt{2}$

Section 5.2

2. (a) No; not the same operation.
   (b) No; zero polynomial (identity) does not belong to P.
   (c) No; not every element of Z^* has an inverse in Z^*.
   (d) Yes
   (e) No; Z is not a subset of $M_2(Z)$.

(f) Yes

(g) No; $\{0, 3, 6\}$ not closed under $+_8$

3. $[\{0\}, +_{12}], [Z_{12}, +_{12}], [\{0, 2, 4, 6, 8, 10\}, +_{12}], [\{0, 4, 8\}, +_{12}] [\{0, 3, 6, 9\}, +_{12}], [\{0, 6\}, +_{12}]$

7. Let $[S, \cdot]$ be a commutative monoid, let $A = \{a \mid a \in S, a^2 = a\}$. Then

$A \subseteq S$; $i \cdot i = i$, so $i \in A$.

For $x, y \in A$, $(x \cdot y)^2 = (x \cdot y)(x \cdot y) = x \cdot x \cdot y \cdot y = x \cdot y$, so $x \cdot y \in A$.

9. Closure: let $x, y \in B_k$. Then $(x \cdot y)^k = x^k \cdot y^k = i \cdot i = i$, so $x \cdot y \in B_k$

Identity: $i^k = i$, so $i \in B_k$

Inverses: for $x \in B_k$, $(x^{-1})^k = (x^k)^{-1} = i^{-1} = i$, so $x^{-1} \in B_k$

11. Closure: let (x, x) and (y, y) belong to A. Then $(x, x) \cdot (y, y) = (x \cdot y, x \cdot y) \in A$

Identity: $(i, i) \in A$

Inverses: for $(x, x) \in A$, $(x, x)^{-1} = (x^{-1}, x^{-1}) \in A$

Section 5.3

1. (a) No; $f(x + y) = 2$, $f(x) + f(y) = 2 + 2 = 4$

(b) No; $f(x + y) = x + y + 1$, $f(x) + f(y) = x + 1 + y + 1$

(c) Yes; $f((x, y) + (p, q)) = f((x + p, y + q)) = x + p + 2(y + q)$,

$f((x, y)) + f((p, q)) = x + 2y + p + 2q$

(d) No; $f(x + y) = |x + y|$, $f(x) + f(y) = |x| + |y|$

(e) Yes; $f(x \cdot y) = |x \cdot y|$, $f(x) \cdot f(y) = |x| \cdot |y|$

(f) Yes; $f(a_n x^n + \cdots + a_1 x + a_0 + b_k x^k + \cdots + b_1 x + b_0)$

$= f(a_n x^n + a_{n-1} x^{n-1} + \cdots + (a_k + b_k) x^k + \cdots + (a_1 + b_1) x + (a_0 + b_0))$

$= a_n + a_{n-1} + \cdots + a_k + b_k + \cdots + a_1 + b_1 + a_0 + b_0$

$= f(a_n x^n + \cdots + a_0) + f(b_k x^k + \cdots + b_0)$

(g) No; $f\left(\begin{bmatrix} a & 0 \\ 0 & b \end{bmatrix}\begin{bmatrix} c & 0 \\ 0 & d \end{bmatrix}\right) = f\left(\begin{bmatrix} ac & 0 \\ 0 & bd \end{bmatrix}\right) = (ac, bd)$

$f\left(\begin{bmatrix} a & 0 \\ 0 & b \end{bmatrix}\right) + f\left(\begin{bmatrix} c & 0 \\ 0 & d \end{bmatrix}\right) = (a, b) + (c, d) = (a + c, b + d)$

(h) No. Take α and β both odd. Then $f(\alpha \circ \beta) = 1$ but $f(\alpha) +_2 f(\beta) = 0 +_2 0 = 0$

None are isomorphisms.

5. Because f is an onto function, $U \neq \emptyset$. To show closure, let $x, y \in U$. Then $f(x \cdot y) = f(x) + f(y) = t + t = t$, so $x \cdot y \in U$.

10. (a) Let $f: G \times H \rightarrow H \times G$ be defined by $f((x, y)) = (y, x)$. Then f is one-to-one and onto, and $f((x_1, y_1) \cdot (x_2, y_2)) = f((x_1 \cdot x_2, y_1 + y_2)) = (y_1 + y_2, x_1 \cdot x_2) = (y_1, x_1) \cdot (y_2, x_2) = f((x_1, y_1)) \cdot f((x_2, y_2))$

(b) $f((x_1, y_1) \cdot (x_2, y_2)) = f((x_1 \cdot x_2, y_1 + y_2)) = x_1 \cdot x_2 = f((x_1, y_1)) \cdot f((x_2, y_2))$

Section 5.4

1. (a) $f(a_2 x^2 + a_1 x + a_0 + b_2 x^2 + b_1 x + b_0)$

$= f((a_2 + b_2) x^2 + (a_1 + b_1) x + (a_0 + b_0)) = a_2 + b_2$

$= f(a_2 x^2 + a_1 x + a_2) + f(b_2 x^2 + b_1 x + b_0)$

Clearly f is onto.

(b) (i) and (iii)

2. (a) $f(Z_{18}) = \{0, 4, 8\}$
   $Z_{18}/f = \{[0], [1], [2]\}$ where

   $[0] = \{0, 3, 6, 9, 12, 15\}$
   $[1] = \{1, 4, 7, 10, 13, 16\}$
   $[2] = \{2, 5, 8, 11, 14, 17\}$

   The isomorphism is

   $[0] \to 0$
   $[1] \to 4$
   $[2] \to 8$

3. (a) $g([1] * [2]) = g([1 +_{18} 2]) = g([3]) = g([0]) = 0$
   (b) $g([1] +_{12} g([2]) = 4 +_{12} 8 = 0$

Section 5.5

2. (a) $(0, 0) + S = \{(0, 0), (1, 2), (2, 0), (0, 2), (1, 0), (2, 2)\} = S$
   $(1, 1) + S = \{(1, 1), (2, 3), (0, 1), (1, 3), (2, 1), (0, 3)\}$
   (b) $(1, 1) + S + (1, 1) + S = (0, 0) + S$

3. (a) $1S = \{1, 2, 4\}$
   $3S = \{3, 6, 5\}$
   (b) $3S \cdot 4S = 3S$

6. (a) $f(\mathbb{Q} \times \mathbb{Q} \times \mathbb{Q}) = \{(y, y) | y \in \mathbb{Q}\}$
   $f((x_1, y_1, z_1) + (x_2, y_2, z_2)) = f((x_1 + x_2, y_1 + y_2, z_1 + z_2))$
   $= (y_1 + y_2, y_1 + y_2) = (y_1, y_1) + (y_2, y_2)$
   $= f((x_1, y_1, z_1)) + f((x_2, y_2, z_2))$
   (b) $K = \{(x, 0, z) | x, z \in \mathbb{Q}\}$
   (c) $(4, 4, 4)$ and $(5, 4, 3)$
   (d) Yes
   (e) $g:(\mathbb{Q} \times \mathbb{Q} \times \mathbb{Q})/K \to f(\mathbb{Q} \times \mathbb{Q} \times \mathbb{Q})$ defined by $g((x, y, z) + K) = (y, y)$

9. This is Exercise 9 of Section 5.3.
   Let i_G and i_H denote the identity elements of G and H. Let f be an isomorphism, $f:G \to H$. Then $f(i_G) = i_H$ by Theorem 5.67, and because f is one-to-one, i_G is the only element mapping to i_H. Thus $K = \{i_G\}$. Now assume that $K = \{i_G\}$; we need to show that the homomorphism f is one-to-one. Let g_1 and g_2 be elements of G with $f(g_1) = f(g_2)$.

   Then

   $f(g_1) \cdot (f(g_2))^{-1} = i_H$
   $f(g_1) \cdot f(g_2^{-1}) = i_H$
   $f(g_1 \cdot g_2^{-1}) = i_H$

   so that

   $g_1 \cdot g_2^{-1} = i_G$

   or

   $g_1 = i_G \cdot g_2 = g_2$

10. Let $x \in A$ and $g \in G$; we want to show that $g^{-1}xg \in A$. But since $x \in A$, $g^{-1}xg = g^{-1}gx = x \in A$. A is normal by Theorem 5.113.

14. Let S be normal in G and let $x \in G$, $s \in S$. Because S is normal, $sx \in Sx = xS$ so that $sx = xs'$ for some $s' \in S$. Then $x^{-1}sx = x^{-1}xs' = s' \in S$.

22. $[G:T] = \dfrac{|G|}{|T|} = \dfrac{|G|}{|S|} \cdot \dfrac{|S|}{|T|} = [G:S] \cdot [S:T]$

Section 6.1

1. (a) $s = 12$; ALLGAULISDIVIDED
   (b) There are no occurrences of the letter E.
   (c) IBM

2. (a) $M^{-1} = \begin{bmatrix} 5 & 24 \\ 19 & 3 \end{bmatrix}$ (remember that this is arithmetic in Z_{26}); decoded message is ATTACK.
   (b) ATMDCK

4. (a) Yes
   (b) 152478; 152748 (error detected); 125748 (no error detected)

10. (a) $n = 5$, $m = 3$, $r = 2$; $m \nleq 2^r - r - 1$, so neither perfect nor single-error correcting.
    (b) $n = 12$, $m = 7$, $r = 5$; $m < 2^r - r - 1$, so single-error correcting but not perfect.
    (c) $n = 15$, $m = 11$, $r = 4$; $m = 2^r - r - 1$, so perfect.

11. (a) $6 \, (32 \leq 2^6 - 6 - 1)$
    (b) $6 \, (36 \leq 2^6 - 6 - 1)$
    (c) $7 \, (60 \leq 2^7 - 7 - 1)$

Section 6.2

3. If only a single error has occurred, the coset leader has weight 1. If the 1 occurs in the kth component, then the syndrome is the kth row of H, which is the binary representation of k. Thus computing the syndrome of X gives the binary representation of the single component in X which should be changed.

7. (a) One possible H is

$$H = \begin{bmatrix} 1 & 1 & 0 & 0 \\ 1 & 0 & 1 & 0 \\ 1 & 0 & 0 & 1 \\ 0 & 1 & 0 & 1 \\ 0 & 0 & 1 & 1 \\ 0 & 1 & 1 & 0 \\ 1 & 1 & 1 & 0 \\ 1 & 1 & 0 & 1 \\ 1 & 0 & 1 & 1 \\ 0 & 1 & 1 & 1 \\ 1 & 1 & 1 & 1 \\ 1 & 0 & 0 & 0 \\ 0 & 1 & 0 & 0 \\ 0 & 0 & 1 & 0 \\ 0 & 0 & 0 & 1 \end{bmatrix}$$

(b) The syndrome for 011000010111001 is 1111. The coset leader is 000000000010000, so the word is decoded as 011000010101001.

Section 7.1

1. (a) 0001111110   (b) *aaacaaaa*   (c) 00100110

2. (a) 110100, 111010   (b) None   (c) $0a_1a_2a_3a_4a_5$

5.

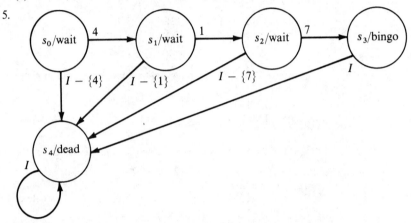

9. Once a state is revisited, behavior will be periodic since there is no choice of paths from a state. The maximum number of inputs which can occur before this happens is $n - 1$ (visiting all n states before repeating). The maximum length of a period is n (output from all n states, with the last state returning to s_o).

Section 7.2

2. (a) $g:0 \to a$ (b) $g:1 \to c$ (c) $g:0 \to a$
$\quad\quad 1 \to a \quad\quad\quad 2 \to b \quad\quad\quad 1 \to c$
$\quad\quad 2 \to b \quad\quad\quad 3 \to b \quad\quad\quad 2 \to b$
$\quad\quad 3 \to c \quad\quad\quad 4 \to c \quad\quad\quad 3 \to a$
$\quad\quad 4 \to c \quad\quad\quad\quad\quad\quad\quad\quad 4 \to d$
$\quad\quad\quad\quad\quad\quad\quad\quad\quad\quad\quad\quad 5 \to b$

3. (a) States of M/g are $[0] = \{0, 1\}, [2] = \{2\}, [3] = \{3, 4\}$.

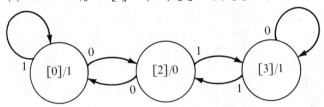

Isomorphism: $[0] \to a$
$\quad\quad\quad\quad\quad [2] \to b$
$\quad\quad\quad\quad\quad [3] \to c$

(c) States of M/g are $[0] = \{0, 3\}$
$[1] = \{1\}$
$[2] = \{2, 5\}$
$[4] = \{4\}$

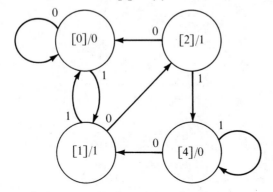

Isomorphism: $[0] \rightarrow a$
$[1] \rightarrow c$
$[2] \rightarrow b$
$[4] \rightarrow d$

Section 7.3

1. (b) $01^* \vee (110)^*$

2. (c) 100^*1

3. (a) Yes   (b) No   (c) No   (d) Yes

6. (a)

	Next state		
M' Present state	Present input		Output
	0	1	
$\{s_0\}$	$\{s_0, s_1\}$	$\{s_1\}$	1
$\{s_1\}$	$\{s_0\}$	$\{s_0, s_1\}$	0
$\{s_0, s_1\}$	$\{s_0, s_1\}$	$\{s_0, s_1\}$	1

$(0^* \vee (0 \vee 1)1^*(0 \vee 1))^*$

Section 8.1

1. (a) s_1 (b) s_2

2. $A = \{0\}, B = \{1, 2, 5\}, C = \{3, 4\}, D = \{6\}$

Present state	Next state		Output
	Present input		
	0	1	
A	C	D	1
B	C	B	0
C	B	A	1
D	C	B	1

5. $A = \{0, 2\}$, $B = \{1, 3\}$, $C = \{4\}$

Present state	Next state			Output
	Present input			
	a	b	c	
A	B	C	A	1
B	C	A	B	0
C	B	A	A	0

8. (a) True   (b) False   (c) True   (d) False

Section 8.2

1. (a) Possible answer:

Present state	Next state		Output
	Present input		
	0	1	
s_0	s_0	s_1	0
s_1	s_2	s_1	0
s_2	s_0	s_3	0
s_3	s_2	s_1	1

	d_1	d_2
s_0	0	0
s_1	0	1
s_2	1	0
s_3	1	1

$x(t)$	$d_1(t)$	$d_2(t)$	$y(t)$	$d_1(t+1)$	$d_2(t+1)$
0	0	0	0	0	0
1	0	0	0	0	1
0	0	1	0	1	0
1	0	1	0	0	1
0	1	0	0	0	0
1	1	0	0	1	1
0	1	1	1	1	0
1	1	1	1	0	1

$$y(t) = d_1 d_2$$
$$d_1(t+1) = x'd_1'd_2 + xd_1d_2' + x'd_1d_2 = x'd_2 + xd_1d_2'$$
$$d_2(t+1) = xd_1'd_2' + xd_1'd_2 + xd_1d_2' + xd_1d_2 = x$$

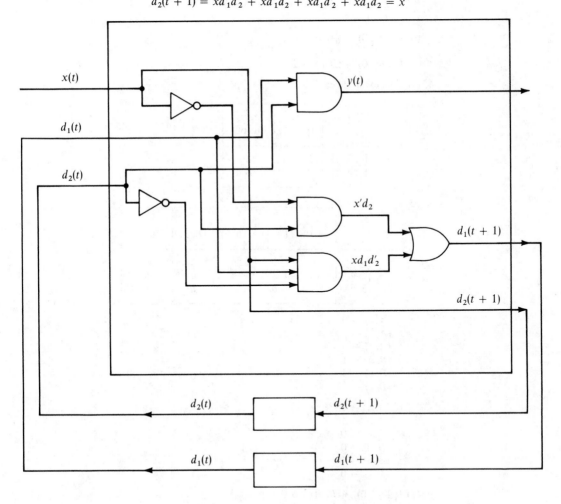

Section 8.3

1. (a) One example is

M_1	Input 0 1	Output
$A = \{0, 1, 3\}$	B A	0
$B = \{2, 4\}$	A A	1

M_2	Input 0 1	Output
$a = \{0, 2\}$	b c	0
$b = \{1, 4\}$	a b	1
$c = \{3\}$	a a	1

$g:(A, a) \to 0$
$(A, b) \to 1$
$(A, c) \to 3$
$(B, a) \to 2$
$(B, b) \to 4$

$g':(0, 0) \to 0$
$(1, 0) \to 1$
$(0, 1) \to 1$
$(1, 1) \to 0$

(b) $((B, b), 0) \to (A, a) \to 0$
$((A, a), 1) \to (A, c) \to 3$
$((A, c), 0) \to (B, a) \to 2$

4. (a) One example is

M_1	Input 0 1	Output
$A = \{0, 2, 3\}$	B A	A
$B = \{1, 4, 5\}$	A B	B

M_2	$(A, 0)$	$(A, 1)$	Input $(B, 0)$	$(B, 1)$	Output
$a = \{0, 1\}$	a	b	a	b	a
$b = \{2, 4\}$	c	c	c	a	b
$c = \{3, 5\}$	b	b	c	a	c

$g:(A, a) \to 0$
$(A, b) \to 2$
$(A, c) \to 3$
$(B, a) \to 1$
$(B, b) \to 4$
$(B, c) \to 5$

$g':(A, a) \to 0$
$(A, b) \to 1$
$(A, c) \to 0$
$(B, a) \to 0$
$(B, b) \to 1$
$(B, c) \to 0$

(b) $((A, b), 0) \to (B, c) \to 5$
$((B, b), 1) \to (B, a) \to 1$
$((A, c), 1) \to (A, b) \to 2$

Section 8.4

4. (a)

i	0	1	2
	01	11	10
S	λ	0	1	00

0	0	1	2	2
1	1	2	1	2
2	2	2	2	2

Table for S_M:

$\circ$	i	0	1	2
i	i	0	1	2
0	0	2	0	2
1	1	2	1	2
2	2	2	2	2

Not a group.

8. The present state of the cascade machine is $([3], 2)$ because $3 = 2 +_6 1$. The next state of the cascade machine is

$$([3 +_6 4], 2 +_6 \beta(4, [3]))$$
$$= ([1], 2 +_6 (1 +_6 4) +_6 (-(3 +_6 4)'))$$
$$= ([1], 2 +_6 5 +_6 (-1))$$
$$= ([1], 2 +_6 5 +_6 5)$$
$$= ([1], 0)$$

and $\gamma([1], 0) = 1$ because $1 = 0 +_6 1$

9. (a) The elements of $Z_6/\{0, 3\}$ are $[0] = \{0, 3\}$, $[1] = \{1, 4\}$ and $[2] = \{2, 5\}$; let $0, 1, 2$ be the fixed coset representatives.

The present state of the cascade machine is $([2], 0)$ because $2 = 0 +_6 2$. The next state of the cascade machine is

$$([2 +_6 5], 0 +_6 \beta(5, [2]))$$
$$= ([1], 0 +_6 (2 +_6 5) + (-(2 +_6 5)'))$$
$$= ([1], 0 +_6 1 +_6 (-1))$$
$$= ([1], 1 +_6 5)$$
$$= ([1], 0)$$

and $\gamma([1], 0) = 1$ because $1 = 0 +_6 1$

Section 9.1

1. (a) halts with final tape $\cdots$

	b	0	0	0	0	0	b		$\cdots$

(b) does not change the tape and moves forever to the left

4. One answer:

State 2 is a final state

$(0, b, b, 2, R)$ blank tape or no more 1's

$(0, 1, 1, 1, R)$ has read odd number of 1's

$(1, 1, 1, 0, R)$ has read even number of 1's

12. $f(n_1, n_2, n_3) = \begin{cases} n_2 + 1 & \text{if } n_2 > 0 \\ \text{undefined} & \text{if } n_2 = 0 \end{cases}$

16. One answer:

$(0, 1, b, 1, R)$ erases one extra 1

$(1, *, b, 3, R)$ $n_1 = 0$

$\left.\begin{array}{l} (1, 1, b, 2, R) \\ (2, 1, 1, 2, R) \\ (2, *, 1, 3, R) \end{array}\right\}$ $n_1 > 0$, replaces * with leftmost 1 of $\bar{n}_1$, halts

19. (a) T may run forever processing a given input string α, and we would be unable to test other strings.
   (b) Let T be a Turing machine which computes the function. Using copies $T_1, T_2, \ldots$ of T, feed input into them, and check to see which have halted with a representation $\bar{m}$ on the tape; any m so represented is in the range set.

Section 9.2

1. For any α, replace any instances of the symbol s with a new symbol. Create a machine T^* which acts like T but replaces any instances of s with the same new symbol and in addition, whenever T reaches a halting configuration, T^* prints s on the tape. Then T^* prints s if and only if T halts on α. Assume that there exists an algorithm P to solve the printing problem. An algorithm to solve the halting problem is: given a (T, α) pair, modify α and create T^*, and then apply P.

Section 9.3

2. In each case all the possible candidates for success can be processed in parallel, that is, an NDTM can "guess" truth assignments (or subgraphs of G or sets from C) and then test to see if any of the guesses produces a solution.

Section 10.1

1. (a) $L(G) = \{a\}$
   (b) $L(G) = \{010101, 010111, 011101, 011111, 110101, 110111, 111101, 111111\}$
   (c) $L(G) = 0(10)^*$
   (d) $L(G) = 0^*1111^*$

4. For example, $G = (V, V_T, S, P)$ where $V = \{0, 1, S, S_1\}$, $V_T = \{0, 1\}$, and P consists of the productions

$S \to \lambda$
$S \to 01$
$S \to 0S_1 1$
$S_1 \to 0S_1 1$
$S_1 \to 01$

8. For example, $G = (V, V_T, S, P)$ where $V = \{0, 1, S, S_1\}$, $V_T = \{0, 1\}$, and P consists of the productions

$$S \rightarrow \lambda$$
$$S \rightarrow S_1$$
$$S_1 \rightarrow 0S_10$$
$$S_1 \rightarrow 0S_11$$
$$S_1 \rightarrow 00$$
$$S_1 \rightarrow 11$$

12.

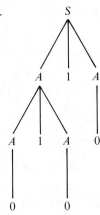 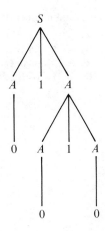

Section 10.2

1. $G = (V, V_T, S, P)$ where $V = \{S, A, B, C, 0, 1\}$

   $V_T = \{0, 1\}$, and P consists of

$S \rightarrow \lambda$	$B \rightarrow 0B$
$S \rightarrow 0$	$B \rightarrow 1B$
$S \rightarrow 0C$	$C \rightarrow 0C$
$S \rightarrow 1A$	$C \rightarrow 1A$
$A \rightarrow 1B$	$C \rightarrow 0$
$A \rightarrow 0C$	
$A \rightarrow 0$	

   $L(G) = (0 \vee 10)^*$

3.

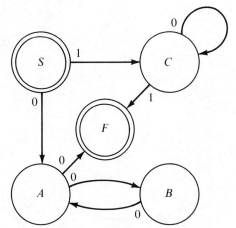

$$L(G) = \lambda \vee 0(00)*0 \vee 10*1$$

5. There are n words of length 1

             $n \cdot n$ words of length 2

                  $\vdots$

          n^k words of length k

So the total number of words is $n + n^2 + \cdots + n^k$

8. (a) $W_0 = \{S\}$

       $W_1 = \{S, a, b, bA\}$

       $W_2 = \{S, a, b, bA, bc\}$

       $W_3 = W_2$ so $cb \notin L(G)$

INDEX

ANSWERS TO PRACTICE PROBLEMS

Note to student: Finish all parts of a practice problem before turning to the answers.

CHAPTER 1

1.2 false, false, false

1.3

A	B	$A \vee B$
T	T	T
T	F	T
F	T	T
F	F	F

1.5 (a) antecedent: a is a positive number.
        consequent: $2a$ is a positive number.
   (b) antecedent: A 2×3 array is to be stored.
        consequent: 6 storage locations are to be used.
   (c) antecedent: Susan will pass her physics course.
        consequent: She is bright and studies hard.
   (d) antecedent: High gasoline mileage.
        consequent: Good combustion.

1.6

A	B	$A \Rightarrow B$
T	T	T
T	F	F
F	T	T
F	F	T

1.7

A	A'
T	F
F	T

1.10 (a)

A	B	$A \Rightarrow B$	$B \Rightarrow A$	$(A \Rightarrow B) \Leftrightarrow (B \Rightarrow A)$
T	T	T	T	T
T	F	F	T	F
F	T	T	F	F
F	F	T	T	T

(b)

A	B	A'	B'	$A \vee A'$	$B \wedge B'$	$(A \vee A') \Rightarrow (B \wedge B')$
T	T	F	F	T	F	F
T	F	F	T	T	F	F
F	T	T	F	T	F	F
F	F	T	T	T	F	F

(c)

A	B	C	B'	$A \wedge B'$	C'	$(A \wedge B') \Rightarrow C'$	$((A \wedge B') \Rightarrow C')'$
T	T	T	F	F	F	T	F
T	T	F	F	F	T	T	F
T	F	T	T	T	F	F	T
T	F	F	T	T	T	T	F
F	T	T	F	F	F	T	F
F	T	F	F	F	T	T	F
F	F	T	T	F	F	T	F
F	F	F	T	F	T	T	F

(d)

A	B	A'	B'	$A \Rightarrow B$	$B' \Rightarrow A'$	$(A \Rightarrow B) \Leftrightarrow (B' \Rightarrow A')$
T	T	F	F	T	T	T
T	F	F	T	F	F	T
F	T	T	F	T	T	T
F	F	T	T	T	T	T

1.12

A	1	A'	$A \vee A'$	$A \vee A' \Leftrightarrow 1$
T	T	F	T	T
F	T	T	T	T

1.15 Possible answers:
   (a) A whale.
   (b) Input given by means of punched cards.

1.18 Hypothesis: x is divisible by 6

   $x = k \cdot 6$ for some integer k

   $2x = 2(k \cdot 6)$

   $2x = (2 \cdot k)6$

   $2x = (k \cdot 2)6$

   $2x = k(2 \cdot 6)$

   $2x = k \cdot 12$

   $2x = k(3 \cdot 4)$

   $2x = (k \cdot 3)4$

   $k \cdot 3$ is an integer

   Conclusion: $2x$ is divisible by 4

1.20 (a) If $2a$ is not a positive number, then a is not a positive number.

(b) If it is false that 6 locations are to be used, then it is false that a 2×3 array is to be stored.

(c) If Susan is not bright or does not study hard, then she will not pass her physics course.

(d) If combustion is not good, then there is not high gasoline mileage.

1.22 (a) If $2a$ is a positive number, then a is a positive number.

(b) If 6 storage locations are to be used, then a 2×3 array is to be stored.

(c) If Susan is bright and studies hard, she will pass her physics course.

(d) If there is good combustion, then there is high gasoline mileage.

1.26 $P(1)$ is $1 = \dfrac{1(1 + 1)}{2}$

Assume $P(k)$, $1 + 2 + 3 + \cdots + k = \dfrac{k(k + 1)}{2}$

To show $P(k + 1)$,

$$1 + 2 + 3 + \cdots + (k + 1) = 1 + 2 + 3 + \cdots + k + (k + 1)$$

$$= \frac{k(k + 1)}{2} + (k + 1) = (k + 1)\left(\frac{k}{2} + 1\right)$$

$$= (k + 1)\left(\frac{k + 2}{2}\right) = \frac{(k + 1)[(k + 1) + 1]}{2}$$

1.29 (a) $\{4, 5, 6, 7\}$

(b) {April, June, Sept., Nov.}

(c) $\{0, 1, 8\}$

1.30 (a) $\{x \mid x = y^2 \text{ for } y \in \{1, 2, 3, 4\}\}$

(b) $\{x \mid x$ is one of the Three Men in a Tub$\}$

(c) $\{x \mid x$ is a prime number$\}$

1.31 $x \in B$

1.33 (a), (b), (d), (e), (h), (i), (l)

1.35 Let $x \in A$. Then $\cos(x/2) = 0$. $\mathrm{Cos}(x/2) = 0$ if and only if $x/2 = \pm \pi/2, \pm 3\pi/2, \pm 5\pi/2, \ldots$ or $x = \pm \pi, \pm 3\pi, \pm 5\pi, \ldots$ and for any multiple of π, the sine function is 0. Thus $x \in B$.

1.36 $\mathscr{P}(A) = \{\varnothing, \{1\}, \{2\}, \{3\}, \{1, 2\}, \{1, 3\}, \{2, 3\}, \{1, 2, 3\}\}$.

1.37 2^n

1.38 By the definition of equality for ordered pairs,

$$2x - y = 7 \quad \text{and} \quad x + y = -1$$

Solving the system of equations, $x = 2$, $y = -3$

1.39 $(3, 3) \quad (3, 4) \quad (4, 3) \quad (4, 4)$

1.46 (a) S is not closed under division.

(c) 0^0 is not defined.

(f) $x^{\#}$ is not unique for, say, $x = 4$ ($2^2 = 4$ and $(-2)^2 = 4$)

1.51 $x \in A$ and $x \in B$

1.54

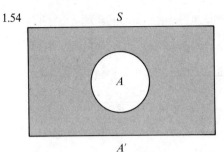

1.55

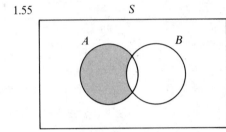

1.57 (a) $\{1, 2, 3, 4, 5, 7, 8, 9, 10\}$
     (b) $\{1, 2, 3\}$
     (c) $\{1, 3, 5, 10\}$

1.59 Show set inclusion in each direction. To show $A \cup \emptyset \subseteq A$, let $x \in A \cup \emptyset$. Then $x \in A$ or $x \in \emptyset$, but since $\emptyset$ has no elements, $x \in A$. To show $A \subseteq A \cup \emptyset$, let $x \in A$. Then $x \in A$ or $x \in \emptyset$, so $x \in A \cup \emptyset$.

1.61 (a) $(C \cap (A \cup B)) \cup ((A \cup B) \cap C') = ((A \cup B) \cap C) \cup ((A \cup B) \cap C')$ 1b
$$= (A \cup B) \cap (C \cup C') \qquad \text{3b}$$
$$= (A \cup B) \cap S \qquad \text{5a}$$
$$= A \cup B \qquad \text{4b}$$
     (b) $(C \cup (A \cap B)) \cap ((A \cap B) \cup C') = A \cap B$

1.63 (a) $A \times B = \{(1, 3), (1, 4), (2, 3), (2, 4)\}$
     (b) $B \times A = \{(3, 1), (3, 2), (4, 1), (4, 2)\}$
     (c) $A^2 = \{(1, 1), (1, 2), (2, 1), (2, 2)\}$
     (d) $A^3 = \{(1, 1, 1), (1, 1, 2), (1, 2, 1), (1, 2, 2), (2, 1, 1), (2, 1, 2), (2, 2, 1), (2, 2, 2)\}$

1.66 $7 \cdot 5 = 35$

1.68 $P(20, 2) = \dfrac{20!}{18!} = 380$

1.69 $6! = 720$

1.72 $C(12, 3) = \dfrac{12!}{3!9!} = 220$

CHAPTER 2

2.6 (a) $(3, 2) \in \rho$ (b) $(2, 4), (2, 6) \in \rho$ (c) $(3, 4), (5, 6) \in \rho$ (d) $(2, 1), (5, 2) \in \rho$

2.7 a subset of $S_1 \times S_2 \times \cdots \times S_n$.

2.8 $\rho + \sigma = \rho \cup \sigma$
$\rho \cdot \sigma = \rho \cap \sigma$
ρ' (here ' denotes an operation on a binary relation) $= \rho'$ (here ' denotes the set operation of complementation)

2.9 (a) $x(\rho + \sigma)y \Leftrightarrow x \leq y$
(b) $x\rho'y \Leftrightarrow x \neq y$
(c) $x\sigma'y \Leftrightarrow x \geq y$
(d) $\rho \cdot \sigma = \varnothing$

2.14 (a) reflexive, symmetric, transitive
(b) reflexive, antisymmetric, transitive
(c) reflexive, symmetric, transitive
(d) antisymmetric
(e) reflexive, symmetric, antisymmetric, transitive
(f) antisymmetric (this property is tricky—recall the truth table for implication), transitive
(g) reflexive, symmetric, transitive
(h) reflexive, symmetric, transitive

2.16 (a) $(1, 1)$ $(1, 2)$ $(2, 2)$ $(1, 3)$ $(3, 3)$ $(1, 6)$
$(6, 6)$ $(1, 12)$ $(12, 12)$ $(1, 18)$ $(18, 18)$
$(2, 6)$ $(2, 12)$ $(2, 18)$ $(3, 6)$ $(3, 12)$
$(3, 18)$ $(6, 12)$ $(6, 18)$
(b) 1, 2, 3
(c) 2, 3

2.18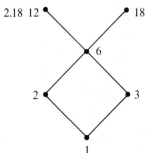

2.19 $y \in S$ is a greatest element if $x \leq y$ for all $x \in S$.
$y \in S$ is a maximal element if there is no $x \in S$ with $y \prec x$.

2.21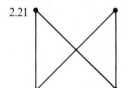

2.25 Let $q \in [x]$. Then $x\rho q$. Because $x\rho z$, by symmetry, $z\rho x$. By transitivity, $z\rho x$ together with $x\rho q$ gives $z\rho q$. Therefore, $q \in [z]$.

2.26 For any $x \in S$, x is in the same subset as itself, so $x\rho x$. If $x\rho y$, x is in the same subset as y, so y is in the same subset as x, or $y\rho x$. If $x\rho y$ and $y\rho z$, then x is in the same subset as y and y is in the same subset as z, so x is in the same subset as z, or $x\rho z$.

2.29 (a) The equivalence classes are sets consisting of lines in the plane with the same slope.
   (b) $[n] = \{n\}$; the equivalence classes are all of the singleton sets of elements of $\mathbb{N}$.
   (c) $[1] = [2] = \{1, 2\}$, $[3] = \{3\}$

2.32 $x \equiv_n y$ if $x - y$ is an integral multiple of n.

2.33 $[0] = \{0, 5, 10, 15, \ldots, -5, -10, -15, \ldots\}$
   $[1] = \{1, 6, 11, 16, \ldots, -4, -9, -14, \ldots\}$
   $[2] = \{2, 7, 12, 17, \ldots, -3, -8, -13, \ldots\}$
   $[3] = \{3, 8, 13, 18, \ldots, -2, -7, -12, \ldots\}$
   $[4] = \{4, 9, 14, 19, \ldots, -1, -6, -11, \ldots\}$

2.35 (a) Not a function; $2 \in S$ has two values associated with it.   (b) Function.
   (c) Not a function; for values 0, 1, 2, 3 of the domain, the corresponding $h(x)$ values fall outside the codomain.
   (d) Not a function; not every member of S has a Social Security number.
   (e) Function (not every value in the codomain need be used).
   (f) Function.   (g) Function
   (h) Not a function; $5 \in N$ has two values associated with it.

2.36 16, ± 3

2.41 (b), (g)

2.44 (e), (g)

2.48 One possibility: $\{(0, 0), (1, 1), (-1, 2), (2, 3), (-2, 4), (3, 5), (-3, 6), \ldots\}$

2.50 $(g \circ f)(x) = g(f(x)) = g((2.3)^2) = g(5.29) = 5$
   $(f \circ g)(x) = f(g(x)) = f(2) = 2^2 = 4$

2.51 Let $(g \circ f)(s_1) = (g \circ f)(s_2)$. Then $g(f(s_1)) = g(f(s_2))$ and because g is injective, $f(s_1) = f(s_2)$. Because f is injective, $s_1 = s_2$.

2.54 (a) $(1, 4, 5) = (4, 5, 1) = (5, 1, 4)$ (b) $\begin{pmatrix} 1 & 2 & 3 & 4 & 5 \\ 1 & 4 & 2 & 5 & 3 \end{pmatrix}$

2.55 (a) $g \circ f = (1, 3, 5, 2, 4) = (3, 5, 2, 4, 1)$, and so on.
   (b) $g \circ f = \begin{pmatrix} 1 & 2 & 3 & 4 & 5 \\ 4 & 2 & 5 & 1 & 3 \end{pmatrix}$

2.56 $(1, 2, 4) \circ (3, 5)$ or $(3, 5) \circ (1, 2, 4)$.

2.57 Let $t \in T$. Then $(f \circ g)(t) = f(g(t)) = f(s) = t$.

2.60 $f^{-1}: \mathbb{R} \rightarrow \mathbb{R}, f^{-1}(x) = (x - 4)/3$

2.66 The middle one.

2.67 One possible picture is given by the figure that follows.

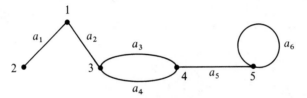

2.68 (a) yes   (b) no   (c) yes   (d) $3, a_5, 5, a_6, 6; 3, a_3, 4, a_4, 5, a_6, 6$
     (e) $3, a_3, 4, a_4, 5, a_5, 3$ (f) a_3 or a_4 or a_5 (g) a_1 or a_2 or a_6 or a_7

2.71 (a)

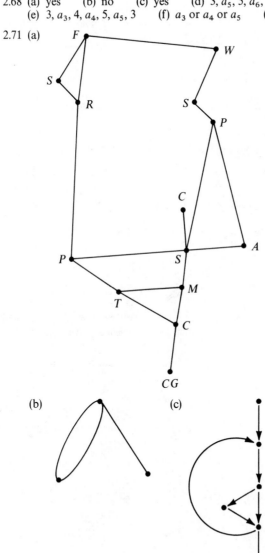

2.73 $A = \begin{bmatrix} 1 & 1 & 0 & 1 \\ 1 & 0 & 1 & 0 \\ 0 & 1 & 0 & 2 \\ 1 & 0 & 2 & 0 \end{bmatrix}$

2.74 $A^2 = \begin{bmatrix} 3 & 1 & 3 & 1 \\ 1 & 2 & 0 & 3 \\ 3 & 0 & 5 & 0 \\ 1 & 3 & 0 & 5 \end{bmatrix}$

2.75 (a) The number of possible paths of length n from n_i to n_j (b) induction

2.77 $R = \begin{bmatrix} 3 & 4 & 3 & 2 \\ 0 & 2 & 3 & 2 \\ 0 & 2 & 2 & 1 \\ 0 & 1 & 2 & 1 \end{bmatrix}$

2.78 There are two distinct nodes, 3 and 4, with $3\rho 4$ and $4\rho 3$.

2.80 (a) No   (b) Yes

2.83 (a) No; four odd nodes.   (b) Yes, no odd nodes.

2.84 No

2.85 (a) No   (b) Yes

2.87 In the figure below, $N = 4$, $A = 5$, $R = 2$. $N - A + R = 4 - 5 + 2 = 1$.

2.89 One solution: there exists a path between n_1 and n_3. Consider n_2 and n_4; no path. Exchange colors 2 and 4 in a section containing n_2. The result is the following figure.

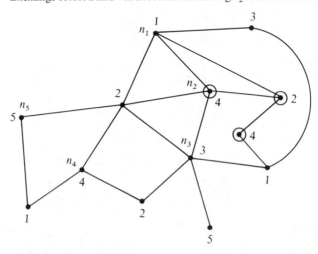

CHAPTER 3

3.1   1a. $x + y = y + x$ 1b. $x \cdot y = y \cdot x$
      2a. $(x + y) + z = x + (y + z)$ 2b. $(x \cdot y) \cdot z = x \cdot (y \cdot z)$
      3a. $x + (y \cdot z) = (x + y) \cdot (x + z)$ 3b. $x \cdot (y + z) = x \cdot y + x \cdot z$
      4a. $x + 0 = x$ 4b. $x \cdot 1 = x$
      5a. $x + x' = 1$ 5b. $x \cdot x' = 0$

3.6   (a) $A \vee A = A$ (b) $A \cup A = A$

3.7   (a) $x + 1 = x + (x + x')$ (5a, complements)
$$\quad\quad\quad = (x + x) + x' \quad \text{(2a, associative property)}$$
$$\quad\quad\quad = x + x' \quad \text{(idempotent property)}$$
$$\quad\quad\quad = 1 \quad \text{(5a, complements)}$$
      (b) $x \cdot 0 = 0$

3.8 $0 + 1 = 1$ (Practice 3.7(a))
      $0 \cdot 1 = 1 \cdot 0$ (1b)
      $\quad\quad = 0$ (Practice 3.7(b))
      Therefore $1 = 0'$.

3.9   Let $X = \begin{bmatrix} a_{11} & a_{12} \\ a_{21} & a_{22} \end{bmatrix}$. Then

$$\begin{bmatrix} a_{11} & a_{12} \\ a_{21} & a_{22} \end{bmatrix} \begin{bmatrix} 1 & 0 \\ 0 & 1 \end{bmatrix} = \begin{bmatrix} a_{11} & a_{12} \\ a_{21} & a_{22} \end{bmatrix} = \begin{bmatrix} 1 & 0 \\ 0 & 1 \end{bmatrix} \begin{bmatrix} a_{11} & a_{12} \\ a_{21} & a_{22} \end{bmatrix}$$

3.10 Let $\begin{bmatrix} a_{11} & 0 \\ 0 & a_{22} \end{bmatrix}$ and $\begin{bmatrix} b_{11} & 0 \\ 0 & b_{22} \end{bmatrix}$ be elements of $M_2^D(\mathbb{Z})$. Then

$$\begin{bmatrix} a_{11} & 0 \\ 0 & a_{22} \end{bmatrix} \begin{bmatrix} b_{11} & 0 \\ 0 & b_{22} \end{bmatrix} = \begin{bmatrix} a_{11}b_{11} & 0 \\ 0 & a_{22}b_{22} \end{bmatrix} \in M_2^D(\mathbb{Z});$$

so the product of two members of $M_2^D(\mathbb{Z})$ exists and is a unique member of $M_2^D(\mathbb{Z})$.

3.11 (a) Let $\begin{bmatrix} a_{11} & 0 \\ 0 & a_{22} \end{bmatrix}$ and $\begin{bmatrix} b_{11} & 0 \\ 0 & b_{22} \end{bmatrix}$ be members of $M_2^D(\mathbb{Z})$. Then

$$\begin{bmatrix} a_{11} & 0 \\ 0 & a_{22} \end{bmatrix} \begin{bmatrix} b_{11} & 0 \\ 0 & b_{22} \end{bmatrix} = \begin{bmatrix} a_{11}b_{11} & 0 \\ 0 & a_{22}b_{22} \end{bmatrix}$$

$$= \begin{bmatrix} b_{11}a_{11} & 0 \\ 0 & b_{22}a_{22} \end{bmatrix} = \begin{bmatrix} b_{11} & 0 \\ 0 & b_{22} \end{bmatrix} \begin{bmatrix} a_{11} & 0 \\ 0 & a_{22} \end{bmatrix}$$

      (b) Not commutative. For example,

$$\begin{bmatrix} 1 & 2 \\ 3 & 4 \end{bmatrix} \begin{bmatrix} 2 & 1 \\ 2 & 2 \end{bmatrix} = \begin{bmatrix} 6 & 5 \\ 14 & 11 \end{bmatrix}$$

$$\begin{bmatrix} 2 & 1 \\ 2 & 2 \end{bmatrix} \begin{bmatrix} 1 & 2 \\ 3 & 4 \end{bmatrix} = \begin{bmatrix} 5 & 8 \\ 8 & 12 \end{bmatrix}$$

3.12 0

3.13 Yes. For any real numbers x and y, $x + y = y + x$

3.14 $X \cdot X^{-1} = X^{-1} \cdot X = I$

3.16 (a) Yes. The product of two positive real numbers is a positive real number.

(b) Yes. For all $x, y, z \in \mathbb{R}^+$, $(x \cdot y) \cdot z = x \cdot (y \cdot z)$

(c) Yes. For all $x, y \in \mathbb{R}^+$, $x \cdot y = y \cdot x$.

(d) If i denotes an identity element, then for any $x \in \mathbb{R}^+$, $x \cdot i = i \cdot x = x$. 1 is an identity element.

(e) For each $x \in \mathbb{R}^+$, there is an element $x^{-1} \in \mathbb{R}^+$ such that $x \cdot x^{-1} = x^{-1} \cdot x = 1$. The inverse of x is $1/x$.

3.17 Answer (e) will change. The element $0 \in \mathbb{R}$ has no inverse since there is no real number y such that $0 \cdot y = y \cdot 0 = 1$.

3.21 (a) $f(-4 + 7) = f(3) = e^3$; top and right

(b) $f(-4) \cdot f(7) = e^{-4} \cdot e^7 = e^3$; left and bottom

3.22 (a) $f(x \cdot y) = f(x)*f(y)$ (b) $f(x') = [f(x)]''$

3.24 (b)   (c)   (d)

$$
\begin{array}{ccc}
B \times B \xrightarrow{\ +\ } B & \quad B \times B \xrightarrow{\ \cdot\ } B & \quad B \xrightarrow{\ '\ } B \\
{\scriptstyle f \times f}\Big\downarrow \quad \Big\downarrow {\scriptstyle f} & {\scriptstyle f \times f}\Big\downarrow \quad \Big\downarrow {\scriptstyle f} & {\scriptstyle f}\Big\downarrow \quad \Big\downarrow {\scriptstyle f} \\
b \times b \xrightarrow{\ \&\ } b & b \times b \xrightarrow{\ *\ } b & b \xrightarrow{\ ''\ } b
\end{array}
$$

3.25 (a) $f(0 + a) = f(a) = \{1\} = \varnothing \cup \{1\} = f(0) \cup f(a)$

(b) $f(a + a') = f(1) = \{1, 2\} = \{1\} \cup \{2\} = f(a) \cup f(a')$

(c) $f(a \cdot a') = f(0) = \varnothing = \{1\} \cap \{2\} = f(a) \cap f(a')$

(d) $f(1') = f(0) = \varnothing = \{1, 2\}' = [f(1)]'$

3.28 $f\left(\begin{bmatrix} a_{11} & 0 \\ 0 & a_{22} \end{bmatrix} \cdot \begin{bmatrix} b_{11} & 0 \\ 0 & b_{22} \end{bmatrix}\right) = f\left(\begin{bmatrix} a_{11}b_{11} & 0 \\ 0 & a_{22}b_{22} \end{bmatrix}\right)$

$= a_{11}b_{11} = f\left(\begin{bmatrix} a_{11} & 0 \\ 0 & a_{22} \end{bmatrix}\right) f\left(\begin{bmatrix} b_{11} & 0 \\ 0 & b_{22} \end{bmatrix}\right)$

CHAPTER 4

4.5 (a) 2^n (b) $2^4 = 16$ (c) 2^{2^n}

4.8 (a)

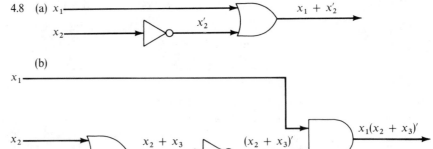

(b)

4.10 (a) $(x_1' + x_2)x_3'$

(b)

x_1	x_2	x_3	$(x_1' + x_2)x_3'$
1	1	1	0
1	1	0	1
1	0	1	0
1	0	0	0
0	1	1	0
0	1	0	1
0	0	1	0
0	0	0	1

4.13 (a) $x_1x_2x_3 + x_1x_2'x_3 + x_1x_2'x_3' + x_1'x_2'x_3 + x_1'x_2'x_3'$

(b)

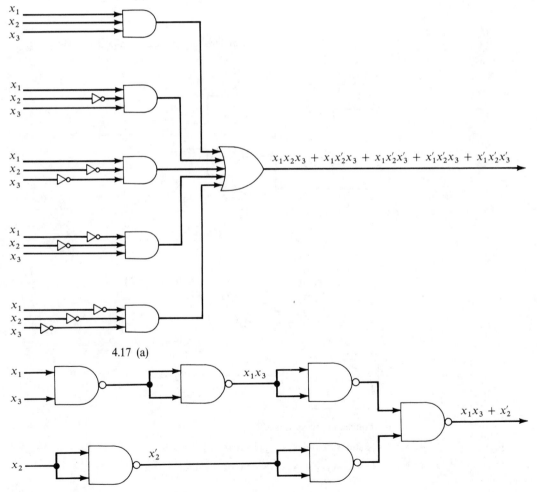

4.17 (a)

4.17 (b) $x_1x_3 + x_2' = ((x_1x_3)'x_2)'$.

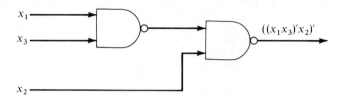

4.19 (a)

x_1	x_2	$f(x_1, x_2)$
1	1	0
1	0	1
0	1	1
0	0	0

(b) One possibility is the canonical sum-of-products form: $x_1x_2' + x_1'x_2$.

(c)

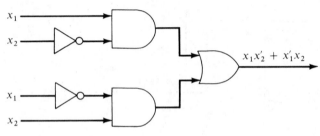

4.21 (a) $x_1x_2 + x_1'x_2 = x_2x_1 + x_2x_1'$
$= x_2(x_1 + x_1')$
$= x_2 \cdot 1$
$= x_2$

(b) $x_1 + x_1'x_2 = x_1 \cdot 1 + x_1'x_2$
$= x_1(1 + x_2) + x_1'x_2$ (See Practice 3.7.)
$= x_1 + x_1x_2 + x_1'x_2$
$= x_1 + x_2(x_1 + x_1')$
$= x_1 + x_2 \cdot 1 = x_1 + x_2$

4.23

	x_1	x_1'
x_2		1
x_2'		1

The reduced expression is x_1'.

4.26 x_1x_3 (4 squares) and $x_1'x_2x_3'$ (2 squares)

4.31 $x_1x_2'x_4 + x_1x_3'x_4 + x_2'x_3'$ (See following figure.)

	x_1x_2	x_1x_2'	$x_1'x_2'$	$x_1'x_2$
x_3x_4		1		
x_3x_4'				
$x_3'x_4'$		1	1	
$x_3'x_4$	1	1	1	

4.34 The reduction table is shown in the following figure.

x_1	x_2	x_3		x_1	x_2	x_3
1	1	1	1	1	1	—
1	1	0	1,2	1	—	0
1	0	0	2,3	—	0	0
0	0	1	4			
0	0	0	3,4	0	0	—

The comparison table is shown in the following figure.

	111	110	100	001	000
11–	✓	✓			
1–0		✓	✓		
–00			✓		✓
00–				✓	✓

Essential terms are $11-$ and $00-$. Either $1-0$ or -00 can be used as the third reduced term. The minimal sum-of-products form is

$$x_1x_2 + x_1'x_2' + x_1x_3'$$

or

$$x_1x_2 + x_1'x_2' + x_2'x_3'$$

CHAPTER 5

5.7 All operations are associative binary operations on their respective sets.

5.8 (a) All (b) All but $[\mathbb{R}^+, +]$; $0, 1, 0, 1, \begin{bmatrix} 0 & 0 \\ 0 & 0 \end{bmatrix}$

5.11 $[\mathbb{R}, +], [\mathbb{R}^+, \cdot], [\mathbb{Z}, +], [M_2(\mathbb{Z}), +]$

5.13 (a) $f(x) + g(x) = g(x) + f(x)$
      $(f(x) + g(x)) + h(x) = f(x) + (g(x) + h(x))$
      (b) The zero polynomial, 0
      (c) $-7x^4 + 2x^3 - 4$

5.15 (a)

$+_5$	0	1	2	3	4
0	0	1	2	3	4
1	1	2	3	4	0
2	2	3	4	0	1
3	3	4	0	1	2
4	4	0	1	2	3

$\cdot_5$	0	1	2	3	4
0	0	0	0	0	0
1	0	1	2	3	4
2	0	2	4	1	3
3	0	3	1	4	2
4	0	4	3	2	1

      (b) 0; 1   (c) 3   (d) All except 0.

5.16 (a)

$\cdot_6$	0	1	2	3	4	5
0	0	0	0	0	0	0
1	0	1	2	3	4	5
2	0	2	4	0	2	4
3	0	3	0	3	0	3
4	0	4	2	0	4	2
5	0	5	4	3	2	1

      (b) 1 and 5

5.18 Let $f, g, h \in S$. Then for any $x \in A$, $((f \circ g) \circ h)(x) = (f \circ g)(h(x)) = f(g(h(x)))$
      and $(f \circ (g \circ h))(x) = f((g \circ h)(x)) = f(g(h(x)))$. Hence, $(f \circ g) \circ h = f \circ (g \circ h)$.

5.20 (a)

$\circ$	α_1	α_2	α_3	α_4	α_5	α_6
α_1	α_1	α_2	α_3	α_4	α_5	α_6
α_2	α_2	α_1	α_6	α_5	α_4	α_3
α_3	α_3	α_5	α_1	α_6	α_2	α_4
α_4	α_4	α_6	α_5	α_1	α_3	α_2
α_5	α_5	α_3	α_4	α_2	α_6	α_1
α_6	α_6	α_4	α_2	α_3	α_1	α_5

      (b) No

5.22 (a) No   (b) No

5.24 (a) $(10, 9)$ (b) $(1, 1)$ (c) $(3, 2)$

5.25 (a) For $(g, h) \in G \times H$, $(1_G, 1_H) \cdot (g, h) = (1_G \circ g, 1_H * h) = (g, h)$, and $(g, h) \cdot (1_G, 1_H) =$
      $(g \circ 1_G, h * 1_H) = (g, h)$.
      (b) $(g, h)^{-1} = (g^{-1}, h^{-1})$ because $(g^{-1}, h^{-1}) \cdot (g, h) = (g^{-1} \circ g, h^{-1} * h) = (1_G, 1_H)$,
      and $(g, h) \cdot (g^{-1}, h^{-1}) = (g \circ g^{-1}, h * h^{-1}) = (1_G, 1_H)$.
      (c) Yes. $(g_1, h_1) \cdot (g_2, h_2) = (g_1 \circ g_2, h_1 * h_2) = (g_2 \circ g_1, h_2 * h_1) = (g_2, h_2) \cdot (g_1, h_1)$.

5.26 $i_1 = i_1 i_2$ because i_2 is an identity.
      $i_1 i_2 = i_2$ because i_1 is an identity.

5.28 Let y and z both be inverses of x. Let i be the identity. Then $y = y \cdot i = y \cdot (x \cdot z) =$
      $(y \cdot x) \cdot z = i \cdot z = z$.

5.31 $7^{-1} = 5, 3^{-1} = 9$; so $10^{-1} = (7 +_{12} 3)^{-1} = 3^{-1} +_{12} 7^{-1} = 9 +_{12} 5 = 2$.

5.33 $z \cdot x = z \cdot y$ implies

$$z^{-1} \cdot (z \cdot x) = z^{-1} \cdot (z \cdot y)$$

$$(z^{-1} \cdot z) \cdot x = (z^{-1} \cdot z) \cdot y$$

$$i \cdot x = i \cdot y$$

$$x = y$$

5.37 $x = 1 +_8 (3)^{-1} = 1 +_8 5 = 6$

5.38

$\circ$	1	a	b	c	d
1	1	a	b	c	d
a	a	b	c	d	1
b	b	c	d	1	a
c	c	d	1	a	b
d	d	1	a	b	c

5.39   (a) $[\mathbb{Z}_{18}, +_{18}]$ (b) S_3 or P_3

5.43   (a) and (c)

5.44   Yes. For $x, y \in A$, $x \cdot y = y \cdot x$ because x and y belong to S. Commutativity is inherited.

5.48   (a) and (b). Note that (c) is not a submonoid by our definition because the set does not contain $(1, 1)$, but $(0, 1)$ is an identity for this set under componentwise multiplication.

5.50   (a) A is closed under.
(b) $i \in A$.
(c) Every $x \in A$ has an inverse element in A.

5.53   (a) Closure holds; $0 \in \{0, 2, 4, 6\}$; $0^{-1} = 0$, $4^{-1} = 4$, 2 and 6 are inverses of each other.
(b) Closure holds; $1 \in \{1, 2, 4\}$; $1^{-1} = 1$, 2 and 4 are inverses of each other.

5.54 $[\{\alpha_1, \alpha_5, \alpha_6\}, \circ], [\{\alpha_1 \alpha_2\}, \circ], [\{\alpha_1, \alpha_3\}, \circ], [\{\alpha_1, \alpha_4\}, \circ]$

5.55   To show f is one-to-one, let α and β belong to A_n and suppose $f(\alpha) = f(\beta)$. Then $\alpha \circ (1, 2) = \beta \circ (1, 2)$. By the cancellation law available in the group S_n, $\alpha = \beta$. To show f is onto, let $\gamma \in O_n$. Then $\gamma \circ (1, 2) \in A_n$ and $f(\gamma \circ (1,2)) = \gamma \circ (1,2) \circ (1,2) = \gamma$.

5.57   For $nz_1, nz_2 \in n\mathbb{Z}$, $nz_1 + nz_2 = n(z_1 + z_2) \in n\mathbb{Z}$, so closure holds; $0 = n \cdot 0 \in n\mathbb{Z}$; for $nz \in n\mathbb{Z}$, $-nz = n(-z) \in n\mathbb{Z}$.

5.60 $f(x) + f(y)$

5.64   (a) Yes. For $x, y \in \mathbb{N}$, $f(x + y) = 2(x + y) = 2x + 2y = f(x) + f(y)$.
(b) No. For $x, y \in \mathbb{N}$, $f(x + y) = 2(x + y) = 2x + 2y$, but $f(x) \cdot f(y) = 2x \cdot 2y = 4xy$. These are not always equal.

(c) Yes. For $\begin{bmatrix} a & b \\ c & d \end{bmatrix}$ and $\begin{bmatrix} e & f \\ g & h \end{bmatrix}$ in $M_2(\mathbb{Z})$,

$$f\left(\begin{bmatrix} a & b \\ c & d \end{bmatrix} + \begin{bmatrix} e & f \\ g & h \end{bmatrix} \right) = f\left(\begin{bmatrix} a + e & b + f \\ c + g & d + h \end{bmatrix} \right) = a + e$$

$$= f\left(\begin{bmatrix} a & b \\ c & d \end{bmatrix} \right) + f\left(\begin{bmatrix} e & f \\ g & h \end{bmatrix} \right)$$

(d) Yes. For $a_2x^2 + a_1x + a_0$ and $b_2x^2 + b_1x + b_0$ in $\mathbb{N}^2[x]$, $f(a_2x^2 + a_1x + a_0 + b_2x^2 + b_1x + b_0) = f((a_2 + b_2)x^2 + (a_1 + b_1)x + (a_0 + b_0))$
$= a_2 + b_2 = f(a_2x^2 + a_1x + a_0) + f(b_2x^2 + b_1x + b_0)$.

5.65   Let $f(x) \in f(S)$. Then $f(x) + f(i_S) = f(x \cdot i_S) = f(x)$ and $f(i_S) + f(x) = f(i_S \cdot x)$
$= f(x)$.

5.66 $f(x^{-1}) + f(x) = f(x^{-1} \cdot x) = f(i_S)$
$f(x) + f(x^{-1}) = f(x \cdot x^{-1}) = f(i_S)$

5.68   Let $f(x)$ and $f(y)$ belong to $f(S)$. Then $f(x) + f(y) = f(x \cdot y) = f(y \cdot x) = f(y) + f(x)$.

5.72   Clearly f is onto. f is also one-to-one: let $f(x) = f(y)$, then $5x = 5y$ and $x = y$. f is a homomorphism: for $x, y \in \mathbb{Z}$, $f(x + y) = 5(x + y) = 5x + 5y = f(x) + f(y)$.

5.73   (a) Composition of bijections is a bijection, and for $x, y \in S$, $(g \circ f)(x \cdot y) = g(f(x \cdot y)) = g(f(x) + f(y)) = (g \circ f)(x) * (g \circ f)(y)$.
   (b) $S \simeq S$ by the identity mapping. If f is an isomorphism from S to T, then f^{-1} is an isomorphism from T to S. If $S \simeq T$ and $T \simeq V$, then by part (a), $S \simeq V$.

5.75   To show that α_g is an onto function, let $y \in G$. Then $g^{-1} \cdot y$ belongs to G and $\alpha_g(g^{-1} \cdot y) = g(g^{-1} \cdot y) = (g \cdot g^{-1})y = y$. To show that α_g is one-to-one, let $\alpha_g(x) = \alpha_g(y)$. Then $g \cdot x = g \cdot y$, and by cancellation, $x = y$.

5.76   (a) For $\alpha_g \in P$, $\alpha_g \circ \alpha_1 = \alpha_{g \cdot 1} = \alpha_g$ and $\alpha_1 \circ \alpha_g = \alpha_{1 \cdot g} = \alpha_g$.
   (b) $\alpha_g \circ \alpha_{g^{-1}} = \alpha_{g \cdot g^{-1}} = \alpha_1$, and $\alpha_{g^{-1}} \circ \alpha_g = \alpha_{g \cdot g^{-1}} = \alpha_1$.

5.77   (a) Let $f(g) = f(h)$. Then $\alpha_g = \alpha_h$ and in particular, $\alpha_g(1) = \alpha_h(1)$, or $g \cdot 1 = h \cdot 1$ and $g = h$.
   (b) For $g, h, \in G$, $f(g \cdot h) = \alpha_{g \cdot h} = \alpha_g \circ \alpha_h = f(g) \circ f(h)$.

5.79   Reflexive: $f(x) = f(x)$. Symmetric: if $f(x) = f(y)$, then $f(y) = f(x)$. Transitive: if $f(x) = f(y)$ and $f(y) = f(z)$, then $f(x) = f(z)$.

5.81 $f(\mathbb{Z}) = \mathbb{Z}_3 = \{0, 1, 2\}$. The distinct members of $\mathbb{Z}/f$ are

$[0] = \{0, 3, 6, 9, \ldots, -3, -6, -9, \ldots\} = [-12]$

$[1] = \{1, 4, 7, 10, \ldots, -2, -5, -8, \ldots\} = [16]$

$[2] = \{2, 5, 8, 11, \ldots, -1, -4, -7, \ldots\} = [8]$

5.82   Let $[x], [y], [z] \in S/f$. Then $([x] * [y]) * [z] = [x \cdot y] * [z] = [(x \cdot y) \cdot z]$ and $[x] * ([y] * [z]) = [x] * [y \cdot z] = [x \cdot (y \cdot z)]$, but $(x \cdot y) \cdot z = x \cdot (y \cdot z)$.

5.84   Computations (b) and (c) are correct.

5.85 $g([x]) = f(x)$

5.86 $g([x] * [y]) = g([x \cdot y]) = f(x \cdot y) = f(x) + f(y) = g[x] + g[y]$.

5.89   For $[x] \in S/f$, $[x] * [i_S] = [x \cdot i_S] = [x]$ and $[i_S] * [x] = [i_S \cdot x] = [x]$.

5.91   For $[x] \in S/f$, $[x^{-1}] \in S/f$ and $[x] * [x^{-1}] = [x \cdot x^{-1}] = [i_S]$; similarly $[x^{-1}] * [x] = [i_S]$.

5.94   (a) $g([2] * [3]) = g([2 + 3]) = g([5]) = g([1]) = 2$.
   (b) $g([2]) +_8 g([3]) = 4 +_8 6 = 2$

5.97   (a) $K = \{0, 3, 6, 9, \ldots, -3, -6, -9, \ldots\} = 3\mathbb{Z}$
   (b) $K = \{0, 2, 4, 6, 8, 10\}$

5.98 $x \in K$ implies $f(x) = i_H$; $y \in K$ implies $f(y) = i_H$. Then $f(x \cdot y^{-1}) = f(x) \cdot f(y^{-1}) = f(x) \cdot [f(y)]^{-1} = i_H \cdot i_H^{-1} = i_H \cdot i_H = i_H$, and $x \cdot y^{-1} \in K$.

5.100 (a) $0 + K = 3\mathbb{Z} = \{0, 3, 6, 9, \ldots, -3, -6, -9, \ldots\}$
$1 + K = 1 + 3\mathbb{Z} = \{1, 4, 7, 10, \ldots, -2, -5, -8, \ldots\}$
$2 + K = 2 + 3\mathbb{Z} = \{2, 5, 8, 11, \ldots, -1, -4, -7, \ldots\}$
(b) $0 + K = \{0, 2, 4, 6, 8, 10\}, 1 + K = \{1, 3, 5, 7, 9, 11\}$

5.101 Let $y \in [x]$. Then $f(y) = f(x) = i_H \cdot f(x)$; $f(y) \cdot [f(x)]^{-1} = i_H$ or $f(y \cdot x^{-1}) = i_H$ and $y \cdot x^{-1} \in K$, or $y \cdot x^{-1} = k$ and $y = kx$. This proves that $[x] \subseteq Kx$. Let $kx \in Kx$. Then $f(kx) = f(k) \cdot f(x) = i_H \cdot f(x) = f(x)$, or $kx \in [x]$ and $Kx \subseteq [x]$. Therefore, $[x] = Kx$.

5.103 (a) $(7 + 3\mathbb{Z}) + (5 + 3\mathbb{Z}) = 12 + 3\mathbb{Z} = 3\mathbb{Z}$
(b) $-(4 + 3\mathbb{Z}) = 2 + 3\mathbb{Z}$ because $(4 + 3\mathbb{Z}) + (2 + 3\mathbb{Z}) = 0 + 3\mathbb{Z}$

5.106 (a) $6 + S = \{6 +_{12} 0, 6 +_{12} 4, 6 +_{12} 8\} = \{6, 10, 2\}$
(b) $(1, 2, 3)S = \{(1, 2, 3) \circ i, (1, 2, 3) \circ (2, 3)\} = \{(1, 2, 3), (1, 2)\}$
$(1, 3, 2)S = \{(1, 3, 2) \circ i, (1, 3, 2) \circ (2, 3)\} = \{(1, 3, 2), (1, 3)\}$
$S(1, 2, 3) = \{i \circ (1, 2, 3), (2, 3) \circ (1, 2, 3)\} = \{(1, 2, 3), (1, 3)\}$
$S(1, 3, 2) = \{i \circ (1, 3, 2), (2, 3) \circ (1, 3, 2)\} = \{(1, 3, 2), (1, 2)\}$

5.107 Reflexive: $x = x \cdot i_G, i_G \in S$. Symmetric: if $x\rho y$ then $y = xs$, so $x = ys^{-1}, s^{-1} \in S$, and $y\rho x$. Transitive: if $x\rho y$ and $y\rho z$, then $y = xs_1$ and $z = ys_2$, so $z = xs_1s_2$, $s_1s_2 \in S$, and $x\rho z$.

5.111 Associativity: $(xS \cdot yS) \cdot zS = (xyS) \cdot zS = (xy)zS = x(yz)S = xS \cdot (yzS) = xS \cdot (yS \cdot zS)$. Identity: $i_GS = S$ because $xS \cdot i_GS = xS = i_GS \cdot xS$. Inverses: $(xS)^{-1} = x^{-1}S$ because $xS \cdot x^{-1}S = i_GS = x^{-1}S \cdot xS$.

5.115 f is clearly onto. For $x, y, \in G$, $f(xy) = xyS = xSyS = f(x)f(y)$.

5.117 f is clearly onto since $s \in S$ has xs as a preimage. If $f(xs_1) = f(xs_2)$, then $s_1 = s_2$ and $xs_1 = xs_2$, so f is one-to-one.

5.120 (a) $|G| = 2 \cdot 3 \cdot 4 = 24$. Possible orders are the divisors of 24—1, 2, 3, 4, 6, 8, 12, 24.
(b) Yes, because G is a finite commutative group.

CHAPTER 6

6.2 $H(X, Y) = 2$

6.3 The answer to this problem is so obvious it's hard to write anything, so we won't.

6.8 (a) (10110)   (b) (10110)

6.12 Each code word added to itself is 0, and any $X +_2 0 = X$.

$(01111) +_2 (10101) = 11010$
$(01111) +_2 (11010) = 10101$
$(10101) +_2 (11010) = 01111$

6.15 Straightforward matrix multiplication.

6.17 00 $\rightarrow$ 00000
01 $\rightarrow$ 01111
11 $\rightarrow$ 11010

6.20 $0000 \rightarrow 0000000$ $1000 \rightarrow 1000110$
     $0001 \rightarrow 0001111$ $1001 \rightarrow 1001001$
     $0010 \rightarrow 0010011$ $1010 \rightarrow 1010101$
     $0011 \rightarrow 0011100$ $1011 \rightarrow 1011010$
     $0100 \rightarrow 0100101$ $1100 \rightarrow 1100011$
     $0101 \rightarrow 0101010$ $1101 \rightarrow 1101100$
     $0110 \rightarrow 0110110$ $1110 \rightarrow 1110000$
     $0111 \rightarrow 0111001$ $1111 \rightarrow 1111111$

6.24 $(11011)H = (001)$
     $(10100)H = (001)$
     $(01110)H = (001)$
     $(00001)H = (001)$

6.27 The syndrome is 010, so it is decoded to $1000100 + 0000010 = 1000110$.

6.30 01100 or 10010

CHAPTER 7

7.3 000110

7.4

	Next state		
	Present input		
Present state	0	1	Output
s_0	s_0	s_3	0
s_1	s_0	s_2	1
s_2	s_3	s_3	1
s_3	s_1	s_3	2

7.5

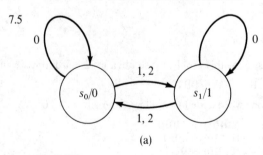

(a)

(b) 01011

7.6 (a) s_1 (b) s_1

7.7 11001011

7.10

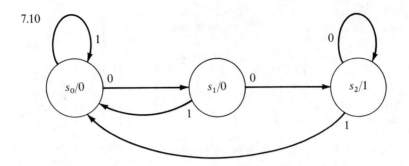

7.13

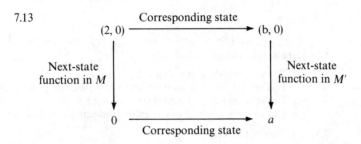

7.14 As α is processed, corresponding states will proceed to corresponding states, and corresponding states have the same output. (A formal proof would use induction on the length of the string α.)

7.16 Bijection

7.17 $g^{-1}:S' \to S$ and g^{-1} is a bijection. For $s' \in S'$ and $i \in I$, let $s' = g(s)$. Then the equation $g(f_s(s, i)) = f'_s(g(s), i)$ can be rewritten as

$f_s(g^{-1}(s'), i) = g^{-1}(f'_s(s', i))$.

The equation $f_0(s) = f'_0(g(s))$ can be rewritten as

$f_0(g^{-1}(s')) = f'_0(s')$.

Therefore, g^{-1} is a homomorphism.

7.19 (a)

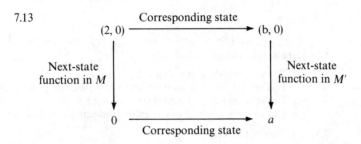

(a)

(b) $h([0]) = a$
    $h([1]) = b$

7.22

Present state	Next state		Output
	Present input		
	0	1	
A	A	A	0
B	C	A	1
C	C	B	1

$g: s_0 \to A$
$s_1 \to B$
$s_2 \to A$
$s_3 \to C$
$s_4 \to C$

7.24 A simulates B by the function

$a \to 0$

$b \to 0$

B cannot simulate A because there is no state in B for state 1 of A to map to; no state in B has an output of 1.

7.29 (a) The set consisting of a single 0.
(b) The set consisting of any number of 0's (including none) followed by 10.
(c) The set consisting of a single 0 or a single 1.
(d) The set consisting of any number (including none) of pairs of 1's.

7.32 (b) does not belong.

7.33 (a) 0   (b) 0*10   (c) $0 \vee 1$ (d) (11)*

7.39

Present state	Next state		Output
	Present input		
	0	1	
$\{s_0\}$	$\{s_0\}$	$\{s_1\}$	0
$\{s_1\}$	$\{s_1\}$	$\{s_0, s_1\}$	1
$\{s_0, s_1\}$	$\{s_0, s_1\}$	$\{s_0, s_1\}$	1

CHAPTER 8

8.2 s_2, s_3

8.4 A state s produces the same output as itself for any input. If s_i produces the same output as s_j, then s_j produces the same output as s_i. Transitivity is equally clear.

8.5 Condition (1) is satisfied because all states in the same class have the same output strings for any input string, including the empty input string. To see that (2) is satisfied, assume s_i and s_j are equivalent states proceeding under the input symbol i to states s_i' and s_j' which are not equivalent. Then there is an input string α such that $f_0(s_i', \alpha) \neq f_0(s_j', \alpha)$. Thus, for the input string $i\alpha$, s_i and s_j produce different output strings, contradicting the equivalence of s_i and s_j.

8.12 (a) Equivalent states of M are $A = \{0, 1, 3\}$, $B = \{2\}$, $C = \{4\}$. The reduced machine is shown in the following figure.

Present state	Next state Present input 0	1	Output
A	B	A	1
B	C	A	0
C	A	A	0

(b) Equivalent states of M are $\{0\}$, $\{1\}$, $\{2\}$, $\{3\}$. M is already minimal.

8.16 Clearly the network behaves properly.

8.17 First, write the state table:

Present state	Next-state Present input 0	1	Output
s_0	s_0	s_1	1
s_1	s_1	s_0	0

The states can be encoded by a single delay element as shown in the figure that follows.

	d
s_0	0
s_1	1

The truth functions are shown in the next figure.

$x(t)$	$d(t)$	$y(t)$	$d(t + 1)$
0	0	1	0
1	0	1	1
0	1	0	1
1	1	0	0

The canonical sum-of-products forms are

$$y(t) = d'$$

$$d(t + 1) = xd' + x'd$$

and the sequential network appears in the figure that follows.

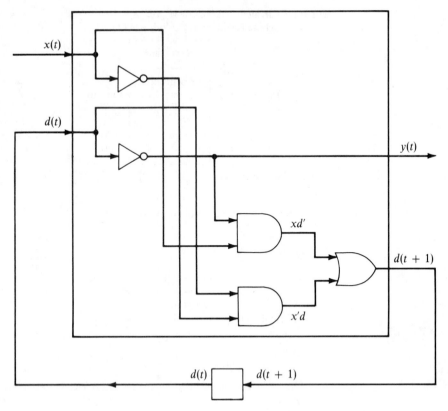

8.20

M_p	Input (0, 0)	(0, 1)	(1, 0)	(1, 1)	Output
(A, a)	(A, b)	(A, a)	(A, b)	(A, a)	$(1, 1)$
(A, b)	(A, a)	(A, b)	(A, a)	(A, b)	$(1, 1)$
(B, a)	(B, b)	(B, a)	(A, b)	(A, a)	$(0, 1)$
(B, b)	(B, a)	(B, b)	(A, a)	(A, b)	$(0, 1)$

8.22 (a) $g(f_p((A, c), 1)) = g(A, b) = 1$ (b) $f_s(g(A, c), 1) = f_s(2, 1) = 1$

8.25 (a) One example is:

M_1	Input 0	1	Output
$A = \{0, 1\}$	A	B	1
$B = \{2, 3\}$	B	A	0

M_2	Input 0	1	Output
$a = \{0, 3\}$	a	a	0
$b = \{1, 2\}$	a	b	1

(b) $g: (A, a) \to 0$ $g': (0, 0) \to 1$
   $\quad (A, b) \to 1$ $\quad (0, 1) \to 0$
   $\quad (B, a) \to 3$ $\quad (1, 0) \to 0$
   $\quad (B, b) \to 2$ $\quad (1, 1) \to 1$

(Other output functions g' and other output assignments can be used.)

8.27

M_s	Input 0	1	Output
(A, a)	(A, b)	(A, a)	(A, a)
(A, b)	(A, a)	(A, b)	(A, b)
(B, a)	(B, a)	(A, b)	(B, a)
(B, b)	(B, b)	(A, b)	(B, b)

8.29 The corresponding state in M_s for 4 is (C, b). Under input 1, $(C, b) \to (C, a)$ and $g(C, a) = 2$.

8.31

M_1	Input 0	1	Output
$A = \{0, 2\}$	B	A	A
$B = \{1, 3\}$	B	A	B

M_2	$(A, 0)$	Input $(A, 1)$	$(B, 0)$	$(B, 1)$	Output
$a = \{0, 1\}$	b	a	a	a	a
$b = \{2, 3\}$	b	b	b	a	b

$$
\begin{array}{ll}
g: (A, a) \to 0 & g': (A, a) \to 1 \\
\quad (A, b) \to 2 & \quad (A, b) \to 0 \\
\quad (B, a) \to 1 & \quad (B, a) \to 1 \\
\quad (B, b) \to 3 & \quad (B, b) \to 0
\end{array}
$$

8.40 (a) $\mathbb{Z}_{12}/\{0, 3, 6, 9\} \simeq \mathbb{Z}_3$
$\{0, 3, 6, 9\}/\{0, 6\} \simeq \mathbb{Z}_2$
$\{0, 6\}/\{0\} \simeq \mathbb{Z}_2$

(b) $\mathbb{Z}_{12}/\{0, 2, 4, 6, 8, 10\} \simeq \mathbb{Z}_2$
$\{0, 2, 4, 6, 8, 10\}/\{0, 4, 8\} \simeq \mathbb{Z}_2$
$\{0, 4, 8\}/\{0\} \simeq \mathbb{Z}_3$
and
$\mathbb{Z}_{12}/\{0, 2, 4, 6, 8, 10\} \simeq \mathbb{Z}_2$
$\{0, 2, 4, 6, 8, 10\}/\{0, 6\} \simeq \mathbb{Z}_3$
$\{0, 6\}/\{0\} \simeq \mathbb{Z}_2$

8.42 (a)

	i	0	1	2
		10		
	11	00		
S	λ	0	1	01
0	0	0	1	1
1	1	0	0	1

(b)

$\circ$	i	0	1	2
i	i	0	1	2
0	0	0	2	2
1	1	0	i	2
2	2	0	0	2

8.44 $g: i \to 0$
$\quad 0 \to 0$
$\quad 1 \to 1$
$\quad 2 \to 1$

8.48 $[g^{*\prime}g] = [g^{*\prime}][g] = [g^*][g] = [g^*g] = [(g^*g)']$. Thus, $g^{*\prime}g = s_1(g^*g)'$ for some $s_1 \in S$ and $(g^{*\prime}g)((g^*g)')^{-1} = s_1 \in S$.

CHAPTER 9

9.3 (a) ...b 0 0 b...
  (b) The machine cycles endlessly over the two nonblank tape squares.
  (c) The machine changes the two nonblank squares to 0 1 and then moves endlessly to the right.

9.6 (a) b X X 1 X X b Machine halts without accepting.
  $\uparrow$
  5

  (b) b X X X X X b Machine halts without accepting.
  $\uparrow$
  2

  (c) b X X X 0 X X b Machine halts without accepting.
  $\uparrow$
  2

9.7 State 3 is the only final state.

  $(0, 1, 1, 0, R)$
  $(0, 0, 0, 1, R)$
  $(1, 0, 0, 2, R)$
  $(1, 1, 1, 0, R)$
  $(2, 0, 0, 2, R)$
  $(2, 1, 1, 0, R)$
  $(2, b, b, 3, R)$

9.8 Change $(2, 1, X, 3, L)$ to $(2, 1, X, 7, L)$ and add $(7, 1, X, 3, L)$.

9.12 One machine that works, together with a description of its actions:

  $(0, 1, 1, 1, R)$ reads first 1

  $(1, b, 1, 6, R)$ $n = 0$, changes to 1 and halts

  $(1, 1, 1, 2, R)$ reads second 1

  $(2, b, b, 6, R)$ $n = 1$, halts

  $(2, 1, 1, 3, R)$ $n \geq 2$

  $\left.\begin{array}{l}(3, 1, 1, 3, R) \\ (3, b, b, 4, L)\end{array}\right\}$ finds right end of $\bar{n}$

  $\left.\begin{array}{l}(4, 1, b, 5, L) \\ (5, 1, b, 6, L)\end{array}\right\}$ erases two 1's from $\bar{n}$ and halts

9.23 For any Turing machine T we can effectively create a machine T^* that acts like T until T reaches a halting configuration, then erases the tape and halts. Then T^* halts with a blank tape when started on a tape containing α if and only if T halts when started on a tape containing α. Assume that there exists an algorithm P to solve this decision problem. We now have an algorithm to solve the halting problem: given a (T, α) pair, create T^* and apply algorithm P.

9.26 $t(n) = 2n + 2$

9.28 Straightforward computation.

CHAPTER 10

10.7 $0S \Rightarrow 00S \Rightarrow 000S \Rightarrow 0000S \Rightarrow 00001$

10.9 $L = \{0^n1 \,|\, n \geq 0\}$.

10.11 For example:
(a) $G = (V, V_T, S, P)$ where $V = \{0, 1, S\}$, $V_T = \{0, 1\}$, and $P = \{S \rightarrow 1, S \rightarrow 0S0\}$.
(b) $G = (V, V_T, S, P)$ where $V = \{0, 1, S, M\}$, $V_T = \{0, 1\}$, and
   $P = \{S \rightarrow 0M0, M \rightarrow 0M0, M \rightarrow 1\}$.

10.20 In G_1: $S \Rightarrow ABA \Rightarrow 00A \Rightarrow 0000A \Rightarrow 00000$
In G_2: $S \Rightarrow 00A \Rightarrow 0000A \Rightarrow 00000$
In G_3: $S \Rightarrow 0A \Rightarrow 00B \Rightarrow 000C \Rightarrow 0000B \Rightarrow 00000$

10.22

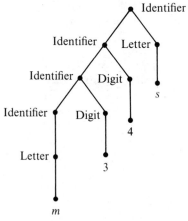

10.25 $G = (V, V_T, S, P)$ where $V = \{S, A, B, L, 0, 1\}$, $V_T = \{0, 1\}$, and P consists of

$S \rightarrow 1A$
$S \rightarrow 0C$
$A \rightarrow 0$
$A \rightarrow 0B$
$A \rightarrow 1C$
$B \rightarrow 1A$
$B \rightarrow 0C$
$C \rightarrow 0C$
$C \rightarrow 1C$
$L(G) = 10(10)^*$

10.27

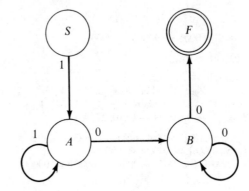

$L(G) = 11*00*0$

10.33 $W_0 = \{S\}$
$W_1 = \{\lambda, ABA\}$
$W_2 = \{\lambda, ABA, 00A, 0BA, AB0\}$
$W_3 = \{\lambda, ABA, 00A, 0BA, AB0, 000, 0B0\}$
$W_4 = W_3$
$0000 \notin L$